Nonlinear Phenomena and Complex Systems

VOLUME 1

The Centre for Nonlinear Physics and Complex Systems (CFNL), Santiago, Chile, and Kluwer Academic Publishers have established this series devoted to nonlinear phenomena and complex systems, which is one of the most fascinating fields of science today, to publish books that cover the essential concepts in this area, as well as the latest developments. As the number of scientists involved in the subject increases continually, so does the number of new questions and results. Nonlinear effects are essential to understand the behaviour of nature, and the methods and ideas introduced to treat them are increasingly used in new applications to a variety of problems ranging from physics to human sciences. Most of the books in this series will be about physical and mathematical aspects of nonlinear science, since these fields report the greatest activity.

Instabilities and Nonequilibrium Structures V

Edited by

Enrique Tirapegui

Facultad de Ciencias Físicas y Matemáticas,
Universidad de Chile,
Santiago, Chile

and

Walter Zeller

Instituto de Física,
Universidad Católica de Valparaíso,
Valparaíso, Chile

KLUWER ACADEMIC PUBLISHERS

DORDRECHT / BOSTON / LONDON

A C.I.P. Catalogue record for this book is available from the Library of Congress.

ISBN-13: 978-94-010-6590-0   e-ISBN-13: 978-94-009-0239-8
DOI: 10.1007/978-94-009-0239-8

Published by Kluwer Academic Publishers,
P.O. Box 17, 3300 AA Dordrecht, The Netherlands.

Kluwer Academic Publishers incorporates
the publishing programmes of
D. Reidel, Martinus Nijhoff, Dr W. Junk and MTP Press.

Sold and distributed in the U.S.A. and Canada
by Kluwer Academic Publishers,
101 Philip Drive, Norwell, MA 02061, U.S.A.

In all other countries, sold and distributed
by Kluwer Academic Publishers Group,
P.O. Box 322, 3300 AH Dordrecht, The Netherlands.

Cover design based on a concept by
David Turkieltaub

TABLE OF CONTENTS

Part III. Stochastic Effects

Part IV. Statistical Mechanics and Applications

Foreword

This book contains the lectures and a selection of the seminars given in the Fifth International Workshop on Instabilities and Nonequilibrium Structures which took place in Santiago, Chile, in December 1993. The Workshop was organized by Facultad de Ciencias Físicas y Matemáticas, Universidad de Chile, Instituto de Física of Universidad Católica de Valparaíso and Centro de Física No Lineal y Sistemas Complejos de Santiago.

This volume is the first of a new series of Kluwer on Nonlinear Phenomena and Complex Systems which will be edited by the Centro de Física No Lineal y Sistemas Complejos de Santiago. We thank Dr. David Larner of Kluwer for his encouragements and support for this project.

LIST OF SPONSORS OF THE WORKSHOP

- Academia Chilena de Ciencias

- Facultad de Ciencias Físicas y Mathemáticas de la Univ. de Chile

- Instituto de Física de la Univ. Católica de Valparaiso

- Centro de Física No Lineal y Sistemas Complejos de Santiago (CFNL)

- CONICYT (Chile)

- Ministère Francais des Affaires Etrangères

- International Centre for Theoretical Physics (Trieste)

- UNESCO

- Fundación Andes (Chile)

- Departamento Técnico de Investigación y de Relaciones Internationa-cionales de la Universidad de Chile

- IDIEM (Fac. Cs. Fís. y Mat., Univ. de Chile)

- CHILGENER S.A.

PREFACE

The articles we present here are intended to give a view of some of the subjects which to-day attract the attention of scientists working in nonlinear physics. All the subjects of current interest in the area are of course not represented, this would be impossible in an expanding area of research as this one, but we feel many of the fascinating new developments are included.

Part I contains two review articles. The first one by D. Walgraef of the Brussels group treats in detail the important problem of the coexistence of locally stable spatio-temporal patterns, of their interactions and of the transitions between them. The study is done in simple representative models and in amplitude equations. In the second article by Abid, Brachet, Debbasch and Nore of the ENS, Paris, the relation of fluid dynamics with nonlinear wave equations through a Madelung transformation is discussed both in the nonrelativistic and in the relativistic case. Special attention is given to the nonlinear Schrödinger equation due to its importance as a dynamical model of superfluids.

Part II is devoted to pattern formation and instabilities. The first contribution by P. Coullet of the Institut Nonlinéaire de Nice proposes a route to complex spatio-temporal behavior which starts with temporal chaos in the core of a defect (a confined object). Lasers are treated in two articles: one by G. Huyet and S. Rica proposes an interesting model for the transverse dynamics and patterns and a second by R. Buceta et al refers to the effect of a periodic injected signal. Front propagation is discussed in non-gradient systems by M. San Miguel et al. together with a discussion of variational properties and mathematical rigorous results on the speed selection mechanism are presented by R. Benguria and M.C. Depassier. Other articles in this Section refer to pattern selection, convection and simulations with a system of hard disks of the Poiseuille flux.

Part III reviews the effects of stochastic perturbation in nonlinear systems and in Part IV we have included articles on nonequilibrium statistical mechanics.

Part V refers to the study of granular matter and its latest developments with an introductory article by E. Guyon and J.P. Troadec.

Finally in Part VI we include a selection of short communications presented to the Workshop.

1

PART I
REVIEW ARTICLES

TRANSITIONS BETWEEN SPATIO-TEMPORAL PATTERNS IN NONEQUILIBRIUM SYSTEMS

D. WALGRAEF †
Center for Nonlinear Phenomena and Complex Systems,
Free University of Brussels, CP 231, Bd du Triomphe,
B-1050 Brussels, Belgium.

ABSTRACT. Physico-chemical systems driven away from thermal equilibrium usually undergo various types of instabilities leading to the formation of spatio-temporal patterns on macroscopic time and space scales. In two and three-dimensional geometries, patterns of different symmetries may be simultaneously stable. The phase stability of these patterns and different transition mechanisms between them are reviewed in the framework of amplitude equations and reduced dynamical models which may describe these systems close to their instability points. On the other hand, many reaction-diffusion systems which present a Turing instability also present a Hopf bifurcation. When these instabilities are close together, the interaction between spatial and temporal modes may lead to complex spatio-temporal patterns. The selection and stability of such patterns as well as their defect behavior are also reviewed in this framework, and the relevance of the results to experimental observations is discussed.

1. Introduction : Selection and Stability of Perfect Patterns

In the vicinity of a pattern forming instability, the dynamics of complex physico-chemical systems may be reduced to much simpler forms via the adiabatic elimination of the stable modes, via multiple scale analysis or normal forms derivation [1] and lead to several types of asymptotic dynamics for the unstable mode σ which plays the role of an order parameter. The simplest one describes the onset of convection in Rayleigh-Benard experiments or pattern forming instabilities of the Turing type in nonlinear chemical kinetics, and is written as :

$$\tau_0 \partial_t \sigma(\vec{r}, t) = [\epsilon - (q_c^2 + \nabla^2)^2]\sigma(\vec{r}, t) - v\sigma^2(\vec{r}, t) - u\sigma^3(\vec{r}, t) \tag{1}$$

where σ is the unstable mode and plays the role of an order parameter, while ϵ is the reduced distance to the instability threshold and is written as $(\lambda - \lambda_c)/\lambda_c$

E. Tirapegui and W. Zeller (eds.), Instabilities and Nonequilibrium Structures V, 5–32.
© *1996 Kluwer Academic Publishers.*

where λ is the bifurcation parameter and λ_c its critical value. For $\epsilon < 0$, the trivial uniform steady state $\sigma = 0$ is stable while for $\epsilon > 0$, patterns of wavenumber q_c are expected to develope. Furthermore, q_c, v and u may be explicitly computed for each particular dynamical model. Let us recall that this equation results from a series expansion in powers of σ, limited to the third powers on assuming that one considers regions of the parameter space near the instability where the amplitude of the slow mode remains small (of the order of $\sqrt{\epsilon}$) and that u is positive. If this is not the case, one has of course to consider higher order nonlinearities. Furthermore, u and v may also be space dependent, but let us consider them first as constants, which is the case for most reaction-diffusion systems (as a consequence of the constant nonlinear kinetic rates) [2] and several convection problems [3].

This equation has a gradient structure and we can define an associate Lyapunov functional that decreases in any dynamics and plays a role similar to the one played by the free energy in equilibrium phase transitions. The corresponding Ginzburg-Landau type of dynamics associated to Eq.(1) may be written as:

$$\tau_0 \partial_t \sigma = -\Gamma \frac{\delta \mathcal{F}}{\delta \sigma} \tag{2}$$

where $\delta / \delta \sigma$ represents the functional derivative and where the potential or Lyapunov functional is given by

$$\mathcal{F} = \frac{1}{\Gamma} \int d\vec{r} [-\frac{\epsilon}{2}\sigma^2 + \frac{1}{2}((q_c^2 + \nabla^2)\sigma)^2 + \frac{v}{3}\sigma^3 + \frac{u}{4}\sigma^4] \tag{3}$$

Each possible planform may be characterized by m pairs of wavevectors $(\{\vec{q}_i, -\vec{q}_i\})$ and, since the selected pattern should in principle correspond to the minimum of this potential, let us consider first structures with $|\vec{q}_i| = q_c$. For such planforms, the order parameterlike variable may be written as :

$$\sigma(\vec{r}) = 2Re \sum_{i=1}^{m} a_i exp(i\vec{q}_i\vec{r}) \tag{4}$$

and on inserting this expression in eq. (1) and neglecting the short space and time variations, one obtains the following evolution equations for the uniform scaled amplitudes A_i ($A_i = a_i/\sqrt{3u}$) :

$$\partial_t A_i = \mu A_i + v\Sigma_j\Sigma_k A_j^* A_k^* \delta(\vec{k}_i + \vec{k}_j + \vec{k}_k) - |A_i|^2 A_i - \Sigma_{i \neq j}\gamma_{ij}|A_j|^2 A_i$$

$$-\Sigma_j\Sigma_k\Sigma_m\lambda_{jkm}A_j^* A_k^* A_m^* \delta(\vec{k}_i + \vec{k}_j + \vec{k}_k\vec{k}_m) \tag{5}$$

where the coefficients γ_{ij} and λ_{jkm} depend on the angles between the corresponding modes. The λ_{jkm} terms only appear in three dimensional systems and correspond to closed circuits built by non coplanar wavevectors with $\vec{k}_i$.

Two classes of patterns have to be considered:

1) the patterns built on m independent pairs of wavevectors. In this case, there is no contribution to the dynamics coming from the quadratic couplings and stationnary amplitudes may appear via a supercritical bifurcation. In the case of eq. (1) $\gamma_{ij} = 2$, and we have :

$$|a_i| = a_m = \begin{cases} 0 & for \quad \lambda < \lambda_c \quad ; \quad \sqrt{\dfrac{\epsilon}{3(2m-1)u}} & for \quad \lambda > \lambda_c \end{cases} \qquad (6)$$

2) the patterns built on triplets of wavevectors satisfying the triangular relation

$$\vec{q}_i + \vec{q}_j + \vec{q}_k = 0$$

These structures correspond, in two-dimensional systems, to honeycomb or triangular lattices for the maxima of the order parameter defined by

$$\sigma_3(\vec{r}) = 2Re\, a_3[expiq_c x + expiq_c \frac{x + \sqrt{3}y}{2} + expiq_c \frac{x - \sqrt{3}y}{2}]$$

with

$$|a_3| = -\frac{v + \sqrt{v^2 + 15u\epsilon}}{15u} \qquad (7)$$

and they appear via a first order like transition at $\epsilon = -\frac{v^2}{15u}$. In three-dimensional systems besides rodlike triangular patterns, we may also have structures built on six pairs of wavevectors forming an octahedron and leading to bcc lattices or filamental structures for the maxima of σ defined by:

$$\sigma_6(\vec{r}) = 2a_6[\cos\frac{q_c}{\sqrt{2}}x\cos\frac{q_c}{\sqrt{2}}y + \cos\frac{q_c}{\sqrt{2}}x\cos\frac{q_c}{\sqrt{2}}z + \cos\frac{q_c}{\sqrt{2}}y\cos\frac{q_c}{\sqrt{2}}z] \qquad (8)$$

with

$$a_6 = -\frac{2v + \sqrt{4v^2 + 33u\epsilon}}{33u} \qquad (9)$$

The physically possible patterns correspond to the stationnary solutions which are stable with respect to arbitrary disturbances of infinitesimal amplitude. For example, for the first class of structures, the linear evolution matrix of homogeneous amplitude disturbances has the following structure:

$$L = - \begin{pmatrix} r_D & r_{ND} & r & r & \cdots \\ r_{ND} & r_D & r & r & \cdots \\ r & r & r_D & r_{ND} & \cdots \\ r & r & r_{ND} & r_D & \cdots \\ \cdots & \cdots & \cdots & \cdots & \cdots \end{pmatrix} \qquad (10)$$

where $r_D = r_{ND} = \epsilon/(2m - 1)$ and $r = 2\epsilon/(2m - 1)$ and since the eigenvalues of this matrix are 0, $2(r - r_D)$ and $-2[r_D + (m - 1)r]$, positive growth rates occur $(2(r - r_D))$ for $m > 1$ implying the instability of these structures. Hence the roll or wall structure is the only stable one in this category. Patterns arising through a subcritical bifurcation are stable in a wide range of the parameters space [4] and the corresponding bifurcation diagram is shown in Figs.1 and 2.

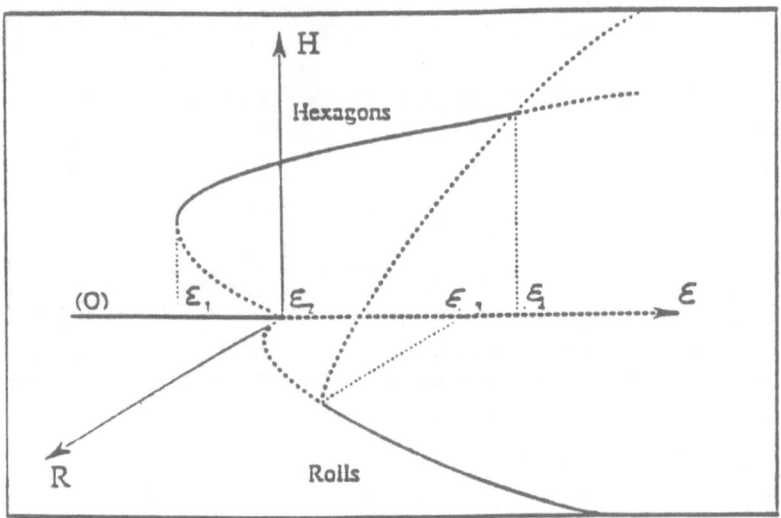

Figure 1 : Bifurcation diagram for the steady states of Eq.(1) for two-dimensional systems. Heavy lines correspond to stable states and dotted lines to unstable states (R stands for the amplitude of rolls and H for hexagons).

When u is negative, one has to consider higher nonlinearities which could favor more complex planforms corresponding for example to quasi-periodic patterns such as Penrose tilings or icosahedral lattices [5].

The dynamics and the bifurcation diagram associated to Eq.(1) are however not universal. Effectively, when v and u are space or gradient dependent, other spatial structures may be selected.

Let us illustrate this on another potential dynamics arising in systems where the dominant nonlinearities couple spatial derivatives of the unstable mode, e.g. the Proctor-Sivashinsky dynamics given by [6]:

$$\tau_0 \partial_t \sigma(\vec{r}, t) = [\epsilon - (q_c^2 + \nabla^2)^2] \sigma(\vec{r}, t) + u\vec{\nabla}(\vec{\nabla}\sigma(\vec{r}, t).(\vec{\nabla}\sigma(\vec{r}, t))^2) \qquad (11)$$

In this case, the only stable structure of the first class described above corresponds to square patterns. The amplitude of the squares with critical wavelengthes is in general determined by the following stationarity condition ($A_i = a_i/\sqrt{3uq_c^4}$) :

$$\epsilon A_i - A_i|A_i|^2 - g_{ij}A_i|A_j|^2 = 0, \qquad i, j = 1, 2 \qquad (12)$$

From Eq.(11) we get $\gamma_{ij} = 2/3$ and the squares of amplitude $|A_i| = |A_j| = \sqrt{3\epsilon/5}$ appear via a supercritical bifurcation or second order transition at $\epsilon = 0$ whereas the rolls are now unstable. In general, as a result of the linear stability analysis, squares will be selected when $\gamma_{ij} < 1$. This inequality is satisfied in the model given by Eq.(11) as a consequence of the dependence of the nonlinear terms on the angles between the interacting wavevectors defining the structure (here $q_i \perp q_j$). Let us also note that in this case hexagonal planforms are marginally stable and could appear in the supercritical region.

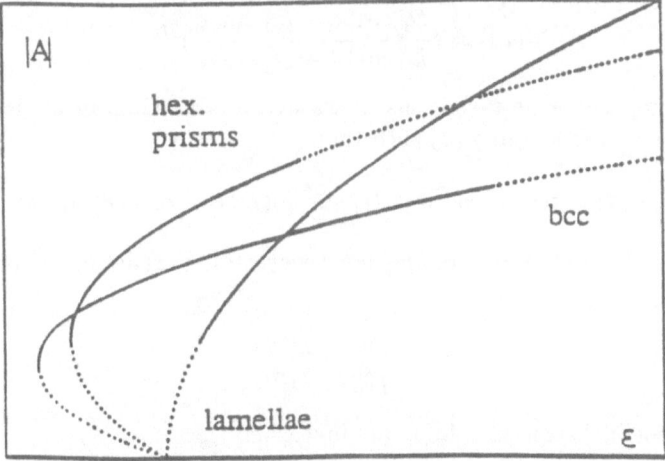

Figure 2 : Bifurcation diagram associated with the dynamics given by Eq.(1) for three-dimensional systems. Heavy lines correspond to stable states and dotted lines to unstable states.

In the Swift-Hohenberg dynamics (cf. Eq.(1)) the preferred wavenumber, according to the Lyapunov functional is q_c, while in the Sivashinsky dynamics (10), the functional associated to a square pattern of wavenumber q is proportional to $-(\epsilon - (q^2 - q_c^2)^2)^2/q^4$ leading to a preferred wavenumber, $\bar{q} = (q_c^4 - \epsilon)^{1/4}$, which decreases for increasing constraints.

Let us also emphasize that a large class of systems where stable inhomogeneous structures appear spontaneously are anisotropic such as for example, liquid crystals, chemically active media and catalytic surfaces, irradiated or stressed materials, systems submitted to external and flow fields, ... In such systems, there is an intrinsic mechanism which raises the orientational degeneracy by inducing preferred directions for the wavevectors characterising the structures. Let us illustrate this effect in the simple case of the Turing instability arising in the Prigogine-Lefever-Nicolis model [7] in a 2D uniaxial medium (the principal axis being parallel to $0x$). The rate equations for the densities of the two reacting species, X and Y, are

$$\dot{X} = A - (B+1)X + X^2Y + D_{\parallel}^x \nabla_x^2 X + D_{\perp}^x \nabla_y^2 X$$
$$\dot{Y} = BX - X^2Y + D_{\parallel}^y \nabla_x^2 Y + D_{\perp}^y \nabla_y^2 Y \tag{13}$$

The homogeneous steady state becomes unstable for $B > B_c = (1 + A\eta)^2$, where

$$\eta(\phi) = \left(\frac{D_{\parallel}^x \cos^2 \phi + D_{\perp}^x \sin^2 \phi}{D_{\parallel}^y \cos^2 \phi + D_{\perp}^y \sin^2 \phi}\right)^{1/2},$$

versus inhomogeneous perturbations of wavevectors making an angle ϕ with the principal axis and of length $q_c(\phi)$ such that

$$q_c^4(\phi) = A/(D_{\parallel}^x \cos^2 \phi + D_{\perp}^x \sin^2 \phi)(D_{\parallel}^y \cos^2 \phi + D_{\perp}^y \sin^2 \phi) \tag{14}$$

Hence the bifurcation occurs first for wavevectors parallel to the principal axis when

$$\frac{D_{\perp}^x}{D_{\perp}^y} > \frac{D_{\parallel}^x}{D_{\parallel}^y}$$

whereas orthogonal structures bifurcate first when

$$\frac{D_{\perp}^x}{D_{\perp}^y} < \frac{D_{\parallel}^x}{D_{\parallel}^y}$$

In this simple example, the anisotropy of the transport coefficients induces preferred directions for the critical wavevector and a first selection mechanism already appears in the linear analysis. However, since nonlinear couplings may favor cellular structures, competition may occur between wall structures, which should dominate close to the instability, and hexagonal or square patterns which should dominate for larger values of the bifurcation parameter, when the nonlinearities become sufficiently important.

This effect also appears in dynamics such as given by Eq.(11) where, in the absence of anisotropy, the nonlinear coupling term tends to select square structures. In the case of anisotropic linear growth rates such as :

$$\epsilon - (q_c^2 + \nabla^2)^2 + A\nabla_y^2 \qquad , \qquad (15)$$

the amplitude of rolls and squares of critical wavevectors are determined in this case by the following :

$$\epsilon A_1 - A_1[|A_1|^2 + g_{12}|A_2|^2] = 0$$
$$(\epsilon - Aq_c^2)A_2 - A_2[|a_2|^2 + g_{12}|A_1|^2] = 0 \qquad (16)$$

The anisotropy A tends to align the critical wavevector parallel to the direction 1. The stability analysis shows that the roll structure induced by the anisotropy and of amplitude $|A_1| = \sqrt{\epsilon}$, $A_2 = 0$ becomes unstable versus transverse modulations at $\epsilon > Aq_c^2/(1-g_{12})$, where one has a direct transition to a square structure. This effect has been observed in liquid crystals submitted to elliptical shears and quantitatively described by this type of asymptotic dynamics [8], as shown in Fig.3.

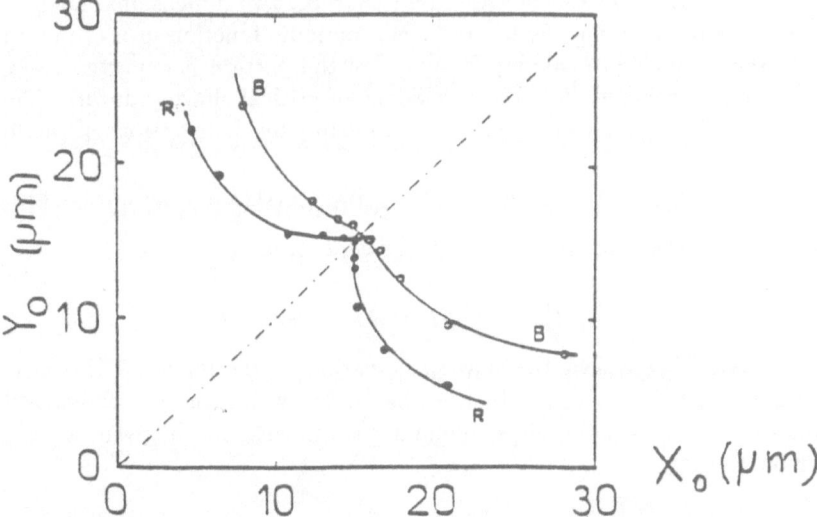

Figure 3 : Threshold curves for roll (R) and bimodal (B) patterns in MBBA liquid crystals submitted to an elliptical shear of intensity $X = X_0 \cos \omega t$ and $Y = Y_0 \sin \omega t$ $(\omega/2\pi = 150Hz)$ [8].

The patterns briefly discribed here occur via continuous symmetry breakings, namely of the translational and rotational symmetries. A consequence of this is the diffusive dynamics of the phase variables which play the role of Goldstone modes. Topological defects may thus be expected, as well as an extreme sensitivity to even small external fields and boundary effects. Since furthermore, the noise intensity is usually so small, in many nonequilibrium phase transitions, that the system may be locked in local minima of the potential for times much longer than any experimental time scale, and selection and transitions may occur via various types of dynamical mechanisms that we will illustrate in the following.

2. Phase Dynamics and Defects of Cellular Patterns

As discussed in the preceding section, the patterns described so far are associated with symmetry breaking instabilities, and the preferred structure corresponds to periodic spatial modulations. However, the real patterns observed in nature are usually much more complicated: they are not perfectly periodic and contain defects, this behavior being also reminiscent of equilibrium phase transitions. As the patterns considered here spontaneously break continuous invariances, continuous families of such patterns are expected to exist and each of these families are characterized by their phases, the dynamics of which being particularly interesting. In fact, the concept of phase dynamics turns out to be extremely fruitful to describe real patterns [9] and will be illustrated here in the most simple cases.

If $X_0(x, y, t)$ is the local value of a physico-chemical quantity associated with a perfectly periodic layered structure in an isotropic two-dimensional medium where y is the coordinate along the layers, it is a periodic function of x of period $2\pi/q_0$. Due to the symmetry breaking, the phase of the pattern is expected to evolve on large time and space scales (the equivalent of critical slowing down). The phase description of the pattern consists then in looking for distributions of the form

$$X(x, y, t) = X_0(x + \phi(X, Y, T), y) + \Xi(\phi_X, \phi_{XX}, \dots, \phi_Y, \dots; x, y)$$

with an equation for the phase ϕ of the layered structure:

$$\partial_T \phi = \Pi(\phi_X, \phi_{XX}, \dots, \phi_Y, \dots) \qquad (17)$$

where X and Y represent the slow space variables, the functions Ξ and Π vanish when ϕ is a constant phase. In the case of the wall structure associated to the dynamics given by (1) the phase dynamics is diffusive and is given, at the leading order by

$$\partial_T \phi = D_x \nabla_X^2 \phi + D_y \nabla_Y^2 \phi \qquad (18)$$

This equation may also be obtained by looking at the amplitude equation for a wall structure of critical wavevector $q_c \vec{1}_x$ and given by

$$\sigma(x, y, t) \propto A(x, y, t)expiq_c x + A^*(x, y, t)exp - iq_c x$$

A and A^* being slowly varying functions of their arguments, their kinetic equations, obtained from (1), may be written as

$$\tau_0 \partial_t A = \epsilon A + (2q_c \nabla_x - i\nabla_y^2)^2 A - |A|^2 A \qquad (19)$$

This equation also possesses a simple family of periodic solutions

$$A_k = \sqrt{\epsilon - k^2} expikx$$

describing patterns of wavevector $(q_c + k)\vec{1}_x$, and on expressing A in amplitude and phase variables, $A = Rexpi(kx + \phi)$, it is easy to see that R has a relaxational dynamics while ϕ has a diffusive dynamics. Once again a space-time scale separation occurs for long ranged perturbations and the amplitude may be adiabatically eliminated from the asymptotic dynamics. It is then a matter of algebra to obtain the phase dynamics of the layered pattern of wavevector $(q_c + k)\vec{1}_x$ which reads

$$\partial_t \phi = D_\| \nabla_x^2 \phi + D_\perp \nabla_y^2 \phi - K\nabla_y^4 \phi$$

with

$$D_\| = \frac{\xi^2}{\tau_0} \cdot \frac{\epsilon - 3\xi^2 k^2}{\epsilon - \xi^2 k^2}, \qquad D_\perp = \frac{\xi^2}{\tau_0} \frac{k}{q_c}, \qquad K = \frac{1}{\tau_0} \qquad (20)$$

and $\xi = 4q_c^2$. $D_\|$ and $D_\perp$ are the parallel and transverse phase diffusion coefficients of the structure. When they are positive, the pattern eventually synchronizes itself as experimentally observed in the relaxation of thermally printed modulations of convective patterns [10]. When one of the diffusion coefficients changes its sign, instability occurs. When $D_\| < 0$, the pattern is unstable versus longitudinal perturbations and this case corresponds to the Eckhaus instability [1], while $D_\perp < 0$ corresponds to the zig-zag instability.

In anisotropic systems, the transverse diffusion coefficient contains a contribution from the term which raises the orientational degeneracy and $D_\perp$ remains finite and positive even near threshold contrary to the isotropic case. For example, with the dynamics

$$\tau_0 \partial_t \sigma(\vec{r}, t) = [\epsilon - (q_c^2 + \nabla^2)^2 + A\nabla_y^2]\sigma(\vec{r}, t) - v\sigma^2(\vec{r}, t) - u\sigma^3(\vec{r}, t) \qquad (21)$$

we have

$$D_\perp = \frac{1}{\tau_0}[\xi^2 \frac{k}{q_c} + A]$$

14

and the zig-zag instability limit is shifted towards $q_\Delta = q_c(1 - \frac{A}{\xi^2})$, as shown in Fig.4.

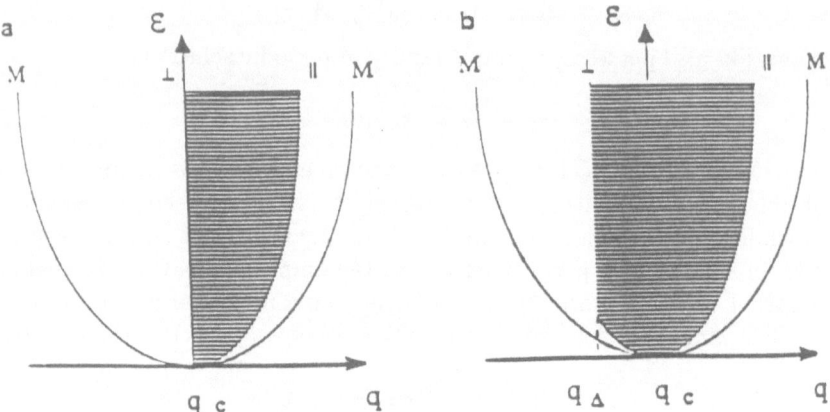

Figure 4 : Stability diagram of layered structures in the isotropic and anisotropic Swift-Hohenberg dynamics. The patterns are stable in the hatched regions limited by Eckhaus and zig-zag instabilities.

For polygonal isotropic structures, the phase is a two-dimensional vector. For instance, in the case of an hexagonal pattern, where the amplitudes equations of the underlying modes may be written as :

$$\partial_t A = \epsilon A + [(\vec{q}_1 + \vec{\nabla})^2]A + vB^*C^* - A|A|^2 - gA(|B|^2 + |C|^2)$$
$$\partial_t B = \epsilon B + [(\vec{q}_2 + \vec{\nabla})^2]B + vA^*C^* - B|B|^2 - gB(|A|^2 + |C|^2)$$
$$\partial_t C = \epsilon C + [(\vec{q}_3 + \vec{\nabla})^2]C + vB^*A^* - C|C|^2 - gC(|B|^2 + |A|^2)$$
$$|\vec{q}_1| = |\vec{q}_2| = |\vec{q}_3| = q = q_c + k \tag{22}$$

After the separation in amplitude and phase and the adiabatic alimination of the relaxing amplitude modes, or the usual multiple scale analysis, one obtains the following phase diffusion equation [11] :

$$\partial_t \vec{\phi} = D_\perp^h \nabla^2 \vec{\phi} + (D_\parallel^h - D_\perp^h)\vec{\nabla}(\vec{\nabla}\vec{\phi}) \tag{23}$$

with

$$D_\perp^h = q_c^2 + 6q_c k - \frac{16q_c^4 k^2}{W_\Delta} \tag{24}$$

and

$$D_{\parallel}^h = 3q_c^2 + 10q_ck - 16q_c^4k^2(\frac{1}{W_\Delta} + \frac{1}{W_\Sigma})$$ (25)

with

$$W_\Delta = 2(1-g)R^2 + 4vR$$ (26)

and

$$W_\Sigma^h = 2(1+2g)R^2 - 2vR$$ (27)

with

$$R = \frac{2}{1+2g}[v + \sqrt{v^2 + (1+2g)(\epsilon - 4q_c^2k^2)}]$$ (28)

It is interesting to compare this dynamics with the equation of motion of the displacement field of an isotropic solid :

$$\rho\frac{\partial^2}{\partial t^2}\vec{u} = \mu\nabla^2\vec{u} + (\lambda + \mu)grad\ div\vec{u}$$ (29)

where μ and λ are the Lamé coefficients. Hence, the diffusion coefficients $D_{\parallel}^h$ and $D_{\perp}^h$ play a role similar to the elasticity coefficients of a solid, but the phase dynamics is diffusive and not propagative. Steady state solutions of the phase dynamics may correspond in both cases to phase singularities associated to dislocations which are now observed in almost all non equilibrium pattern forming systems.

Furthermore, he stability analysis shows that the hexagonal planforms are only stable in the a closed domain defined by $D_{\parallel}^h > 0$ and $D_{\perp}^h > 0$ as shown on the phase diagram represented in Fig.5.

These stability limits have been tested numerically as well as the relaxation of unstable hexagonal patterns towards patterns with stable wavelengthes, as shown in the figures 6 to 8. These results are in agreement with an other recent study of this problem [12].

Figure 5 : Stability diagram associated with the dynamics given by Eq.(1) for roll and hexagonal patterns, for g=2 and v=0.2.

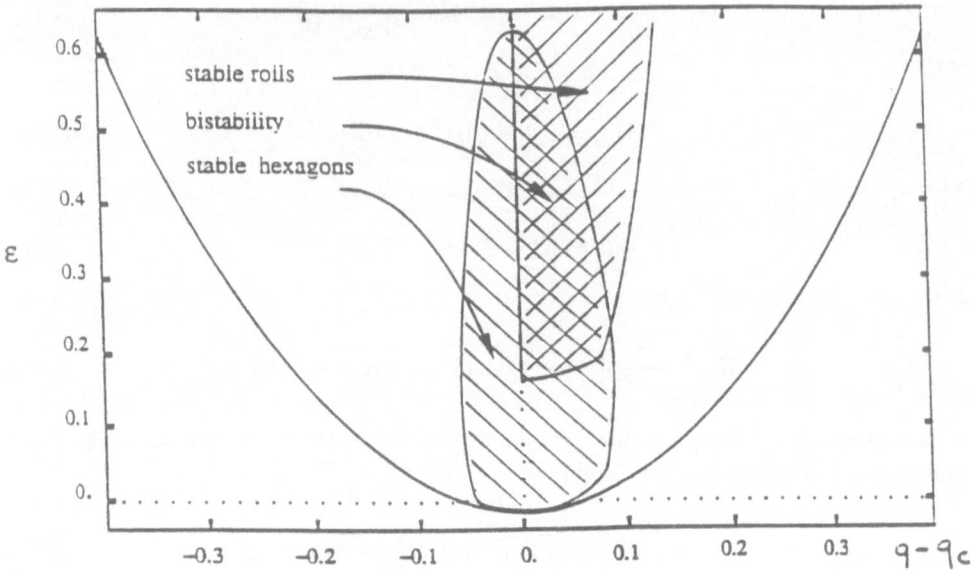

Figure 6 : Numerical tests of the stability of hexagonal patterns described by Eq.(1) for g=2 and v=0.2.

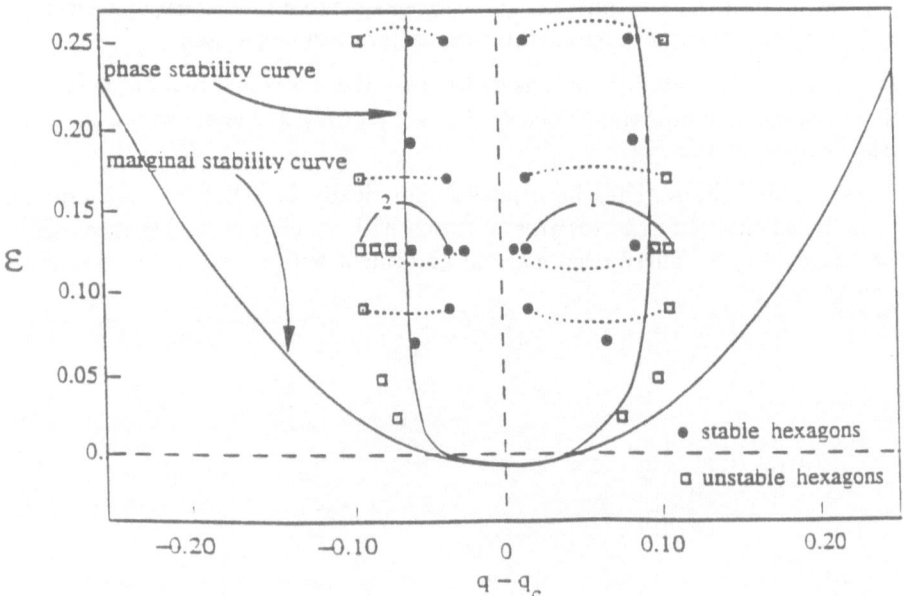

Figure 7 : Snapshots of the evolution of unstable hexagonal planforms towards stable ones ($\lambda_{initial} < \lambda_{s-}$, test 1 of fig.6) .

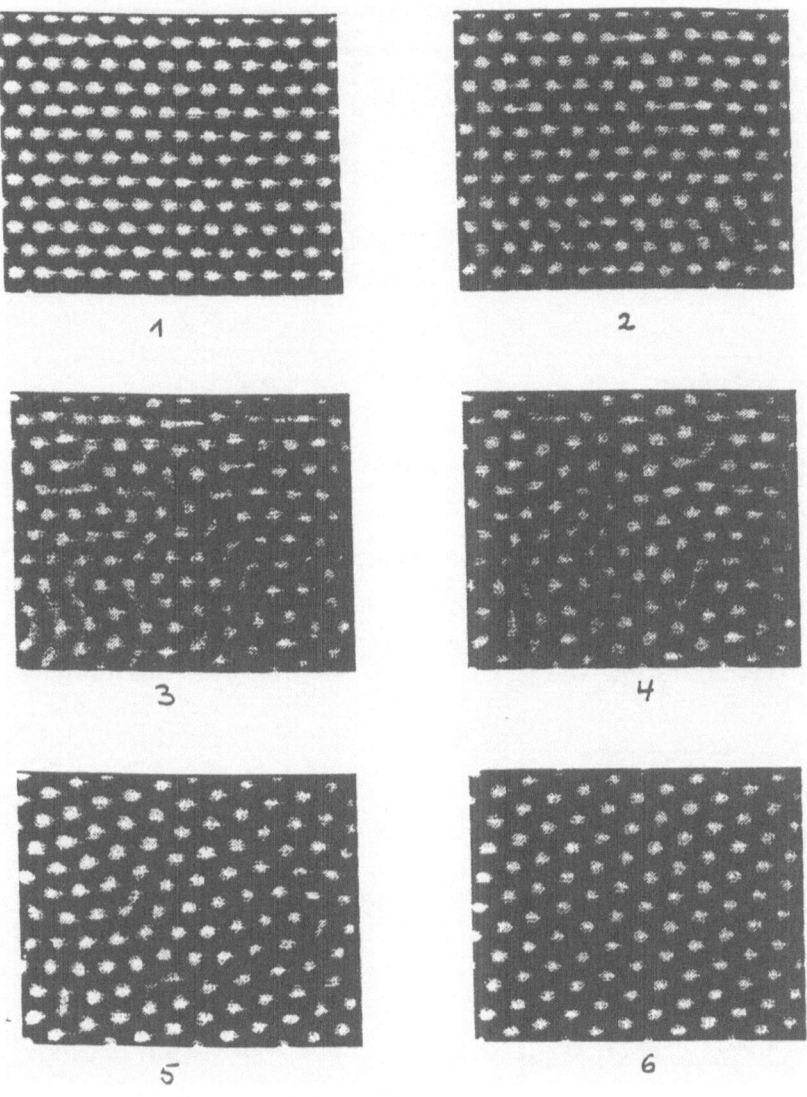

18

Figure 8 : Snapshots of the evolution of unstable hexagonal planforms towards stable ones ($\lambda_{initial} > \lambda_{s-}$, test 2 of fig.6) .

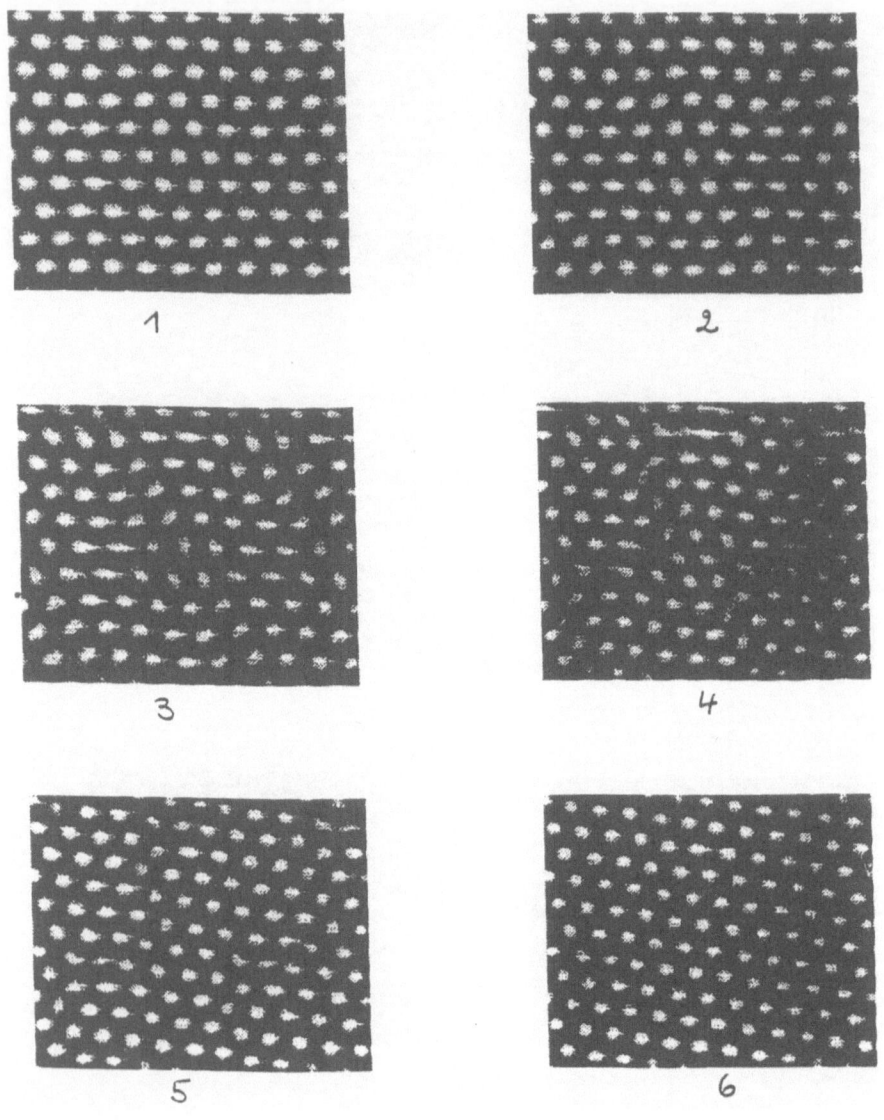

Figure 9 : Transition from zig-zag unstable rolls to an hexagonal pattern in the dynamics (1).

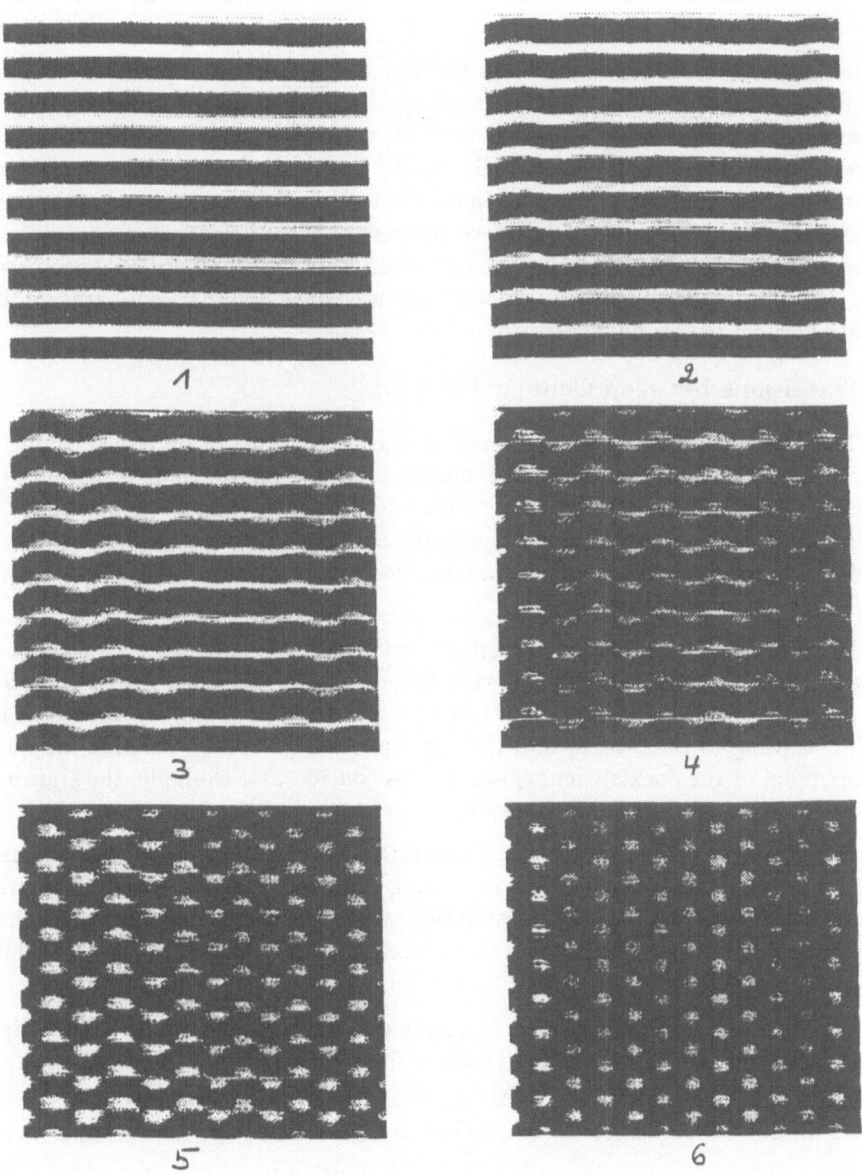

The existence of Goldstone modes or the phase degeneracy reveals that all the symmetries are not lifted by the transition to spatial patterns, and that, beyond the instability, different solutions are equally stable and equally probable. Defects are links between such different solutions that may develope in the system. In cellular patterns, besides the dislocationlike or codimension two defects, the most common ones are wall-like or codimension one defects.

Let us come back, for example, to the case of the hexagonal planforms. Besides the global phase symmetry which only affects the orientation of the texture, one has to take into account the $A \to A$, $B \to -B$, $C \to -C$ symmetry. Effectively, the patterns defined by (R, R, R) and (R, -R, -R) are equivalent and the wall defects, or anti-phase-boundaries, which connect these two structures are such that their core has the structure $(\sqrt{\epsilon}, 0, 0)$. Hence the core of these defects correspond to rolls embedded between anti-phase hexagonal patterns. In three-dimensional geometries, b.c.c. antiphases also exist, and the walls or anti-phase-boundaries separating them should have an hexagonal structure.

3. Transitions between Cellular Patterns

The various transition mechanisms between structures of different symmetries may be illustrated on the dynamical models described above. Such transitions may be continuous and correspond to secondary bifurcations as in the transitions from roll to bimodal structures in liquid crystal instabilities [8]. In this case, the secondary mode grows continuously from zero and there is a mere exchange of stability between unimodal and bimodal patterns .

Discrete transitions, as the transitions between hexagonal or b.c.c. lattices and rolls or walls structures, occur also very frequently. In this case, the structures are simultaneously stable in some range of the parameters space with an hysteresis loop and, according to the system and the experimental set-up, a transition may occur in any point of the coexistence region. Let us consider, for example, the transitions between stripes, or rolls, and hexagons.

In the purely deterministic case, a transition from hexagons to rolls occurs, in the dynamics given by Eq.(1), on increasing the constraint, at $\epsilon = 4v^2/3u$, where the hexagonal solution looses its stability, while, on decreasing the constraint, a transition from rolls to hexagons takes place at $\epsilon = v^2/3u$, where the roll solution looses its stability.

In the presence of fluctuations, since the dynamics (1) is potential, both transitions should occur when the potential takes the same value for stripes and hexagons, namely $\epsilon = \epsilon_s = \frac{4v^2}{3u} \frac{\sqrt{6}-1}{5(\sqrt{6}-2)^2}$. However, the noise intensity is usually too small, in nonequilibrium systems, to be able to induce the transition.

Nevertheless, ϵ_s gives the value of the bifurcation parameter for which a front separating the two structures remains stationnary, i.e. when the two structures may coexist in the system. Effectively, if the two structures appear simultaneously in the system, a front separating them will move towards the rolls for $\epsilon < \epsilon_s$ and towards the hexagons for $\epsilon > \epsilon_s$. This effect is important in systems with frustration, where , for example, astqble structure is incompatible with the boundary conditions, or when a structure is forced at the boundaries. This is the case in convective instabilities, where haxagonal planforms are usually incompatible with the geometry of the system, and where rolls are forced at the boundaries, even in regimes where the hexagons are the only stable pattern [13]. In this case, the first structure to appear, on increasing the bifurcation parameter, corresponds to an hexagonal planform with a boundary layer of rolls. Then, for $\epsilon \geq \epsilon_s$, the rolls invade the sample in a forced nucleation process [14]. Defects such as dislocations or antiphase boundaries can also trigger the transition from hexagons to rolls in a similar wettinglike phenomenon where the roll structure which forms the core of the defect expands. for $\epsilon \geq \epsilon_s$, at the expenses of the surrounding hexagonal structure [16].

Finally, as shown in Fig.9, transitions from rolls to hexagons may occur through the zig-zag instability of the rolls. This may happen by the lowering of the wavenumber of the rolls in the whole coexistence region between rolls and hexagons (cf. Fig.4)

4. Systems with Competing Turing-Hopf Instabilities

During the last years, the interest for pattern formation in nonlinear chemical systems strongly increased due, on one side, to the experimental observation of genuine Turing patterns [15, 16], and, on the other side, to the possible interpretation of these results by reaction-diffusion models [2]. In particular, the analytical and numerical analysis of different dynamical models presenting a Turing instability confirmed the generic aspects of pattern formation in the vicinity of such an instability [17, 18, 19, 20]. Furthermore, it was shown that specific aspects imposed by experimental requirements, such as the spatial variations of the control parameters induced by the feeding processes, can also be described in the framework of the amplitude equation formalism [21]. These variations allow for example the coexistence of structures of different symmetries in the same reactor and also the existence of localized structures embedded in structures of other types [22]. They also affect the transitions between different structures by selecting their orientation and symmetries [23].

An interesting aspect of many reaction-diffusion systems is that they present both Turing and Hopf instabilities. When the diffusion coefficients of the activator and the inhibitor are very different, these instabilities are usually well separated and there is no interaction between spatial and temporal modes. However, on varying these diffusion coefficients, the Turing and Hopf instability thresholds may be moved close together in such a way that complex spatio-temporal behaviors may arise as a result of the interaction between spatial and temporal modes. This situation may also be realized experimentally in the so-called CIMA reaction performed in gel reactors

[24]. Effectively, it is the formation of a complex of reduced mobility between iodide and starch that allows the separation between Turing and Hopf bifurcations in this reaction. As a result, on acting on this complex. the two instability thresholds can be brought close together in order to realize a codimension two point. Although degenerate bifurcations have been studied from the theoretical point of view by various authors (e.g. [25, 26, 27, 28, 29]), and experimentally in the framework of convection in binary mixtures and in Taylor-Couette instabilities [30, 31], such a situation has been encountered and analyzed only recently in nonlinear chemical systems . The most interesting findings are perhaps the observation of localized oscillations embedded in Turing patterns and vice-versa. This led, for example, to the characterization of what has been called the "chemical flip-flop" [32]:

From the theoretical point of view, the coupled amplitude equations describing interacting Turing and Hopf bifurcations may be easily derived, either from symmetry arguments or from multiple scale analysis performed on kinetic models . The stability of the possible structures (pure Turing, pure Hopf, mixed modes) may then be explicitly analyzed and various experimental observations may be recovered in the numerical analysis of these equations [33]. Furthermore, new types of complex spatio-temporal behaviors induced by the phase instability of mixed modes have been observed in this framework [33].

I will review here the different aspects of pattern formation, selection and stability as wel l as the defect behavior in reaction-diffusion systems presenting interacting Turing and Hopf bifurcations in the framework of amplitude equations for one and two-dimensional geometries. Besides the generic or universal properties of the spatio-temporal behavior of such systems, I will also discuss non generic properties specific to chemical systems operating in laterally fed batch reactors.

4.1. AMPLITUDE EQUATIONS FOR INTERACTING TURING-HOPF MODES.

The general form of the amplitude equations for structures constructed on m pairs of unstable modes close to a pattern forming instability of the Turing type was presented in a preceeding section (eq.). On the other hand, it is well known that, close to a Hopf bifurcation, the evolution equations for the complex amplitudes H of the unstable modes take the form of the complex Ginzburg-Landau equation :

$$\partial_t H = \epsilon H + (1 + i\alpha)\Delta H - (1 + i\beta)|H|^2 H \tag{30}$$

In a codimension two situation, where the Hopf and Turing instabilities are close together, spatial and temporal modes both evolve on long space and time scales and are thus supposed to interact. Their interactions need to respect the symmetries of the problem, i.e. time and space translations and space rotation. As a result the corresponding coupled amplitude equations may be written, for two-dimensional systems, as :

$$\partial_t H = \epsilon H + (1 + i\alpha)\Delta H - (1 + i\beta)|H|^2 H - (\rho + i\delta)\Sigma_i |A_i|^2 H \qquad (31a)$$

$$\partial_t A_i = \mu A_i + (2q_c \vec{k}_i.\vec{\nabla})^2 A_i + v\Sigma_j \Sigma_k A_j^* A_k^* \delta(\vec{k}_i + \vec{k}_j + \vec{k}_k) - |A_i|^2 A_i -$$

$$\Sigma_{i \neq j} \gamma_{ij} |A_j|^2 A_i - \lambda |H|^2 A_i \qquad (31b)$$

4.2. PATTERN SELECTION FOR CODIMENSION TWO TURING-HOPF BIFURCATIONS.

4.2.1. One-dimensional geometries.

The amplitude equations (31) (with $\gamma_{ij} = \gamma$ and $v > 0$) admit three types of uniform steady states corresponding to perfect structures :
1) pure Hopf patterns ($H = \sqrt{\epsilon} exp - i\beta\epsilon t, A_i = 0$), which exist for $\epsilon > 0$ and are stable for $\mu - \lambda\epsilon < 0$;
2) pure Turing patterns ($H = 0, A_i \neq 0$) corresponding to wall structures ($H = 0, A_1 = \sqrt{\mu} exp i\phi, A_{i1} = 0$), which exist for $\mu > 0$ and are stable for $\epsilon - \rho\mu < 0$,
3) mixed modes ($H = \sqrt{\epsilon - \rho\mu} exp - i[\beta(\epsilon - \rho\mu) + \delta(\mu - \lambda\epsilon)]t, A_1 = \sqrt{\mu - \lambda\epsilon} exp i\phi$, $A_{i1} = 0$), which exist in the range $\lambda\epsilon < \mu < \epsilon/\rho$. and are only stable when $\lambda\rho < 1$.

The corresponding phase diagrams is represented in fig.10:

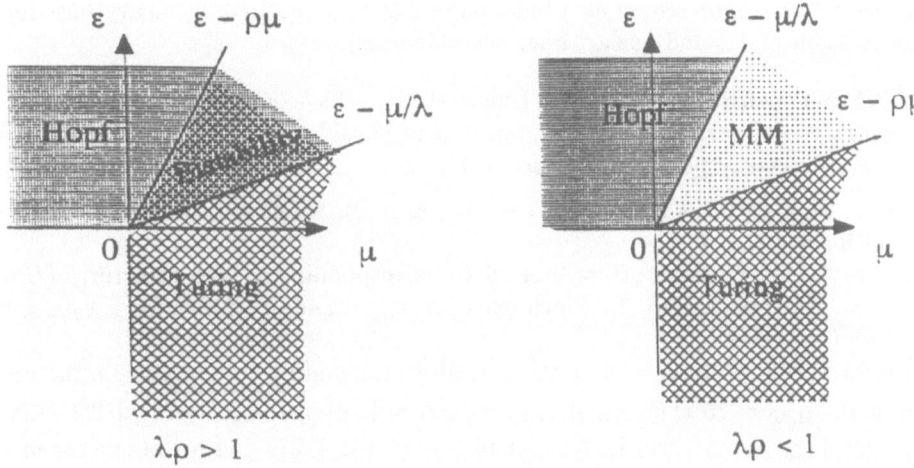

Figure 10 : Stability ranges of pure Turing, pure Hopf and mixed mode solutions for codimension two Turing-Hopf bifurcations in one-dimensional systems.

24

and the associated bifuration diagrams are drawn in fig.11.

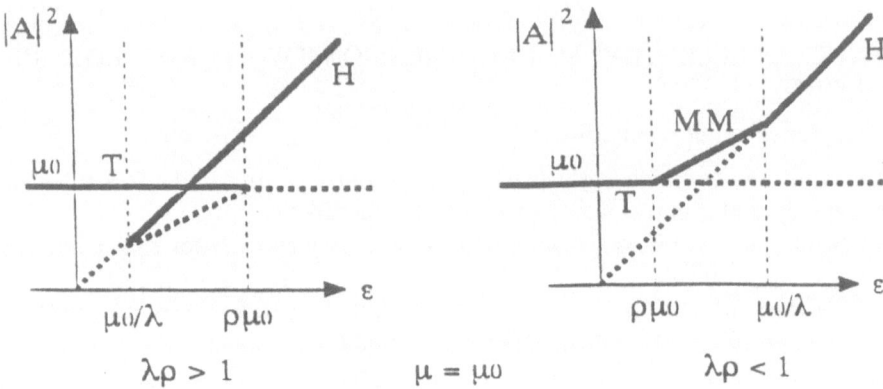

Figure 11 : Bifurcation diagram for pure Turing, pure Hopf and mixed mode solutions of eq.(31) in one-dimensional geometries (the square of the amplitude modulus of each mode is represented as a function of ϵ at fixed positive μ ; heavy lines represent stable states and dashed lines unstable ones).

4.2.2. *Two-dimensional geometries.* Due to the possible existence of hexagonal planforms in two-dimensional systems, one has to consider now five types of uniform steady states corresponding to perfect structures :

1) pure Hopf patterns ($H = \sqrt{\epsilon}exp - i\beta\epsilon t, A_i = 0$), which exist for $\epsilon > 0$ and are stable for $\mu - \lambda\epsilon < 0$;
2) pure Turing patterns ($H = 0, A_i \neq 0$) corresponding to wall structures ($H = 0, A_1 = \sqrt{\mu}expi\phi, A_{i\neq1} = 0$), which exist for $\mu > 0$ and are stable for $\epsilon - \rho\mu < 0$, and $v^2/(1-\gamma)^2$,
3) pure Turing patterns ($H = 0, A_i \neq 0$) corresponding to hexagonal structures ($H = 0, A_{1,2,3} = Rexpi\phi_{1,2,3}, A_{i\neq1,2,3} = 0, R = [v + \sqrt{v^2 + 4(1 + 2\gamma)\mu}]/2(1 + 2\gamma), \phi_1 + \phi_2 + \phi_3 = 2\pi$) , which exist for $m > -v^2/4(1 + 2\gamma)$ and are stable for $m < 4v^2/(1-\gamma)^2$ and for $\epsilon - 3\rho R^2 = \epsilon - \rho\mu^* < 0$,
4) mixed Hopf-wall modes ($H = \sqrt{(\epsilon - \rho\mu)}exp - i[\beta(\epsilon - \rho\mu) + \delta(\mu - \lambda\epsilon)]t$, $A1 = \sqrt{(\mu - \lambda\epsilon)}expi\phi, A_{i\neq1} = 0$), which exist in the range $\lambda\epsilon < \mu < \epsilon/\rho$, and are only stable when $\lambda\rho < 1$.

5) mixed Hopf-hexagon modes $(A_{1,2,3} = R_M expi\phi_{1,2,3}$, $A_{i\neq 1,2,3} = 0$, $R_M = [v + \sqrt{v^2 + 4(1 + 2\gamma - 3\lambda\rho)(\mu - \lambda\epsilon)}]/2(1 + 2\gamma - 3\lambda\rho)$, $\phi_1 + \phi_2 + \phi_3 = 2\pi$, $H = \sqrt{(\epsilon - 3\rho R_M^2)}exp - i[\beta\epsilon + 3(\delta - \beta\rho)R_M^2]t)$, which exist, when $\lambda\rho < 1$, in the range $\frac{3\rho}{2(1+2\gamma)^2}[2(1 + 2\gamma)\mu + v^2 + v\sqrt{v^2 + 4(1 + 2\gamma)\mu}] < \epsilon < \frac{\mu}{\lambda} + \frac{v^2}{4\lambda(1+2\gamma-3\lambda\rho)}$, and are only stable for $\lambda\epsilon - \frac{v^2}{4\lambda(1+2\gamma-3\lambda\rho)} < \mu < \lambda\epsilon + \frac{v^2}{(1-\gamma)^2}\frac{(4+2\gamma-3\lambda\rho)(2+4\gamma-3\lambda\rho)}{4\lambda(1+2\gamma-3\lambda\rho)}$.

The corresponding phase diagram looks like :

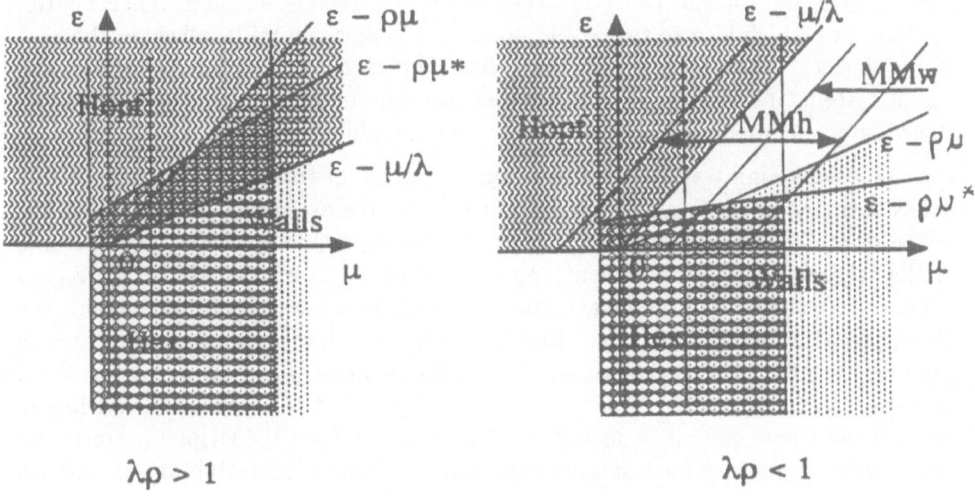

$\lambda\rho > 1$ $\lambda\rho < 1$

Figure 12 : Stability ranges of pure Turing (walls or hexagons), pure Hopf and mixed mode solutions (MMw : Hopf-walls, MMh : Hopf-hexagons) for codimension two Turing-Hopf bifurcations in two-dimensional systems.

4.3. DEFECTS, DEFECT BIFURCATIONS AND LOCALIZED STRUCTURES.

4.3.1. *One-dimensional geometries.*

Besides the usual defects of pure Hopf or Turing structures like wave sources (or one-dimensional spirals) or anti-phase boundaries, new types of defects and defect behavior may be observed, due to the proximity of the Turing and Hopf bifurcations. When $\lambda\rho > 1$, wall or stripe structures and limit cycle oscillations may be simultaneously stable. Domains of each of these structures may thus coexist in the system and are separated by fronts which move or not, according to the value of the control parameters of the system.

Line defects, allowed by the symmetries of the order parameterlike variables, may also appear, and, due to the proximity of the two bifurcations, bifurcations of the defect core should occur, triggering the transition between the different structures, via a mechanism discussed in another context by Coullet et al. [34].

Let us discuss the two possible cases here :

- in the region where limit cycle oscillations are the only stable structures ($\epsilon > 0$, $\mu - \lambda\epsilon < 0$), a typical line defect is a hole or a stable front separating domains where the amplitude of the oscillations takes the equivalent opposite values $+\sqrt{\epsilon}$ and $-\sqrt{\epsilon}$. In the region where $\mu < 0$, the core of the defect, corresponding to a zero amplitude for both the Turing and Hopf modes, is stable and plays the role of a wave source, emiting alternatively left and right traveling waves with a π dephasing, as a one-dimensional spiral. However, when $\mu > 0$. it becomes unstable versus spatial modulations which lead to the development of a wall structure localized in the core of the defect. When μ increases, the defect core increases and finally invades the whole system, inducing the transition between the oscillations and a steady wall structure for μ smaller than $\lambda\epsilon$, i.e. before the instability of uniform oscillations.

- the corresponding transition mechanism should be observed starting from a line defect in a striped pattern : in the region where stripes are stable ($\mu > 0$, $\epsilon - \rho\mu < 0$), the typical line defect is a stable front separating domains where the amplitude of the stripes takes the equivalent opposite values $+\sqrt{\mu}$ and $-\sqrt{\mu}$. In the region where $\epsilon < 0$, the core of the defect, corresponding to a zero amplitude for both the Turing and Hopf modes, is stable. However, when $\epsilon > 0$, it becomes unstable versus temporal oscillations which lead to the development of reentrant waves localized in the core of the defect. When ϵ increases, the defect core increases and finally invades the whole system, inducing the transition between the striped pattern and limit cycle oscillations for ϵ smaller than $\rho\mu$, i.e. before the instability of uniform striped structures.

Several defects among the ones described above have been observed in numerical simulations of dynamical models of the reaction-diffusion type describing nonlinear chemical systems [35] and some examples are given in fig.13. Defect mediated transitions between patterns of different symmetries have been observed in convection experiments [13, 36] . Similarly, the defect mediated transitions between temporal and spatial patterns, as described above, should be observable in chemical systems displaying Turing and Hopf instabilities.

4.3.2. Two-dimensional geometries.

In this case, the same situations as in one-dimensional systems also arise, but new possibilities occur, associated with the existence of hexagonal planforms. For example, the typical line defect of hexagonal patterns corresponds to a change of sign of the amplitudes of two of the underlying modes, leaving a roll or wall structure in the core of this defect.

When $\lambda\rho > 1$, there is a tristability domain where walls, hexagons and oscillations are simultaneously stable. Coexisting domains corresponding to the three structure should thus be possible as well as the generalization of the defect behavior described in the preceeding paragraph.

Figure 13 : One-dimensional Turing-Hopf domains for the Brusselator model in a region of the parameter space where the pure Turing and Hopf modes are stable whereas the mixed mode is unstable [35]. The graphs are space-time plots with time running upwards vertically and space horizontally : (a)(up-left) pinned front between Turing and Hopf domains, the Turing structure emits waves in the Hopf region; (b)(up-right) Turing structure invading the Hopf domain; (c)(down-left) stable Turing droplet embedded in a Hopf domain; (d)(down-right) stable Hopf droplet embedded in a Turing structure.

When $\lambda\rho > 1$, mixed modes may exist in a well defined parameter range (see above). These mixed modes may be amplitude stable, but nevertheless undergo a phase instability of the Benjamin-Feir type leading to a form of defect mediated turbulence [37]. In their phase stability regime, these mixed modes are expected to display topological defects. Since there is no direct coupling between the phases of the Hopf and Turing modes which are only coupled through the modes amplitudes, the topological defect should be of the (n, m) type, where n is the topological charge of a singularity in the phase field of the Hopf mode, and m is the topological charge of a singularity in the phase field of the Turing mode. Hence the elementary defects of these patterns are: (a) the (1,0) defect which corresponds to a spiral wave in the underlying Hopf mode (cf. fig.14); (b) in the case of mixed Hopf-walls modes, the (0,1) defect, which corresponds to a dislocation in the underlying Turing mode ; (c) in the case of mixed Hopf-hexagons modes, the corresponding defect is a dislocation in the underlying hexagonal pattern (due to the coupling of the individual phases through the amplitudes of the modes, such dislocations may of course radiate traveling waves in the Hopf mode, as in the case of standing wave patterns [38]).

Figure 14a : Spiral (or (1,0)) defect of a mixed Hopf-wall mode (the amplitude ratio of the Turing and Hopf modes is 1.35, $q_c = 1$, $k_{spiral} = 0.35$)

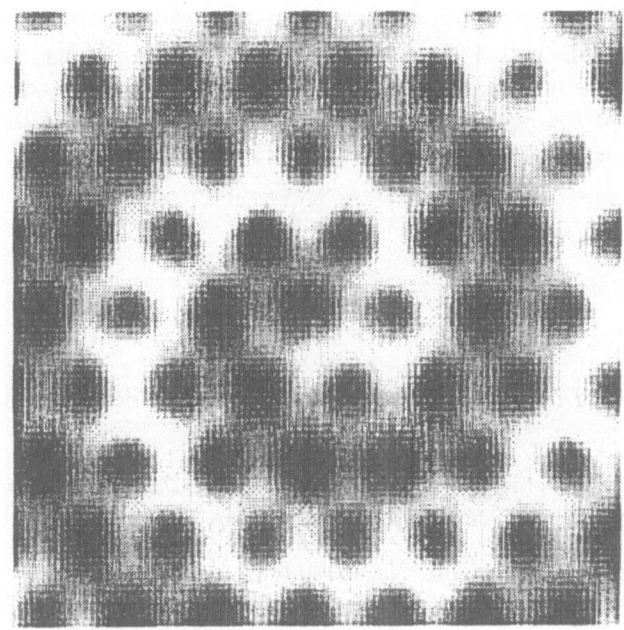

Figure 14b : Spiral (or $(1,0)$) defect of a mixed Hopf-hexagon mode (the amplitude ratio of the Turing and Hopf modes is 0.75, $q_c = 1$, $k_{spiral} = 0.35$)

5. Conclusions.

Different transition mechanisms between nonequilibrium patterns of different symmetries have been described in the framework of weakly nonlinear analysis around instability points. Some of them are reminiscent of the nucleation phenomena of phase transition, when the corresponding amplitude equations have a variational structure. Others are related to specific boundary effects or experimental set-ups. Defects may also trigger transitions between different spatio-temporal patterns through bifurcations in the defect core. This phenomenon leads to a rich variety of possibilities, especially in the case of competing instabilities and non variational dynamics, and a few of them were described here. They should be observable in nonlinear reaction-diffusion systems with neighbouring Turing and Hopf instabilities, a situation which can now be experimentally realized in the chemical systems where genuine Turing patterns have been observed.

Acknowledgments.

Fruitful discussions with P.Borckmans, P.Coullet, P.De Kepper, G.Dewel and A.De Wit are gratefully acknowledged. This work has been supported by the grant CI1*-CT92-0006 of the Science Programme of the Commission of the European Community. DW is Director of Research at the National Fund for Scientific Research of Belgium.

References

[1] Manneville P., *Dissipative Structures and Weak Turbulence*, Academic Press, Boston, 1990.

[2] Baras F. and Walgraef D. eds., *Nonequilibrium Chemical Dynamics : from experiment to microscopic simulation*, Physica A **188** , 1992.

[3] Coullet P. and Huerre P. eds., *New Trends in nonlinear dynamics and pattern forming phenomena : The geometry of non-equilibrium*, Plenum, New York, 1990.

[4] De Wit A., Dewel G., Borckmans P. and Walgraef D., Physica **D 61**, 289-296, (1992).

[5] Walgraef D., Borckmans P. and Dewel G., Nature **318**, 606, (1985).

[6] Gertsberg V.L. and Sivashinsky C.I., Prog.Theor.Phys., **66**, 1219, (1981).

[7] Nicolis G. and Prigogine I., *Self-Organization in Non Equilibrium Systems*, Wiley, New York, 1977.

[8] Guazzelli E., Dewel G., Borckmans P., Walgraef D., Physica **D35**, 220, (1989).

[9] Brand H.R. in *Patterns, Defects and Materials Instabilities*, D.Walgraef and N.M.Ghoniem eds., Kluwer Academic Publishers, Dordrecht, 1990, pp.25-34.

[10] Croquette V.and Schosseler F., J.Physique (Paris), **43**, 1183, (1982).

[11] Lauzeral J., Metens S. and Walgraef D., Europhys.Lett. **24**, 707, (1993).

[12] Sushchik M.M. and Tsimring L.S., "The Eckhaus Instability in Hexagonal Patterns" , preprint; 1993.

[13] Ciliberto S., Pampaloni E. and Perez-Garcia C., Phys.Rev.Lett., **61**, 1198, (1988).

[14] Walgraef D. in *Instabilities and Nonequilibrium Structures III*, E. Tirapegui and W. Zeller eds., Kluwer, Dordrecht, 1991, pp.269-282.

[15] V. Castets, E. Dulos, J. Boissonade, P. De Kepper, Phys. Rev. Lett. **64**, 2953-2956 (1990)

[16] Q. Ouyang, H. Swinney, Nature **352**, 610, (1991).

[17] V.Dufiet and J.Boissonade, Physica **A188**. 158-171 (1992).

[18] I.Lengyel and I.R.Epstein, J.Am.Chem.Soc. **251**, 650-652 (1991)

[19] P.Borckmans, G.Dewel, A.De Wit and D.Walgraef, Entropie **164**, 83 (1991).

[20] A. De Wit, G. Dewel, P.Borckmans and D. Walgraef, Physica **D61**, 289-296, (1992).

[21] G.Dewel and P.Borckmans, Phys.Lett . **A138**, 189-192 (1989).

[22] P.Borckmans, A.De Wit and G.Dewel, Physica **A188**, 137-157 (1992).

[23] A.De Wit, P.Borckmans and G.Dewel, in *Instabilities and Nonequilibrium Structures IV*, E.Tirapegui and W.Zeller eds., Kluwer, Dordrecht, 247-258 (1993).

[24] J.J.Perraud, K.Agladze, E.Dulos and P.De Kepper, Physica **A188** , 1-16 (1992).

[25] P.Holmes, Ann.N.Y.A.S. **357**, 473-488 (1980).

[26] J.Guckenheimer, in *Dynamical Systems and Turbulence*, D.Rand and L.Young eds., Springer, Berlin, 1981.

[27] J.Guckenheimer and P.Holmes, *Nonlinear Oscillations, Dynamical Systems and Bifurcations of Vector Fields*, Springer, Berlin, 1983.

[28] H.R.Brand, P.C.Hohenberg and V.Steinberg, Phys.Rev. **A30**, 2548-2561 (1984).

[29] W.Zimmerman, D.Ambruster, L.Kramer and W.Kuang, Europhys.Lett. **6**, 505, (1988).

[30] I.Rehberg and G.Ahlers, Phys.Rev.Lett. **55**. 500-503 (1985).

[31] T.Mullin and T.J.Price, Nature **340**, 294-296 (1988).

[32] J.J.Perraud, A.De Wit, E.Dulos, P.De Kepper, G.Dewel and P.Borckmans, Phys.Rev.Lett. **71** , 1272-1275 (1993).

[33] A.De Wit, *"Brisure de symetrie spatiale et dynamique spatio-temporelle dans les systemes reaction-diffusion"*, PhD Thesis. Free University of Brussels, 1993.

[34] P.Coullet, L.Gil and D.Repaux, in *Instabilities and Nonequilibrium Structures II*, E.Tirapegui and D.Villarroel eds., Kluwer, Dordrecht, 189-205 (1989).

[35] P.Borckmans, G.Dewel, A.De Wit and D.Walgraef, *"Turing Bifurcations and Pattern Selection*, to appear in *Chemical Waves and Patterns*, R.Kapral and K.Showalter eds., Kluwer, Dordrecht.

[36] S.Ciliberto, P.Coullet, J.Lega, E.Pampaloni and C.Perez-Garcia, Phys. Rev. Lett. **65**, 2370, (1990).

[37] A.De Wit, G.Dewel and P.Borckmans, Phys.Rev., **E48**, 4191, (1993).

[38] J.Lega, in *Patterns, Defects and Materials Instabilities*, D.Walgraef and N.M.Ghoniem eds., Kluwer Academic Publishers, Dordrecht, 1990, pp.7-24.

Galilean and Relativistic Nonlinear Wave equations: an Hydrodynamical Tool?

Malek ABID[1], *Marc BRACHET*[2], *Fabrice DEBBASCH*[3] *and Caroline NORE*[2]

[1] Institut de Recherche sur les Phénomènes Hors Equilibre.
UMR CNRS et Université d'Aix-Marseille I, service 252, Centre St-Jérôme,
13397 Marseille Cedex 20, France
[2] Laboratoire de Physique Statistique
CNRS URA 1306, ENS Ulm
24 Rue Lhomond, 75231 Paris Cedex 05, France
[3] Laboratoire de Radioastronomie
ENS Ulm, 24 Rue Lhomond, 75231 Paris Cedex 05, France
Observatoire de Paris
Section de Meudon, place J.-Jansen
F-92195 Meudon Cedex, France

The connexion of nonlinear wave equations with the dynamics of barotropic fluids by Madelung's transformation is reviewed in the case of fluids with arbitrary equations of state. Numerical simulations of the Nonlinear Schrödinger Equation (NLSE) reproducing the instabilities of non rotating and rotating cylindrical jets are presented. It is shown that NLSE, a dynamical model of superflows, reproduces many flow features usually obtained in the context of Euler or Navier-Stokes equations.

A generalization to relativistic wave equations and superfluids is reviewed. The Galilean limit is shown to be NLSE. A model for a relativistic self-gravitating superfluid is obtained by minimally coupling the wave equation to Einstein's gravity. Applications corresponding to static stars and isotropic cosmologies are discussed.

1 Introduction

It has been known for some time that Madelung's transformation maps the (defocusing) Nonlinear Schrödinger Equation (NLSE) into hydrodynamical equations for a compressible fluid with dispersion. In the early 80's [1], E. Spiegel emphasized that the dynamics of irrotational barotropic fluids, with arbitrary equations of state, can be linked to NLSE-type equations with suitable nonlinearities. Furthermore NLSE contains topological defects that are known to follow Eulerian dynamics in the incompressible limit [2, 3, 4]. These topological defects correspond to the quantum vortices of superfluid Helium [5]. In this context, NLSE is the correct dynamical equation of motion for superfluids [6].

More recently [7, 8, 9, 10, 11], numerical simulations of NLSE have been used to probe its ability to reproduce typical fluid dynamical phenomena. One of the motivations behind this recent surge of interest is the possibility of understanding the details of intricate dynamical mechanisms, such as vortex nucleation [7] and vortex-sound interaction [12], in superfluid Helium.

E. Tirapegui and W. Zeller (eds.), Instabilities and Nonequilibrium Structures V, 33–52.
© *1996 Kluwer Academic Publishers.*

A special relativistic generalization of the NLSE dynamics, using the nonlinear Klein Gordon equation (NLKGE), has been studied by J.C. Neu [3, 13] with emphasis on the derivation of equations of motion for vortices, without taking into account the acoustic sector of the dynamics. In a general relativistic framework, static solutions of this wave equation describing boson stars have already been considered by various authors [14, 15], but without a Madelung-like correspondence to usual hydrodynamics. Such a correspondence was recently given in [16].

The purpose of this paper is to synthesize some of the recently obtained results. It is organized as follows. Section 2 is devoted to the fluid dynamical representation of NLSE. Numerical simulations of NLSE in the context of free-shear flows instabilities are presented in section 3. Section 4 concerns the special relativistic generalization of the material presented in section 2. Section 5 contains applications to general relativistic self-gravitating fluids. And section 6 is our conclusion.

2 Fluid dynamical representation of NLSE

In this section, we present the hydrodynamical form of NLSE with an arbitrary nonlinearity, corresponding to a barotropic fluid with an arbitrary equation of state. Basic hydrodynamical features such that acoustic propagation and time independent solutions are also discussed.

2.1 Formal correspondence

Perhaps the most direct way to understand the scope and generality of the connexion between nonlinear waves and fluid dynamics is to consider the following action [1] :

$$\mathcal{A} = 2\alpha \int dt \left\{ d^3x \left(\frac{i}{2} \left(\overline{\psi} \frac{\partial \psi}{\partial t} - \psi \frac{\partial \overline{\psi}}{\partial t} \right) \right) - \mathcal{F} \right\} \tag{1}$$

with

$$\mathcal{F} = \int d^3x \left(\alpha |\nabla \psi|^2 + f(|\psi|^2) \right) \tag{2}$$

where $\psi(\boldsymbol{x}, t)$ is a complex wave field and $\overline{\psi}$ its complex conjugate, α is a positive real constant and f is a polynomial in $|\psi|^2 \equiv \overline{\psi}\psi$ with real coefficients :

$$f(|\psi|^2) = -\Omega|\psi|^2 + \frac{\beta}{2}|\psi|^4 + f_3|\psi|^6 + \ldots + f_n|\psi|^{2n} \tag{3}$$

The NLSE is the Euler-Lagrange equation of motion for ψ corresponding to (1), it reads

$$\frac{\partial \psi}{\partial t} = -i \frac{\delta \mathcal{F}}{\delta \overline{\psi}},$$

or

$$\frac{\partial \psi}{\partial t} = i(\alpha \nabla^2 \psi - \psi f'(|\psi|^2)) \tag{4}$$

Madelung's transformation [5, 1]

$$\psi = \sqrt{\rho}\exp\left(i\frac{\varphi}{2\alpha}\right) \tag{5}$$

maps the nonlinear wave dynamics of ψ into equations of motion for a fluid of density ρ and velocity $\boldsymbol{v} = \nabla\varphi$. Indeed with the help of (5), (1) can be written

$$\mathcal{A} = -\int dt d^3x \left(\rho\frac{\partial\varphi}{\partial t} + \frac{1}{2}\rho(\nabla\varphi)^2 + 2\alpha f(\rho) + \frac{1}{2}(2\alpha\nabla(\sqrt{\rho}))^2\right) \tag{6}$$

and the corresponding Euler-Lagrange equations of motion read

$$\frac{\partial\rho}{\partial t} + \nabla\cdot(\rho\boldsymbol{v}) = 0 \tag{7}$$

$$\frac{\partial\varphi}{\partial t} + \frac{1}{2}(\nabla\varphi)^2 + 2\alpha f'(\rho) - 2\alpha^2\frac{\Delta\sqrt{\rho}}{\sqrt{\rho}} = 0 \tag{8}$$

Without the last term of (8) (the so-called "quantum pressure" term), these equations are the continuity and Bernoulli equations [17] for an isentropic, compressible, irrotational fluid.

It is possible to use this identification to define the corresponding "thermodynamical functions". Being isentropic, the fluid is barotropic, and there is thus only one independent thermodynamical variable. The Bernoulli equation readily gives the fluid's enthalpy *per unit mass* as

$$h = 2\alpha f'(\rho). \tag{9}$$

On the other hand, the $\frac{1}{2}\rho(\nabla\varphi)^2$ term of (6) is seen to correspond to kinetic energy. Thus the fluid's internal energy *per unit mass* is given by

$$e = \frac{2\alpha f(\rho)}{\rho}. \tag{10}$$

The general thermodynamical identity

$$h = e + p/\rho, \tag{11}$$

gives the fluid's pressure

$$p = 2\alpha(\rho f'(\rho) - f(\rho)). \tag{12}$$

The physical dimensions of the variables used in (2) and (3) are fixed by the following considerations. Madelung's transformation (5) imposes that $[|\psi|^2] = [\rho] = M L^{-3}$ and $[\alpha] = L^2 T^{-1}$. Using (10), one gets $[f(\rho)/\rho] = T^{-1}$ and thus, from (3), $[\Omega] = T^{-1}$, $[\beta] = T^{-1}\rho^{-1}$ and $[f_i] = T^{-1}\rho^{1-i}$. Note that, in the case of a Bose condensate of particles of mass m, α has the value $\hbar/2m$ [6].

2.2 Acoustic regime

Dispersion relation The nature of the extra quantum pressure term in (8) is best understood by looking at the dispersion relation corresponding to acoustic waves propagating around a constant density level ρ_0. Setting $\rho = \rho_0 + \delta\rho$ (with $f'(\rho_0) = 0$), $\nabla\varphi = \delta u$ in (7) and in the gradient of (8), one gets (keeping only the linear terms) :

$$\partial_t \delta\rho + \rho_0 \nabla \delta u = 0$$

$$\partial_t \delta u + 2\alpha f''(\rho_0)\nabla \delta\rho - 2\alpha^2 \Delta \frac{\nabla \delta\rho}{2\rho_0} = 0$$

or

$$\partial_t{}^2 \delta\rho = 2\alpha\rho_0 f''(\rho_0)\Delta\delta\rho - \alpha^2 \Delta^2 \delta\rho.$$

The dispersion relation for an acoustic wave $\delta\rho = \epsilon(\exp(i(\omega t - \mathbf{k}\cdot\mathbf{x})) + c.c.)$ (with $\epsilon \ll 1$) is thus

$$\omega = \sqrt{2\alpha\rho_0 f''(\rho_0)\mathbf{k}^2 + \alpha^2 \mathbf{k}^4} \tag{13}$$

This relation shows that the quantum pressure has a dispersive effect that becomes important for large wave numbers. For small wavenumbers, one recovers the usual propagation, with a sound velocity given by

$$c = \left(\frac{\partial p}{\partial \rho}\right)^{\frac{1}{2}} = \sqrt{2\alpha\rho_0 f''(\rho_0)}.$$

The length scale $\xi = \sqrt{\alpha/(\rho_0 f''(\rho_0))}$ at which dispersion becomes noticeable is known as the "coherence length".

Note that, in the context of superfluid helium modelling, it has been suggested [18] to replace (2) by a non local term of the form

$$\mathcal{F}_{nl} = \int d^3x\,\alpha|\nabla\psi|^2 + \int d^3x_1 d^3x_2 \frac{1}{2}|\psi|^2(\mathbf{x_1})V(|\mathbf{x_1}-\mathbf{x_2}|)|\psi|^2(\mathbf{x_2}).$$

It is easy to check that with such a term, one gets the following dispersion relation

$$\omega = \sqrt{2\alpha\rho_0\hat{V}(\mathbf{k})\mathbf{k}^2 + \alpha^2 \mathbf{k}^4}$$

where $\hat{V}(\mathbf{k})$ is the fourier transform

$$\hat{V}(\mathbf{k}) = \int d^3x\,e^{i\mathbf{k}\cdot\mathbf{x}}V(|\mathbf{x}|).$$

The function V can then be chosen such that the dispersion relation fits the one experimentally known for helium. Let us remark here that the same goal can be achieved in a local framework by Taylor expanding the function $\hat{V}(\mathbf{k})$. This amounts to add to (2) dispersive terms of the form :

$$\mathcal{F}_d = \int d^3x\,|\psi|^2 g(\Delta)|\psi|^2,$$

where $g(\Delta)$ is a polynomial in the laplacian operator. We shall not consider such dispersive terms in the rest of this paper.

Nonlinear acoustics The description given by linear acoustic can be somewhat improved by including the dominant nonlinear effects. Such an equation was derived in [9].

Numerical simulations of NLSE in one space dimension using a standard Fourier pseudo-spectral method [19] can be used to study the acoustic regime triggered by an initial disturbance of the form :

$$\psi(x) = 1 + ae^{-\frac{x^2}{l^2}}.$$

A simulation result is displayed in Figures 1. The nonlinear effect present in these Figs can be distinguished from a purely linear dispersive effect by the scale of the generated wave-trains. In the nonlinear case, the scale of the wave-train is much smaller than the scale of the initial perturbation. The pulses travel at supersonic speed. The propagation speed can be explained [9] as a nonlinear renormalization of sound velocity.

The data presented in Figs 1 also show that the shocks that would have appeared under compressible Euler dynamics (i.e. following Eq. (8) without the last term in r.h.s.) have been regularized by the dispersion. There is no evidence of finite-time singularity in our numerics. The spectrum of the solution (data not shown) is well resolved, with a conspicuous exponential tail.

2.3 Time independent solutions

Further insight on the connexion between wave and fluid dynamics can be obtained by considering stationary solutions of the equations of motion. Indeed, by inspection of (1), time independent solutions of NLSE (4), are also solutions of the Real Ginzburg-Landau Equation (RGLE)

$$\frac{\partial \psi}{\partial t} = -\frac{\delta \mathcal{F}}{\delta \overline{\psi}} = (\alpha \nabla^2 \psi - \psi f'(|\psi|^2)). \tag{14}$$

They are thus extrema of the free energy $\mathcal{F}$.

The simplest solution of this type corresponds to a constant density fluid at rest. In this simple case, ψ is constant in space and (14) reads

$$f'(|\psi|^2) = -\Omega + \beta|\psi|^2 + 3f_3|\psi|^4 + \ldots + nf_n|\psi|^{2n-2} = 0. \tag{15}$$

This equation, for given values of the coefficients β and $f_i, i = 3, \ldots, n$ relates the fluid density $|\psi|^2$ to the value of Ω. Note that the Ω term of f does not play a crucial role in the NLSE dynamics. Indeed, it could be removed from the Bernoulli equation (8) by the change of variable $\varphi \to \varphi + 2\alpha\Omega t$ that amounts to a change of phase $\psi \to \psi e^{i\Omega t}$ in NLSE (4). We will however, by convention, not perform these changes of variable, in order that stationary solutions of (14) coincide with stationary solutions of (4). The Ω term of f will thus be fixed by the fluid's density through (15).

38

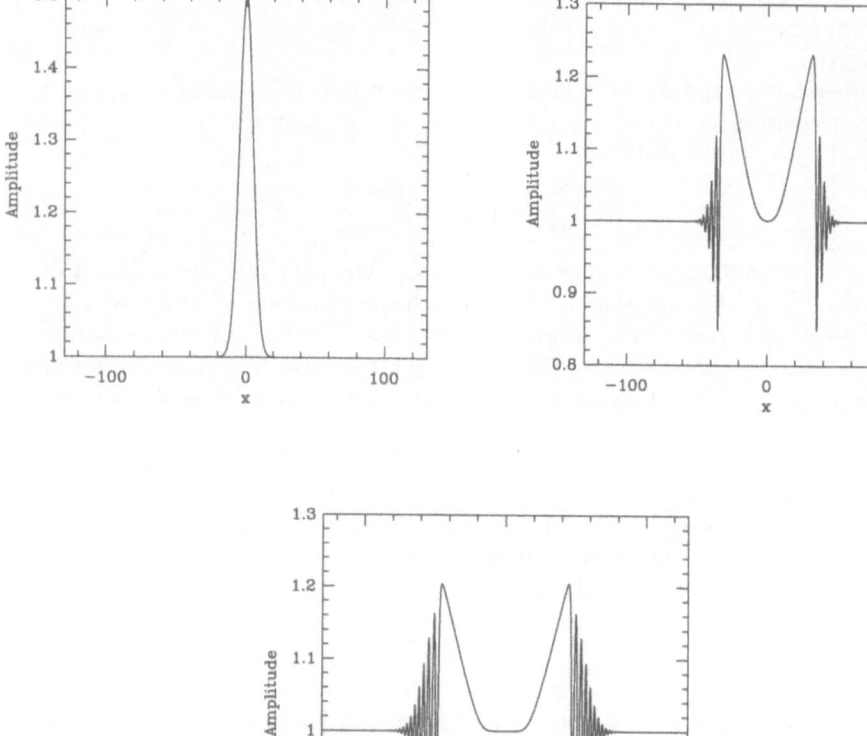

Fig. 1. Numerical integration of NLSE in 1D with an initial perturbation of amplitude ($a = 0.5$) and large width ($l = 10 \ \xi$): (a) amplitude of the initial data. (b) amplitude of the solution at $t = 20$. (c) amplitude of the solution at $t = 30$.

Note that scales significantly smaller than the length scale on the initial data have been generated through nonlinear effects.

Another important type of time-independent solutions of NLSE are the vortex solutions. Madelung's transformation is singular when $\rho = 0$ (i.e. when both $\Re(\psi) = 0$ and $\Im(\psi) = 0$, see Figure 2 (a)). As two conditions are required, the singularities generically happen on points in two dimensions and lines in three dimensions. The circulation of v around such a generic singularity is $\pm 4\pi\alpha$. These topological defects are known in the context of superfluidity as "quantum vortices" [5]. Solutions of (14) with cylindrical symmetry can be obtained numerically [20]. The profile of a vortex is shown on Figure 2 (b). The density admits a horizontal asymptote near the core (see Figure 3 (c)) while the velocity diverges as the inverse of the core distance (see Figure 3 (d)). Then the momentum density ρv is a regular quantity shown on Figure 4 (f).

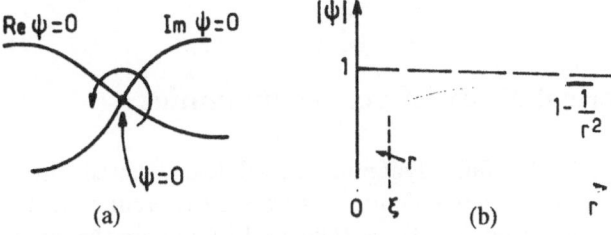

Fig. 2. Nodal point of the condensate wave function and vortex profile.

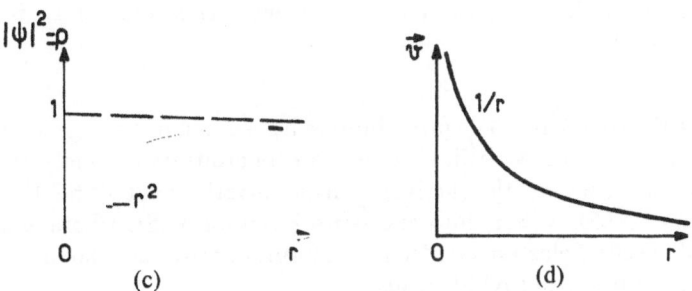

Fig. 3. Vortex density and velocity profile.

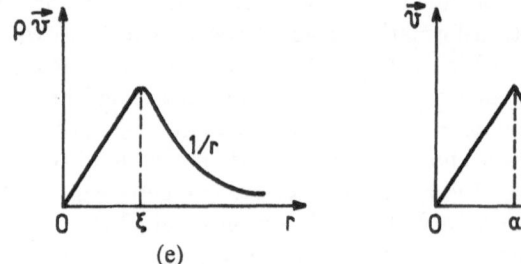

Fig. 4. Quantum vortex momentum versus classical vortex momentum (vorticity patch of radius a).

3 Numerical study of vortex dynamics

Under compressible fluid dynamics, an arbitrary chosen initial condition will generally lead to a regime dominated by acoustic radiation. In order to study vortex dynamics using NLSE, we thus need to prepare the initial data in such a way that the acoustic emission is as small as possible.

3.1 Preparation method

We know that the RGLE (14) starting with an initial data containing a nodal line converges towards the exact time independent vortex solution ψ_v described at the end of section (2.3).

The ARGLE procedure The procedure we have developed is a generalization of this property of RGLE. Our aim is to prepare an arbitrary assembly of moving vortices. To do so, we use the (active) Galilean invariance of NLSE that maps any solution of NLSE $\psi(\boldsymbol{x}, t)$ into another solution of NLSE whose associated velocity and density fields are Galilean transforms of those associated to ψ. The Galilean transformation of NLSE reads:

$$\psi(\boldsymbol{x}, t) \rightarrow \psi(\boldsymbol{x} - \boldsymbol{u}t, t) \, \exp\left(i\left(\frac{\boldsymbol{u}}{2\alpha} \cdot \boldsymbol{x} - \frac{u^2}{4\alpha}t\right)\right),$$

($\boldsymbol{u}$ being the constant velocity of the boost). Thus, the initial solution $\tilde{\psi}(\boldsymbol{x}) = \psi_v(\boldsymbol{x}) \exp(i\frac{\boldsymbol{u}}{2\alpha} \cdot \boldsymbol{x})$ corresponds to a vortex translating with velocity $\boldsymbol{u}$. It can be directly obtained as a stationary solution of the following equation

$$\frac{\partial \psi}{\partial t} = \alpha \nabla^2 \psi - \psi f'(|\psi|^2) - i\boldsymbol{u} \cdot \nabla \psi - \frac{u^2}{4\alpha}\psi \tag{16}$$

that we will call the Advective Real Ginzburg-Landau Equation (ARGLE). This equation corresponds to an extremum of the following modified free energy:

$$\mathcal{F} = \int d^3x \left(\alpha \left| \nabla \psi - i \frac{u}{2\alpha} \psi \right|^2 + f(|\psi|^2) \right). \tag{17}$$

Our preparation method consists in using (16) and (17) with a given *space dependent* divergence less velocity field $u(x)$. Using the Madelung transformation, (17) reads :

$$\mathcal{F} = \frac{1}{2\alpha} \int d^3x \left(\frac{1}{2} (2\alpha \nabla \sqrt{\rho})^2 + 2\alpha f(\rho) + \frac{1}{2} \rho (\nabla \varphi - u(x))^2 \right).$$

The last term in the r.h.s will be minimum if the potential velocity $\nabla \varphi$ is as close as possible to the imposed advective velocity $u(x)$.

Initial conditions for ARGLE For a constant u, the absolute minimum of $\mathcal{F}$ is given by $\psi = \sqrt{\Omega/\beta} \exp(iu \cdot x/(2\alpha))$ which corresponds to a fluid moving with the imposed velocity u. Another stationary solution of ARGLE is $\psi_0 = \sqrt{\Omega/\beta} (1 - (u^2/4\alpha\Omega))^{1/2}$ which corresponds to a fluid with zero velocity. The linear stability study of ψ_0 can be obtained in term of normal modes using $\psi = \psi_0 (1 + u_0 \exp(\sigma t + ikx) + v_0 \exp(\sigma t - ikx))$. We find :

$$\sigma = -\Omega \left(1 - \frac{u^2}{4\alpha\Omega} \right) - \alpha k^2 \pm \sqrt{u^2 k^2 + \Omega^2 \left(1 - \frac{u^2}{4\alpha\Omega} \right)^2}, \tag{18}$$

Note that this stability computation is formally identical to the one leading to the Eckhaus instability of convective rolls [21]. Therefore, in the stable Eckhaus region $(u/c = u/(\sqrt{2\alpha\Omega(1 - (u^2/4\alpha\Omega))}) < 1)$, the stationary solution ψ_0 corresponds to a local minimum of $\mathcal{F}$. In general, with a variable $u(x)$, we also expect several branches of stable solutions for (16) corresponding to local minima of $\mathcal{F}$.

3.2 Numerical results for 3D free-shear flows

To carry out the numerical integration, we use standard pseudo-spectral codes [19]. In the lateral directions, we use sine-cosine transforms to implement free-slip boundary conditions away from the flow's center. Our codes were validated by cross-checking with general periodic codes and linear theory [10].

Round jet The jet's profile is that studied by [22] :

$$u(r) = (U_0/2)[1 + \tanh(R(1 - r/R)/(2\theta))],$$

where U_0 is the centerline velocity on the x jet's axis, $r = \sqrt{y^2 + z^2}$ is the radial coordinate, θ is the momentum thickness and R is the jet radius. The initial data for ARGLE, ψ_0, is obtained by interpolating between a fluid moving with the velocity U_0 at the center of the jet to a motionless fluid far from the jet :

$$\psi_0 = \sqrt{\frac{\Omega}{\beta}} \left[\frac{1}{2} \left(1 - th \left(\frac{r-R}{\theta} \right) \right) \exp \left(i \frac{U_0 x}{2\alpha} \right) + \frac{1}{2} \left(1 + th \left(\frac{r-R}{\theta} \right) \right) \right] \quad (19)$$

The ARGLE converged solution consists in a periodic array of vortex rings of radius R, separated by a length $l = 4\pi\alpha/U_0$. Under NLSE dynamics, this ARGLE converged solution yields a moving periodic array of vortex rings, with little acoustic emission. Adding a small non-axisymmetric perturbation, we obtain a behavior similar to that experimentally observed by Kambe[23] (see Figs 5). The motion of each azimuthally perturbed vortex ring is a rigid rotation in the direction opposite to that of peripheral fluid rotation around the filament.

Swirling jet We have also studied a swirling jet [24] :

$$u(r) = U_0 \exp(-(r/R)^2),$$
$$v(r) = -\frac{z\,R}{r^2} q \left(1 - \exp(-(r/R)^2) \right),$$
$$w(r) = \frac{y\,R}{r^2} q \left(1 - \exp(-(r/R)^2) \right),$$

where q is the rotation rate. The initial data for ARGLE, ψ_0, is obtained by multiplying a scalar field corresponding to the jet, as described above (19), with a field containing as many axial defect filaments as there are quantized quanta $4\pi\alpha$ in the global circulation $2\pi qR$.

The ARGLE converged solution consists in locked-up helices. Under NLSE dynamics, the helices undergo a cork-screw like motion, with little acoustic emission. The dynamics corresponding to an helix with a (small) random perturbation is rich and complex and includes reconnection phenomena (see Figs 6). These reconnection events and the corresponding rapid spreading of the rotating jet can be related to vortex breakdown observed in turbulent flows [25].

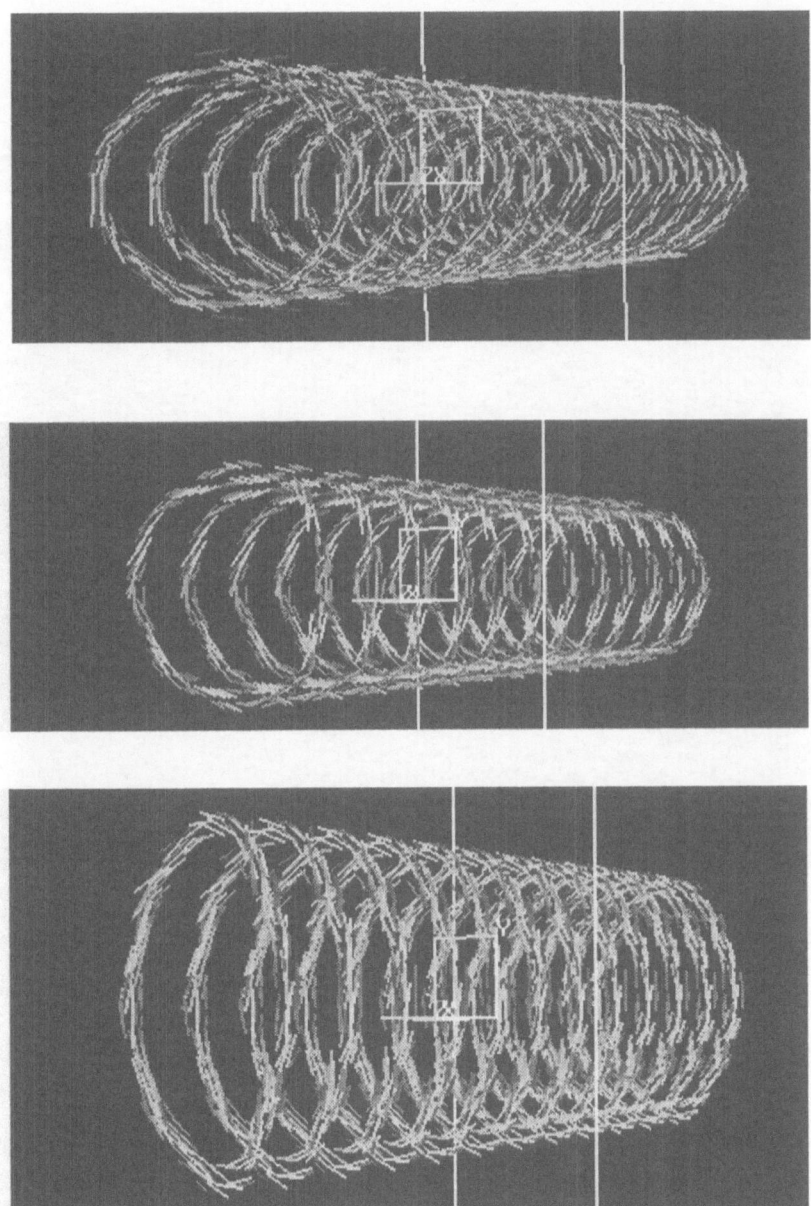

Fig. 5. Azimuthal perturbation of the jet ($m = 2$ azimuthal wave number) : slantwise views at different times (the phenomenon is time-periodical). The jet parameters are : $U_0 = 1$, $\theta = 0.16$ and $R = 1$ with a vortex core size $\xi = 0.07$.

44

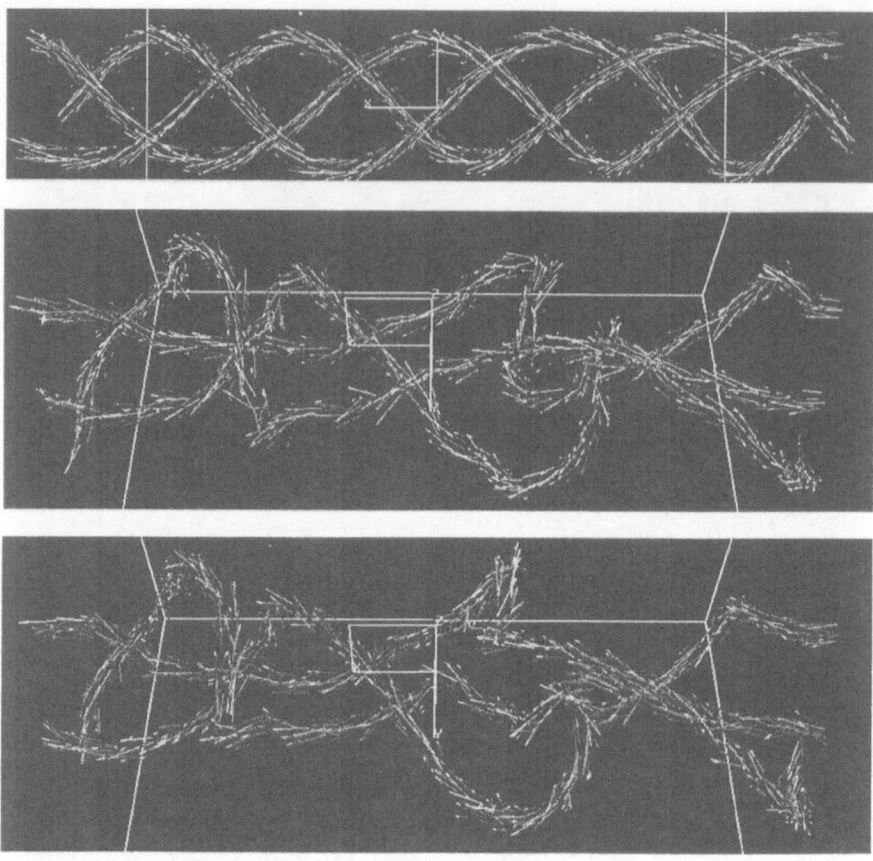

Fig. 6. Swirling jet flow : a) slantwise view of the initial data for the NLSE, b) Reconnection event : before, d) after the reconnection. The jet parameters are the same as in Figure 5 with $q = 0.8$ and $\xi = 0.047$. The amplitude of the random perturbation is 0.1.

4 Fluid dynamical representation of NLKGE in flat space-time

This section contains the special relativistic generalization of the material presented in section 2.

Notations: Throughout the two last sections, the space-time coordinates will be written $x^\mu = (ct, \boldsymbol{x})$, corresponding to the Minkovskian metric given by $diag(1, -1, -1, -1)$.

4.1 Formal correspondence

Let us now consider the following action [16, 26]:

$$\mathcal{A}_r = \frac{1}{c} \int d^4 x L$$

$$L = 2\alpha^2 \Phi_\nu \Phi^{\nu*} - 2\alpha f(|\Phi|^2) - \frac{c^2}{2}|\Phi|^2 \tag{20}$$

where Φ is a complex scalar field and α and f have already been defined in section (2.1). Extremizing $\mathcal{A}_r$ with respect to Φ^*, we obtain the NLKGE:

$$2\alpha^2 \Phi_\mu^{\ \mu} + 2\alpha \Phi f'(|\Phi|^2) + \frac{c^2}{2}\Phi = 0 \tag{21}$$

Equations (20) and (21) are the special relativistic generalizations of (1) and (4). As a matter of fact, if one introduces a new field ψ by the relation:

$$\psi = \Phi \exp\left(i\,\frac{c^2}{2\alpha}t\right) \tag{22}$$

Equations (20) and (21) give back (1) and (4) (for ψ) if one lets c tend to infinity.

As in the Galilean case, a formal correspondence between (21) and special relativistic potential flows of barotropic fluids can be achieved by using the Madelung transform:

$$\Phi = \sqrt{\rho} \exp\left(i\,\frac{\sigma}{2\alpha}\right) \tag{23}$$

If one defines [16] the velocity of the fluid u, the scalar particle density n and the enthalpy density w by

$$u_\mu = -\frac{\sigma_\mu}{(\sigma_\mu \sigma^\mu)^{1/2}} \tag{24}$$

$$n = \frac{\rho}{mc}(\sigma_\mu \sigma^\mu)^{1/2} \tag{25}$$

$$w = \rho(\sigma_\mu \sigma^\mu) \tag{26}$$

the conserved current j associated to the phase invariance of $\mathcal{A}_r$ takes, after a convenient normalization, the usual hydrodynamical form:

$$j_\mu = n\,u_\mu \tag{27}$$

and one also obtains from (24), (25), (26) the normal special relativistic condition for potential flows:

$$mc\sigma_\mu = -\frac{w}{n}u_\mu \tag{28}$$

It should be noted at this point that, for the preceding identifications to be meaningful, the current j has to be time-like in the space-time region under consideration. This is in particular the case if the relativistic effect only constitutes corrections to the main Galilean phenomena. This point is further discussed in [16].

The equation of motion (21) delivers the following relations between n, w and ρ, separating real and imaginary parts:

$$n = \frac{\sqrt{\rho}}{mc}\left(\rho c^2 + 4\alpha\rho f'(\rho) + 4\alpha^2\sqrt{\rho}\left(\sqrt{\rho}\right)_\mu^\mu\right)^{1/2} \tag{29}$$

$$w = \rho c^2 + 4\alpha\rho f'(\rho) + 4\alpha^2\sqrt{\rho}\left(\sqrt{\rho}\right)_\mu^\mu \tag{30}$$

so that the enthalpy per particle H reads

$$H = mc\left(c^2 + 4\alpha f'(\rho) + 4\alpha^2\frac{\left(\sqrt{\rho}\right)_\mu^\mu}{\sqrt{\rho}}\right)^{1/2} \tag{31}$$

If quantum pressure terms were absent, i.e. for ordinary barotropic fluids, it would be possible to deduce automatically from (29), (30) and (31) the correct expression for the pressure p, using the standard thermodynamical relation (with vanishing entropy):

$$dp = n\,dH \tag{32}$$

This would give:

$$p = 2\alpha(\rho f'(\rho) - f(\rho)) \tag{33}$$

However, if the dispersive terms are taken into account, it is no longer possible to eliminate ρ from (29) and (30) to obtain the enthalpy density as a functional of the particle density and its derivatives. It is convenient to retain (33) as the definition of p and to further define the internal energy density ε by the usual relation:

$$\varepsilon = w - p \tag{34}$$

The definitions (29), (30), (31), (34) neglecting the dispersive quantum pressure term give valid thermodynamical relations. However we keep this dispersive term in order for (28) to hold exactly.

Let us now rapidly investigate how the correct Galilean thermodynamics can be recovered from the preceding results. The Galilean particle density is equivalent to n as c tends to infinity [16]. From (29), we therefore deduce that the Galilean particle density is represented by ρ/m. The Galilean counterpart of H is obtained for c tending to infinity. (31) gives immediately that:

$$H \sim mc^2 + 2\alpha mf'(\rho) - 2\alpha^2 m\frac{\Delta\sqrt{\rho}}{\sqrt{\rho}} \tag{35}$$

The first term represents a rest-mass energy, the third one is clearly a dispersive term, not usually included in the Galilean definition of thermodynamical quantities. Recalling that the thermodynamical enthalpy (31) is defined *per particle*, whereas the Galilean one (9) is defined *per unit mass*, the second term coincides with (9). Similarly, the Galilean equivalent of ε/n is:

$$\frac{\varepsilon}{n} \sim mc^2 + 2\alpha m \frac{f(\rho)}{\rho} - 2\alpha^2 m \frac{\Delta\sqrt{\rho}}{\sqrt{\rho}} \qquad (36)$$

In the same way as for H, the second term, divided by m, coincides with the Galilean internal energy *per unit mass* (10). Finally, using the preceding results, (28) implies that, in the Galilean limit,

$$v = \nabla\sigma = \nabla\varphi \qquad (37)$$

as it should be. One can also investigate how the Galilean limiting procedure works on the stress-energy tensor as a whole. This has been done in [16] and we refer the reader to this article for further insights on the question.

4.2 Acoustic phenomena

Dispersion relation In section (2.2), acoustic phenomena were introduced to help in the understanding of the extra quantum pressure dispersive terms present in $\mathcal{A}$. The corresponding relativistic dispersion relation is obtained [16] by linearization of (21) around (22) with $f'(|\Phi_0|^2) = 0$. It reads

$$\left(\frac{\omega^2}{c^2} - k^2\right)\left(k^2 - \frac{\omega^2}{c^2} + \frac{2}{\alpha}|\Phi_0|^2 f''(|\Phi_0|^2)\right) + \left(\frac{\omega}{\alpha}\right)^2 = 0 \qquad (38)$$

Letting c tend to infinity, one obtains (13).

Nonlinear acoustics If one takes as the definition of an acoustic wave that the corresponding density perturbation and 3-velocity are both small quantities of the same order, conventionally chosen to be the first, it can be easily shown [26] that there actually exist an infinity of different acoustic sectors, both for the NLKGE and the NLSE. Each sector is characterized by a real number η, $0 \leq \eta \leq 1$, such that the phase σ is of order η, and the 4D gradient, when acting on ρ and σ, is of order $1 - \eta$. The scaling studied in section (2.2) corresponds to $\eta = 1$, and is the one most commonly worked with in hydrodynamical literature. However, that this is clearly not the only interesting one comes from the fact that the 1D NLKGE and NLSE admit soliton solutions which, in the acoustic regime, correspond to $\eta = 1/2$.

It turns out that it is possible, for all sectors for which $\eta \neq 1$, to obtain [26] a variational principle involving the single field σ, capable of delivering a non-linear equation for this field alone which completely describes the wave propagation at any desired perturbation order and the associated conserved quantities as well. We will not dwell into further details here.

4.3 Time-independent solutions

Let us now elaborate on the special relativistic generalization of the RGLE and on some particular 'static solutions' that can be found with its help. Since time is naturally not a Lorenz scalar, the concept of a static solution and the developments that follow presuppose the choice of a particular inertial frame *ab initio*.

The NLSE and the NLKGE can be put under a hamiltonian form which naturally simplifies the obtaining of the associated RGLE. Let $\mathcal{H}(p, q)$ be the Hamiltonian under consideration. The equations of motion generated by $\mathcal{H}$ take the well-known form:

$$\frac{\partial \mathcal{H}}{\partial p} = \dot{q}$$

$$\frac{\partial \mathcal{H}}{\partial q} = -\dot{p} \tag{39}$$

and the corresponding RGLE simply read:

$$\frac{\partial \mathcal{H}}{\partial p} = -\dot{p}$$

$$\frac{\partial \mathcal{H}}{\partial q} = -\dot{q} \tag{40}$$

Starting from the Lagrangian density L, considered as a function of $\delta = \rho^{1/2}$ and σ:

$$L = -\delta^2 \sigma_t - \frac{\delta^2}{2}(\nabla \sigma)^2 + \frac{\delta^2}{2c^2}\sigma_t^2 - 2\alpha f(\delta^2) - 2\alpha^2(\nabla \delta)^2 + \frac{2\alpha^2}{c^2}\delta_t^2 \tag{41}$$

one finds the following expression for the conjugate momenta (with respect to time) associated to these quantities:

$$\Pi_\delta = \frac{4\alpha^2}{c^2}\delta_t$$

$$\Pi_\sigma = \frac{\delta^2 \sigma_t}{c^2} - \delta^2 \tag{42}$$

The Hamiltonian $\mathcal{H} = pq_t - L$ reads then:

$$\mathcal{H} = \frac{c^4}{8\alpha^2}\Pi_\delta^2 + \frac{c^4}{2\delta^2}\Pi_\sigma^2 + c^3 \Pi_\sigma + \frac{\delta^2}{2}(\nabla \sigma)^2$$

$$+ 2\alpha^2(\nabla \delta)^2 + 2\alpha f(\delta^2) + \frac{\delta^2 c^2}{2} \tag{43}$$

and the RGLE are given by (40) with $p = (\Pi_\delta, \Pi_\sigma)$ and $q = (\delta, \sigma)$. In particular, it is then easy to verify that the Galilean limit of these equations coincide with equation (16) in section (2.3) and that the static special relativistic solutions also minimize the free energy $\mathcal{F}$ introduced in section (2.1).

One can wonder what the special relativistic equivalent of the vortices introduced in section (2.3) are [16]. Let us say that a solution of the NLKGE is a vortex if there exists an inertial frame in which it is static and has cylindrical symmetry (with the density vanishing on the symmetry axis). Using (21), this definition automatically implies that a solution of the NLKGE is a vortex if, and only if, it is also a vortex for the NLSE. Let now r and θ be polar coordinates around the axis. The phase σ of a vortex solution is an integral multiple of $2\alpha\theta$ and the circulation I of the 4-gradient of σ around a closed space-like contour which 'surrounds' the vortex takes the form:

$$I = 4\pi\alpha\, q \tag{44}$$

where q is a positive or negative integer. Moreover, the enthalpy per particle H reads:

$$H = mc\left(1 - \left(\frac{r_{min}}{r}\right)^2\right)^{1/2} \tag{45}$$

where r_{min} is related to q and α by:

$$r_{min} = q\frac{2\alpha}{c} \tag{46}$$

This clearly shows that, in the special relativistic case, no hydrodynamical representation of the vortex is possible for r smaller than r_{min}. In particular, the special relativistic vortex cannot be interpreted as a line-like distribution of vorticity.

5 General relativistic self-gravitating fluids

This section is devoted to applications of the material developed in section 4 to self-gravitating fluids. This is achieved by extending the formalism to general relativity.

If the fluid is a Bose condensate of particles of rest-mass m, two characteristic length-scales can be defined: the Compton length $\hbar/mc$ and the gravitational radius $2Gm/c^2$. These two length-scales would be equal for a particle with a Planck mass $\sqrt{\hbar c/2G}$. At such scales, a quantum theory of gravity would be required for consistency.

For the temptative applications that we consider in this section, the quantum wave-length is much smaller than the gravitational radius and the theory presented here should be of some relevance.

5.1 Fundamentals

The correct action $\mathcal{A}_g$ describing the minimal coupling between the complex scalar field Φ and Einstein's gravitational field reads:

$$\mathcal{A}_g = \frac{1}{c}\int \sqrt{-g}\, d^4x \left(L - \frac{c^4}{16\pi G}R\right) \tag{47}$$

where G is Newton's gravitational constant, g stands for the determinant of the metric tensor and R for the scalar curvature of the metric-compatible connection. Variation of $\mathcal{A}_g$ with respect to Φ gives the curved space-time generalization of equation (21):

$$2\alpha^2\nabla_\mu\nabla^\mu\Phi + 2\alpha\Phi f'(|\Phi|^2) + \frac{c^2}{2}\Phi = 0 \qquad (48)$$

and the variation with respect to the metric furnishes Einstein equations [27]. One could also consider more general coupling between gravitation and the scalar field, as for example the so-called conform coupling [28]. This will not be done here.

5.2 Spherically symmetric 'soliton stars'

A first interesting application of the formalism just presented lies in the study of spherically symmetric solutions to the equations of motion. This has been done by various authors [14, 15] with a pure semi-classical field theory point of view, their principal aim being to describe yet unobserved astrophysical objects usually known as boson stars. In the light of section (4.1), it is rather clear that the structure of this type of stars and, more generally, of any star made out of a barotropic, possibly dispersive fluid (in potential motion), should admit a rather simple description in hydrodynamical language. As a matter of fact, it is possible [26] to derive for these stars an equivalent of the well-known Tollmann-Oppenheimer-Volkoff (TOV) equation. This derivation and the existing supplementary relations between the different unknown functions which are necessary to its solution are fully discussed in [16]. The standard Newtonian limit of the TOV-like system can be shown [16] to be a NLSE equation coupled to a gravitational potential obeying a Poisson equation with a source term $-4\pi\rho G$. This Newtonian self-gravitational system has been considered independently by S. Rica [29].

5.3 Homogeneous isotropic cosmological models

An interesting application of the relativistic formalism is the study of the equations governing the evolution of an isotropic 'toy-universe'. They are presented in [16], where it is shown that they give back, in suitable regimes, a standard Friedman-Robertson-Walker cosmology as well as Linde's chaotic inflation model.

6 Conclusion

The formalisms we have reviewed in this paper make it possible to deal with both Newtonian and Einsteinian perfect barotropic fluids with nonlinear wave equations.

In the case of simple Galilean free shear-flows, the numerical computations we have presented show the ability of NLSE to capture subtle hydrodynamical mechanisms such as vortex reconnection. Furthermore, in the case of superfluids modeled as weakly interacting Bose condensates, NLSE-type descriptions present the interest of naturally containing quantum vortices. This is an important advantage over the more conventional description of superflows in term of classical Euler equations supplemented with a quantification condition on the velocity circulation. As we have demonstrated, NLSE-type descriptions can be tuned to accommodate arbitrary equations of state and dispersion relations. It seems to us that they thus have a great potential for the quantitative explanation of phenomena in real superfluids such as Helium He II, provided that the temperature is low enough for the normal part of the flow to be neglected.

We have shown that the Galilean results can be quite simply extended to special relativity, thereby obtaining a natural description of relativistic barotropic fluids. Such a description is a necessary step to obtain a general relativistic consistent theory of a self-gravitating superfluid. The temptative results concerning self-gravitating bodies presented in section 5 show that this is perhaps the simplest way to introduce a non trivial fluid into general relativity. We are confident that careful studies of such toy models can shed new light on fundamental cosmological problems.

References

1. E. A. Spiegel. Fluid dynamical form of the linear and nonlinear schrödinger equations. *Physica D*, 1:236, 1980.
2. A. L. Fetter. Vortices in an imperfect bose gas iv. translational velocity. *Phys. Rev.*, 151:100, 1966.
3. J. C. Neu. Vortices in complex scalar fields. *Physica D*, 43:385, 1990.
4. F. Lund. Defect dynamics for the nonlinear schrödinger equation derived from a variational principle. *Phys.Rev.Lett.*, A 159:245, 1991.
5. R. J. Donnelly. *Quantized Vortices in Helium II.* Cambridge Univ. Press, 1991.
6. P. Nozières and D. Pines. *The Theory of Quantum Liquids.* Adv. Book Classics, Addison Wesley, 1990.
7. Y. Pomeau T. Frisch and S. Rica. Transition to dissipation in a model of superflow. *Phys.Rev.Lett.*, 69:1644, 1992.
8. Y. Pomeau and S. Rica. Model of superflow with rotons. *Phys. Rev. Lett.*, 71,2:247, 1993.
9. C. Nore, M. Brachet, and S. Fauve. Numerical study of hydrodynamics using the nonlinear schrödinger equation. *Physica D*, 65:154–162, 1993.
10. C. Nore, M. Abid, and M. Brachet. Simulation numérique d'écoulements cisaillés tridimensionnels à l'aide de l'équation de schrödinger non linéaire. *C.R.Acad.Sci. Paris*, 319 II(7):733, 1994.
11. J. Koplik and H. Levine. Vortex reconnection in superfluid helium. *Phys. Rev. Lett.*, 71:1375–1378, 1993.
12. C. Nore, M. Brachet, E. Cerda, and E. Tirapegui. Scattering of first sound by superfluid vortices. *Phys. Rev. Lett.*, 72(16):2593–2596, 1994.
13. J. C. Neu. Vortex dynamics of the nonlinear wave equation. *Physica D*, 43:407, 1990.
14. R. Friedberg, T.D. Lee, and Y. Pang. Mini-soliton stars. *Phys. Rev. D*, 35:3640, 1987.

15. N. Straumann. *Fermion and Boson stars in Relativistic Gravity Research.* Springer Verlag, Berlin, j. ehlers and g. schfer edition, 1992.

16. F. Debbasch and M. E. Brachet. Relativistic hydrodynamics of semiclassical quantum fluids. *to appear in Physica D*, 1995.

17. L. Landau and E. Lifchitz. *Fluid Mechanics*, volume 6. Pergamon Press, 1980.

18. T. Frisch, S. Rica, P. Coullet, and J. M. Gilli. Spiral waves in liquid crystal. *Phys. Rev. Lett.*, 72(10):1471, 1994.

19. D. Gottlieb and S. A. Orszag. *Numerical Analysis of Spectral Methods.* SIAM, Philadelphia, 1977.

20. M. P. Kawatra and R. K. Pathria. Quantized vortices in imperfect bose gas. *Phys. Rev.*, 151:1, 1966.

21. W. Eckhaus. *Studies in Nonlinear Stability Theory.* Springer, Berlin, 1965.

22. M. Abid and M. Brachet. Numerical characterisation of the dynamics of vortex f ilaments in round jets. *Phys. Fluids A*, 5(11):2582–2584, November 1993.

23. T. Kambe and T. Takao. Motion of distorted vortex rings. *J.Phys.Soc.Jap.*, 31(2):591–599, 1971.

24. G. K. Batchelor. Axial flow in trailing line vortices. *J. Fluid Mech.*, 20:645–658, 1964.

25. O. Cadot, S. Douady, and Y. Couder. Characterization of the low-pressure filaments in a three-dimensional turbulent shear flow. *Phys. Fluids*, 7(3):630, 1995.

26. F. Debbasch and M. E. Brachet. Nonlinear acoustics in a special relativistic superfluid. *submitted to Physica D*, 1995.

27. L. Landau and E. Lifchitz. *The Classical Theory of Fields*, volume 2. Pergamon Press, 1980.

28. R. Wald. *General Relativity.* Univ. of Chicago press, chicago edition, 1984.

29. S. Rica. *Défauts et Structures dans les Systèmes hors d'équilibre.* PhD thesis, Institut Non Linéaire de Nice, 1993.

PART II
INSTABILITIES AND
PATTERN FORMATION

Chaotically induced defect diffusion

P. Coullet and K. Emilsson

INLN, UMR CNRS 129, 1361 Route des Lucioles,

06560 Valbonne, France

Defects in systems driven far from equilibrium display interesting features that are consequences of the underlying non-variational dynamics [1]. We are interested in this paper in the bifurcation of defects, in the sense that non trivial topological changes occur in the defects core as a control parameter is varied. In non-equilibrium systems, such bifurcations can lead to states of rich time dependent behaviour.

A classical example of a defect bifurcation occurs in magnetic systems in the presence of an anisotropy [2] [3] [5]. In such a system, regions of opposite magnetization are separated with a so called Ising wall where the magnetization goes through zero at its core. When the anisotropy is decreased an Ising wall undergoes a transition towards a Bloch wall, at the core of which the magnetization rotates. A non-equilibrium analog of this bifurcation is observed in a system of parametrically forced coupled oscillators [4]. When the frequency of the forcing is close to being twice the natural frequency of the oscillators, the forcing plays the role of the anisotropy, the oscillators being locked on two symmetrically opposite equilibrium states. Solutions which spatially join these two states are equivalent to Bloch and Ising walls. The transition from Ising to Bloch breaks the chiral symmetry of the system, as the Bloch wall can have right or left rotation. Non-variational effects associate with this transition a movement of the wall with a velocity which is proportional to the chirality of the Bloch wall [4]. A similar effect should be observed in magnets submitted to a rotating magnetic field [5].

In this paper we propose a scenario in which the core of a defect undergoes a transition

55

E. Tirapegui and W. Zeller (eds.), Instabilities and Nonequilibrium Structures V, 55–62.
© 1996 Kluwer Academic Publishers.

to chaos through a cascade of period doubling bifurcations. In the chaotic regime this state is characterized by the diffusion of the position of the defect, which behaves as if it was subjected to stochastic perturbations. The mechanism of the instability is related to an instability of the bulk which will eventually lead to turbulent behaviour. These complex space-time behaviours manifest themselves first in the core of defects. Since the size of defect cores is finite, the instability leads to scenarios which are typical of finite dimensional dynamical systems. We suggest that such a phenomenon could be a generic behaviour for defects in non-equilibrium systems.

Non-equilibrium analogs of magnetic walls arise naturally in parametrically forced self-oscillatory systems [6]. Such systems are described, near the onset of oscillation, by an equation which describes the slow time dependence of the oscillation amplitude A:

$$\partial_t A = (\mu + i\nu)A - (\beta_r + i\beta_i)|A|^2 A + (\alpha_r + i\alpha_i)\nabla^2 A + \gamma \bar{A} \tag{1}$$

where μ represents the distance from the oscillation threshold and ν the detuning, β_r and β_i the coefficients which measures the nonlinear saturation and the amplitude dependence of the frequency respectively, α_r and α_i the diffusion and the dispersion of waves packets respectively and γ a coefficient which is directly related to the strength of the forcing. Typical systems which are described by such an amplitude equation include the Faraday experiment [6], lasers with injected signals and the periodic forcing of a chemical reaction such as the Belousov-Zhabotinksy reaction.

Without the forcing ($\gamma = 0$), Eq.(1) has a simple solution:

$$A = \sqrt{\frac{\mu}{\beta_r}} \exp -\frac{i\beta_i \mu t}{\beta_r} \tag{2}$$

which describes the basic homogeneous oscillatory pattern. This state is stable with regard to small phase perturbations as long as $\alpha_r \beta_r - \alpha_i \beta_i > 0$ [7] [8]. In spatially extended systems this instability leads to a turbulent state with a continuous creation and annihilation of defects pair [9].

It is common belief, supported by accurate numerical simulations, that the transition to complex spatio-temporal behaviour in extended systems does not follow the classical scenario of transition to chaos, although it may do so when the system is spatially very constrained. In a sense, the transition to chaos involving few degrees of stochasticity, can be understood as a finite size effect. In an extended system, a smooth transition to spatio-temporal chaos involves a number of degrees of stochasticity characterized by a number of positive Lyapounov exponents which should diverge as the size of the systems tends to infinity [10]].

The presence of the forcing in Eq.(1), stabilizes a spatially homogeneous state, which we will call the locked state. Putting $A = R_0 \exp i\phi$, R_0 is the solution of the equation:

$$\gamma^2 = (R_0^2 - \mu)^2 + (\nu - \beta_i R_0^2)^2 \tag{3}$$

where we have put $\alpha_r = \beta_r = 1$ without loss of generality. Solving for the phase yields two different locked states, differing by a phase difference of π. These states are the analogs of magnetic domains with opposite orientation. The system described by Eq.(1) has wall solutions, joining two different locked states. In the core of the wall, the order parameter A can either rotate or go through zero, depending on the value of γ. For small values of γ the wall becomes Bloch like, and moves due to non-variational effects [4].

In this paper we describe another transition which occurs when lowering γ, but in a situation such that $1 + \alpha_i \beta_i < 0$, i.e the basic homogeneous state is unstable in the limit where γ tends to zero, though the presence of the forcing postpones this instability. We study Eq.(1) in one space dimension in a situation such that the locked states exist, and we have a Ising wall solution between these two states (In all what follows we will have $\alpha_i = -1.3$, $\beta_i = 1.5$ and $\nu = 1.55$)(see figure 1.). As we now lower γ, we observe a bifurcation occurring at the core of the wall, which can be clearly identified as a Hopf bifurcation of the core. This can be seen for example by calculating the mobility of the defect, given by:

$$M(t) = \int |\partial_x A(t)|^2 dx \tag{4}$$

and plotting it in $(M(t), M(t + \tau))$ space (see figure 2). In this case the defect is on average symmetric, and thus its average position remains fixed. For a lower value of γ ,

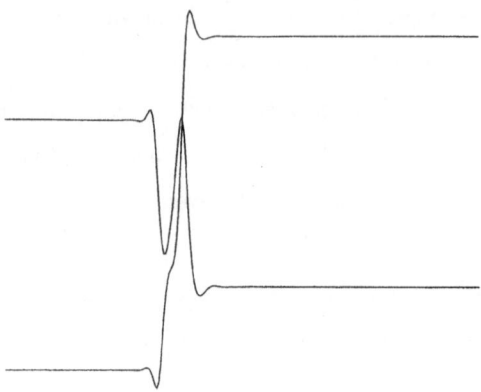

Figure 1: A snapshot of the real and the imaginary part of A as a function of x. Oscillatory but not moving state, $\gamma = 0.509$.

another transition occurs, which corresponds to the Ising-Bloch parity breaking transition. The time average of the solution exhibits a small even part. This manifests itself by the movement of the defect, either in the right or left direction, with a constant velocity. This is followed by a cascade of period doubling bifurcations as γ is lowered (see figure 2.a-d), eventually leading to a chaotic attractor (see figure 2.e).

The defect then moves in a given direction, with a chaotically modulated velocity. Both directions of propagation are observed, which corresponds to two different symmetrical attractors. When the amplitude of the forcing is lowered even more, the two attractors merge through what we can identify as a shilnikov homoclinic bifurcation (see figure 2.f) [11]]. For these value, the average parity is restored, but the defect undergoes a diffusive motion (see figure 3.). The numerical simulations corresponding to each point of figure 3. were performed on a Cray, using a simulation with 256 points, using 3000 time steps per sample, the average being taken over 1000 such samples.

Below some critical value of γ, the defect, which can be seen as a turbulent drop, expands indefinitely through 2 fronts which propagates at a given finite velocity [12] as

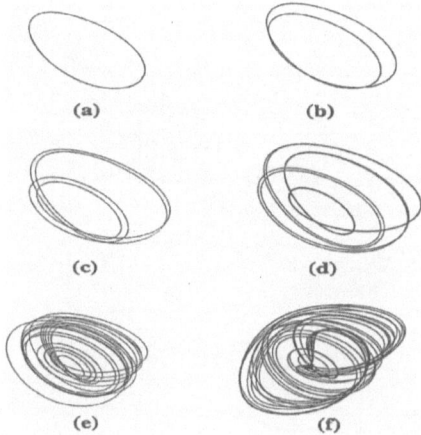

(a)   (b)

(c)   (d)

(e)   (f)

Figure 2: Plot of $(M(t), M(t + \tau))$ for different values of γ (0.455,0.445,0.428,0.422,0.415 and 0.370) with dx=0.51, dt=0.033 on 256 points. Scales are arbitrary.

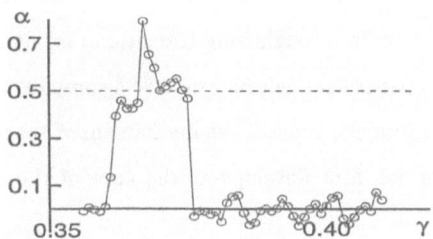

Figure 3: Fit values of α for the dispersion of the defect position $P(t)$ in $P(t) = (< P(t)^2 > - < P(t) >^2)^{\frac{1}{2}} = Ct^\alpha$ for various value of γ. The dashed line corresponds to Brownian motion.(dx=1.0, dt=0.3).

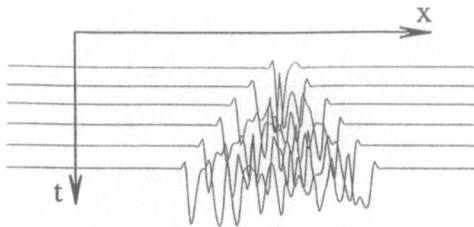

Figure 4: Expansion of turbulent drop for sucessive time intervals ($\gamma = 0.13$)

it was anticipated under general considerations [13] (see figure 4.). Numerical simulation have also been performed in two space dimensions, where we studied the behaviour of a domain wall. In the chaotic regime the wall exhibits behaviour which is somewhat reminiscent of those observed in a roughening transition (see figure 5.).

We have shown in this paper that defects can have a complex behaviour, not accounted for by usual gradient flow models. Indeed, the mechanism of the motion is related to time dependent instabilities which first develops at the core of the defect. These instabili-

Figure 5: Evolution of a chaotic Ising domain wall in two dimensions (256 by 256 points, $\gamma = 0.36$)

ties, known to lead to complex spatio-temporal behaviour in infinite systems, manifests themselves first as finite dimensional chaos, due to the finite size imposed by the topological confinement of the defect. As one approaches the onset of the transition toward a turbulent state, the defects behave in a more and more erratic way, and as expected from general arguments, increases in size. Although we have illustrated this phenomenon on the concrete example of the parametric forcing of oscillators, we are convinced of its robustness. As a conclusion, chaotic behaviours are relevant in the transition towards turbulence in extended systems. They appear in the core of defects, and manifest typical scenarios toward chaos. There are many experimental candidates in which there is hope to observe this phenomenon, including chemical, optical and fluid experiments.

Acknowledgment:

This work has been partially supported by the EEC grants SC1*CT90 0325 and SC1*CT91 0683. The numerical work have been performed on a CRAY which belongs to the "Institut Méditerranéen de Technology (IMT)" in Marseille.

References

[1] F. Busse, L. Kramer, eds., Nonlinear Evolution of Spatio-Temporal Structures in Dissipative Continuous Systems, (Plenum Press, New York, 1990).

[2] Bulaevski L.N. and Ginzburg V.L., Sov. Phys. JETP 18, 530 (1964).

[3] J. Lajzerowicz, J.J. Niez, Le Journal de Physique Lettres 40, L-165 (1979).

[4] P. Coullet, J. Lega, B. Houchmanzadeh, J. Lajzerowicz, Phys. Rev. Lett 65, 1352 (1990).

[5] P. Coullet, J. Lega, Y. Pomeau, Europhys. Lett. 15, 221 (1991).

[6] S. Douady and S. Fauve, Eur. Phys. Lett. 6, 221 (1988).

[7] T.B Benjamin, J.E Feir, J. Fluid Mech. 27 , 417 (1967).

[8] A.C. Newell, Lect. Appl. Math. 15, 157 (1974).

[9] P. Coullet, L. Gil, J. Lega, T, Phys. Rev. Lett. 62, 1619 (1989).

[10] P. Manneville, Liapunov exponents for the Kuramoto-Shivashinsky model in Macroscopic modeling of turbulent flows, Lect. Notes in Physics 230, Springer (1985).

[11] A. Arneodo , P. Coullet and C. Tresser, J. Stat. Phys. 27, 171 (1982).

[12] The same phenomenon was also reported by B. Houchmanzadeh, Doctoral thesis, Grenoble (1992).

[13] Y. Pomeau, Physica D23, 3 (1986).

Pattern formation and phase turbulence in the transverse section of lasers

GUILLAUME HUYET AND SERGIO RICA

Institut Non Linéaire de Nice, UMR 129 CNRS-UNSA,
1361 Route des Lucioles, 06560 Valbonne, France.

The transverse pattern dynamics in a laser with large Fresnel number are often governed by two relatively independent but simultaneous instabilities. One leads to a "turbulent" state (uncorrelated in space and time) through modulation, principally, of the phase of the field while the other yields a periodic modulation in space and time. The measured electric field amplitude is modulated by both effects. As a consequence the laser intensity is locally chaotic on short time scales but the time-average intensity patterns retains the global symmetry of the system. We compare our studies of the Maxwell-Bloch equations with recent experimental results.

During the last 30 years, nonequilibrium structures have been the subject of intensive research in fluid mechanics, chemistry and biology [1]. Patterns have been observed in lasers since their invention [2]; but systematic studies resumed recently [3]. Evidence of spatio-temporal dynamics have been investigated experimentally [4, 5, 6] and theoretically [7, 8, 9] in lasers and in passive systems [10], when the Fresnel number and/or the pump was increased.

The aim of this paper is to provide a theoretical model that describes the evolution and characteristics of patterns which have been observed in recent experiments [5]. The interpretative framework may be extended to any other wave phenomena. Assuming that the transverse dynamics involves only a single longitudinal mode of the Maxwell-Bloch (MB) equations, we show that the homogeneous steady-state solution may be destabilized by two generic instabilities. The first is a long wavelength instability which is related to the phase invariance of the electromagnetic field and is described by a Kuramoto-Shivasinsky type equation (KS) [11]. The second is a short

E. Tirapegui and W. Zeller (eds.), Instabilities and Nonequilibrium Structures V, 63–73.
© *1996 Kluwer Academic Publishers.*

wavelength instability which corresponds to a Hopf bifurcation which selects a well-defined wavelength and is described by a complex Swift-Hohenberg equation (CSH) [12].

The KS field becomes very irregular in space and time for a large set of parameter values. This observed state is often refered to as "phase turbulence" [11, 13]. However, the time averaged structures it creates are spatially homogeneous, as shown in FIG.3. By contrast, the CSH equation generates periodic structures such as rolls, hexagons, or squares, depending on the nonlinearities -and/or the boundary conditions. As a result of these two instabilities, the laser intensity is disorganized in space and in time due to the turbulent behavior of the KS dynamics, yet the time-averaged pattern recovers the spatially periodic structure as selected by the CSH equation. The fact that "turbulent" patterns retain underlying symmetry on average, has been predicted in [14] and experimentally studied in Faraday instability [15] or in Rayleigh-Bénard convection [16].

In this paper, we will first summarize the main features of a recent experiment [5], and we will then analyse the MB equation, showing that the dynamics is described by KS+CSH. We will follow with a discussion of the dynamics of such combined amplitude equations. An example of an experimental transverse pattern displayed by a Fabry-Pérot CO_2 laser with an intracavity lens resulting in a Fresnel number of 60 is shown in FIG. 1. The time averaged pattern FIG.1-a observed on an image plate appears as an ordered structure of concentric rings, squares, and/or more complex structures. Near the lasing threshold, the intensity at a point in the pattern measured by a fast HgCdTe detector is periodically modulated at $150kHz$, with a DC signal larger than the AC peak to peak modulation. This frequency is very close to the the relaxation oscillation frequency of the single transverse mode laser. For higher pump values, the temporal signal becomes chaotic with a broadband power spectrum (FIG. 1-d). Spatial correlation of such laser patterns are presented in [5], where it is found that, for large patterns, correlations are lost at distances smaller than the overall scale of the pattern. We begin with the MB equations [8] in an dimensionless form, for the slowly varying envelopes of electric field $E(\mathbf{x}, t)$ [$\mathbf{x} = (x, y)$], polarization $P(\mathbf{x}, t)$, and population inversion $D(\mathbf{x}, t)$:

$$\partial_t E = -\kappa \left[(1 - i\delta) - i\frac{a}{2}\nabla^2 \right] E - \kappa CP; \tag{1}$$

$$\partial_t P = -\gamma_\perp \left[DE + (1 + i\delta)P \right]; \tag{2}$$

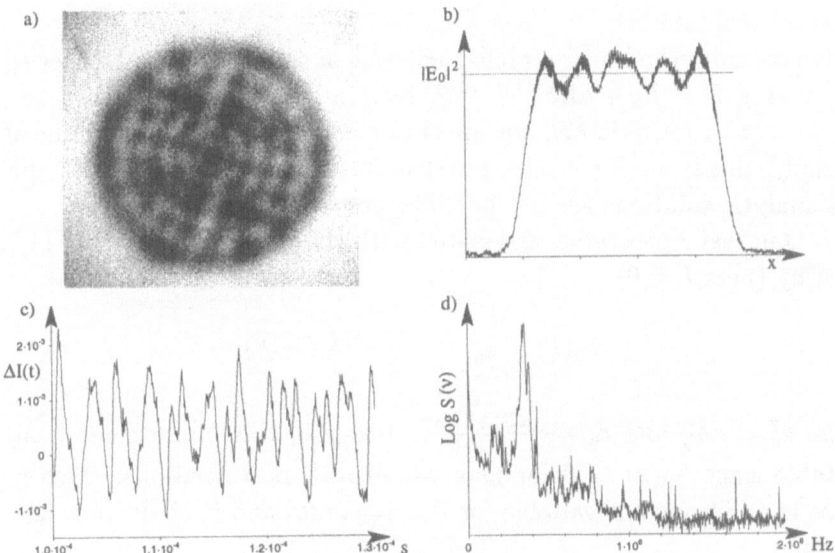

Figure 1: a) The average pattern of a CO_2 laser on a thermal plate. b) Average intensity vs. a transverse coordinate measured by a HgCdTe detector. We may notice that it is a constant (and far from lasing threshold) and modulated almost by a periodic structure. c) Shows the AC component of the light intensity vs. time. d) Its power spectrum, which distinguishes continuous background characteristic of a chaotic signal, and also reveals a narrow band around a frequency between 350-400kHz and a small one between $150 - 200kHz$.

$$\partial_t D = -\gamma_\| \left[-\frac{1}{2}(E^* P + E P^*) + D - 1 \right] ; \tag{3}$$

where $\kappa, \gamma_\perp$ and $\gamma_\|$ are the decays rates for E, P and D, respectively. For modelling a CO_2 laser we take $\frac{\gamma_\perp}{\kappa} \approx 5$, $\frac{\gamma_\|}{\kappa} \approx 0.01$ and $\kappa \approx 2 \times 10^7 s^{-1}$. The pump strength is denoted by C (between 2 and 3 in our experiment), δ is the cavity detuning; a is inversely proportional to the Fresnel number; and ∇^2 is the transverse Laplacian.

To simulate the experimental profile presented in in FIG. 1-b, we consider a homogeneous solution of the MB equations (1,2,3) of the form

$$|E_0|^2 = C - (1 + \delta^2); \quad P_0 = -(1 - i\delta)\frac{E_0}{C}; \quad D_0 = \frac{1 + \delta^2}{C}. \tag{4}$$

The phase of E_0 is not fixed, since MB equations (1,2,3) are invariant under a global phase change $E \rightarrow Ee^{i\phi_0}$, together with $P \rightarrow Pe^{i\phi_0}$, for ϕ_0 an arbitrary real number.

We consider the linear stability analysis of solution (4) in Fourier space, (*i.e.* taking $E = E_0 + \delta E e^{\lambda(k)t - ik\cdot x}$, *etc.*, in equations (1,2,3) and keeping only first order terms in δE, *etc.* we obtain equations which determine $\lambda(k)$). Formally, this is an *Eigenvalue* problem for a 5×5 real matrix. It appears that analytic solutions are not possible, yet we note that we have:

i) One real *Eigenvalue*, associated with the phase invariance of (1,2,3) , given by (near $k \approx 0$):

$$\lambda_0(k) = k_0^2 k^2 - \frac{c_0}{2} k^4 + \mathcal{O}(k^6);$$ (5)

where $k_0^2 \approx \kappa \delta \frac{a}{2}$ and $c_0 \approx \kappa \frac{C(1+\delta^2)a^2}{2|E_0|^2}$. If $\delta > 0$, this phase mode is always unstable since $\lambda_0(k) > 0$ for long wavelength perturbations. This global phase is an important variable for the dynamics and it obeys to a KS type equation.

ii) Two complex conjuguated *Eigenvalues* with an imaginary part which can be estimated in the long wavelength limit:

$$\lim_{k \to 0} Im \left(\lambda_1^{\pm} \right) \approx \sqrt{2\kappa\gamma_{\parallel} |E_0|^2 / (1 + \delta^2)}.$$

In the same limit, their real part is small and negative (roughly given by $-\gamma_{\parallel} \left[1 + |E_0|^2 / (1 + \delta^2) \right] / 2$), and becomes positive for $\delta = \delta_c$ (at $|k| = q_0$), as shown in FIG.2. When $\gamma_{\perp}$ is much larger than $\gamma_{\parallel}$ and κ,

$$\left(\frac{a q_0^2}{2} \right)^2 \approx \frac{\gamma_{\perp} \gamma_{\parallel} |E_0|^2}{6\kappa^2}.$$

The Taylor expansion of this *Eigenvalue* as a function of k^2 around q_0^2 gives:

$$\lambda_1(k) = (\mu + i\Omega) - iv(q_0^2 - k^2) - \alpha(q_0^2 - k^2)^2 + \ldots,$$ (6)

where μ, Ω and v are real coefficients while α is a complex quantity. All these quantities are directly related with the *Eigenvalues* of the full linear operator discussed above [17]. In the neighborhood of the bifurcation, the dynamics is governed by a complex field which obeys a CSH equation. We note that the frequency of the Hopf bifurcation roughly given by relaxation oscillation frequency, as experimentally observed for low pump values.

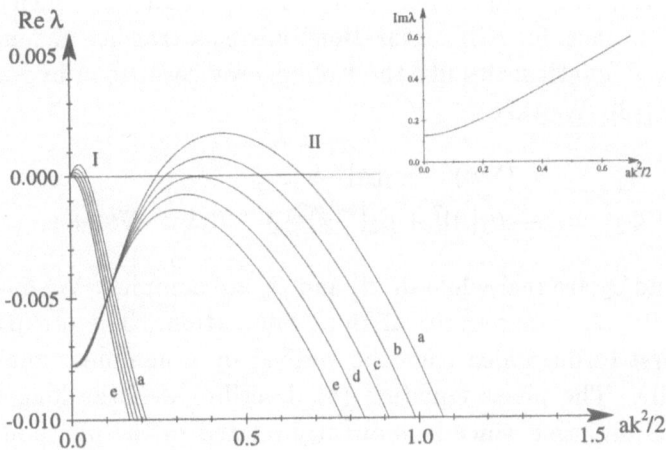

Figure 2: $\lambda_0(k)/\kappa$ (I) and $Re\ \lambda_1(k)/\kappa$ (II) vs. $ak^2/2$ for $C = 2$, $\frac{\gamma_\perp}{\kappa} = 5$, $\frac{\gamma_\|}{\kappa} = 0.01$, and values of δ: a) 0.05, b) 0.04, c) 0.03, d) 0.02 and e) 0.01. The threshold for the Hopf bifurcation is $\delta_c = 0.0305$ at $aq_0^2/2 = 0.3650$. Inset shows the frequency $Im\ \lambda_1(k)/\kappa$.

iii) Two complex *Eigenvalues* with very large negative real parts, roughly given by $\lambda_2^\pm \approx -\gamma_\perp(1 \pm i\delta)$. The associated variables in the phase space do not play an important role in the dynamics.

Consequently, the further dynamics will be governed by a real field ϕ and a complex one A. The linear dynamics (in Fourier space) is given by

$$\partial_t\phi_k = \lambda_0(k)\phi_k \quad \text{and} \quad \partial_t A_k = \lambda_1(k)A_k.$$

The physical variables are functions of ϕ and A. Taking the electric field amplitude $|E|$ as an example we have:

$$\frac{|E|}{|E_0|} = 1 - \frac{a}{4}(1 + \frac{1+\delta^2}{|E_0|^2})\nabla^2\phi - i\left(Ae^{i\Omega t} - A^*e^{-i\Omega t}\right) + \mathcal{O}(|A|^2, (\nabla\phi)^2). \quad (7)$$

The linear problem and the symmetry argument (invariance under the transformation $A \to Ae^{i\psi_0}$ and $\phi \to \phi + \phi_0$) can be used to obtain, directly, the general form of the equations:

$$\partial_t\phi = -k_0^2\nabla^2\phi - \frac{c_0}{2}\nabla^4\phi + \Phi_1; \quad (8)$$

$$\partial_t A = \mu A - iv(q_0^2 + \nabla^2)A - \alpha(q_0^2 + \nabla^2)^2 A + \Phi_2; \quad (9)$$

where $\Phi_{1/2}$ is a real/complex polynomial in A, A^* and $\nabla\phi$, which ensures the saturation. In fact, for KS, "saturation" means a transfer of "energy" from large to small length scales like the Kolmogorov cascade in hydrodynamical turbulence [13]. We take

$$
\begin{aligned}
\Phi_1 &= c_1(\nabla\phi)^2 + c_2|A|^2 + \cdots, \\
\Phi_2 &= -(\beta_1|A|^2 + \beta_2(\nabla\phi)^2)A + \beta_3\nabla\phi\cdot\nabla A + \cdots,
\end{aligned}
$$

where c_1 and c_2 are real while β_1, β_2 and β_3 are complex. We have assumed $Re(\beta_1) > 0$, *i.e.* a supercritical Hopf bifurcation. These expansions are only the first terms which could be verified by a nonlinear analysis or experimentally. The phase equation (8) describes well the long wavelength behavior of any laser, since it is directly related to the phase invariance of the electromagnetic field.

The behavior of the field depends on the signs of k_0^2 and μ, as a consequence four different cases could occur:

i) If $k_0^2 < 0$ and $\mu < 0$, (*i.e.* for lasers $\delta < 0$) the steady-state solution is stable.

ii) If $k_0^2 > 0$ and $\mu < 0$, (*i.e.* $0 < \delta < \delta_c$), there is an input of "energy" at the long wavelengths scales given by $0 < k^2 < 2k_0^2/c_0$, with a maximum rate

$$
\lambda_0(k_m) = k_0^4/2c_0 \approx \kappa\delta^2\frac{|E_0|^2}{4C}
$$

at k_m and, with $k_m = k_0/\sqrt{c_0}$ or

$$
\frac{ak_m^2}{2} = \frac{\delta|E_0|^2}{2C}.
$$

The phase field ϕ, remains essentially chaotic, and has a temporal correlation time function $e^{-t/\tau}$, with τ of the order of $1/\lambda_0(k_m) \approx 10^{-3}sec$, which agrees with the experimental value [5]. However, the temporal mean values ($\langle\nabla^2\phi\rangle$ and $\langle(\nabla\phi)^2\rangle$) become constant in space (the former is zero), as we see (FIG. 3) in a direct simulation of the KS equation (8). Consequently, the signal $|E(\mathbf{x}, t)|^2$ (7) displays a complex behavior in time, but the time averaged pattern appears homogeneous in space.

iii) If $k_0^2 < 0$ and $\mu > 0$ equation (8) is purely diffusive and $\nabla\phi$ goes to zero. Equation (9) creates a periodic structure which oscillates in time. This case does not occur in lasers but it is usefull considering it in order in order to analyze the dynamics of the following case.

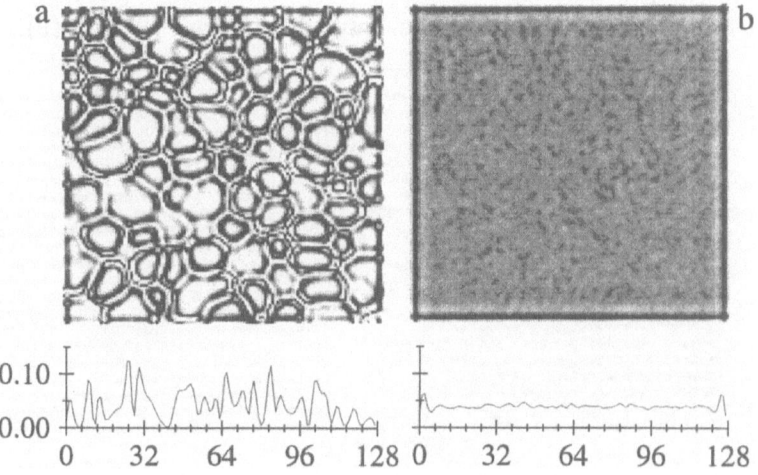

Figure 3: Numerical simulations of the KS equation for $k_0 = 0.5$, $c_0 = c_1 = 1$ and $c_2 = 0$ in a 128×128 grid with Neumann boundary conditions. a) The instantaneous value of $(\nabla \phi)^2$ and its cross-section in one direction. The center of each cell is a source and emits waves which collide and form shocks. Some cells "eat" their neighbors and become bigger and bigger until a breakdown occurs in the smallest cell. This fast dynamic is very similar to the convection on the solar surface. b) The temporal average $\langle (\nabla \phi)^2 \rangle$ over 2000 time units and its cross-section.

iv) If $k_0^2 > 0$ and $\mu > 0$ ($\delta > \delta_c$) -the most interesting case- then the two instabilities occur; ϕ has a dynamics analogous to the numerical simulations of KS in FIG. 3, and dismantles the lamellar structure created by CSH. Subsequently, A shows a complex spatio-temporal behavior. If we seek a stationary solution of the form $A = |A|e^{iq_0 x}$ (in which all the coefficients are taken as reals and $\beta_3 = 0$) we find that A depends on the "stochastic variable" ϕ as ($\beta_2^{(r)} \equiv Re\ \beta_2$):

$$|A|^2 \sim \mu - \beta_2^{(r)}(\nabla \phi)^2. \tag{10}$$

The instantaneous value of $A = |A|e^{iq_0 x}$ looks more disorganized than the mean value $\langle |A|e^{iq_0 x} \rangle$ because of its explicit dependence on the turbulent field $\nabla \phi$. The mean value recovers the periodic structure:

$$\langle A \rangle \sim \left(\sqrt{\mu} - \frac{\beta_2^{(r)}}{2\sqrt{\mu}} \left\langle (\nabla \phi)^2 \right\rangle \right) e^{iq_0 x} \sim C^{te.}\ e^{iq_0 x} ,$$

since $\langle (\nabla \phi)^2 \rangle$ is constant. We sketch our analytical arguments through a direct numerical simulation of the coupled equations (8) and (9).

a) 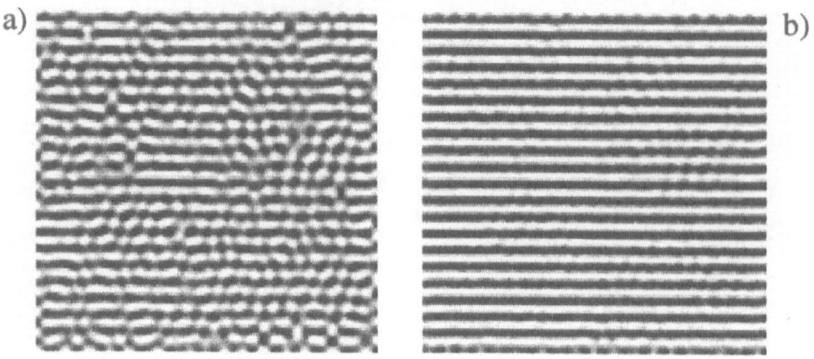 b)

Figure 4: Numerical simulations of the model with all the coefficients real. a) The instantaneous value of $Re\ A$. The initial conditions are rolls parallel to the x axis. b) The temporal average $\langle A \rangle$.

If we take all the coefficient different from zero, equation (9) generates traveling rolls, and thus $\langle A \rangle = 0$. If, we assume that the field selects a stationary pattern it is obvious that $\langle A \rangle = 0$, but $\langle |A|^2 \rangle$ remains non-zero. The problem of crystal selection remains and it is likely the boundary conditions, such as the cylindrical tube of the laser, the spherical mirrors and imperfections, play an important role as in [4].

From this analysis, we see that the steady state solution of the laser is stable below resonance ($\delta < 0$) as it is very well known [18]. As soon as the laser is tuned above resonance, the phase instability occurs. But in the case of CO_2 lasers, a second instability occurs for a small value of the detuning ($\delta \simeq 0.03$). Consequently, one must take into account the two bifurcations in order to describe the dynamics of the observed patterns for most positive detunings. But, if one analyze in 1D the stability of the traveling waves solutions given by $E = E_0 e^{(i\omega_t t - k_t x)}$ [9], the same instabilities occur and can be understood with equations (8,9) by changing only ϕ into $\phi - k_t x$.

An interesting consequence appears when we take into account the effect of the boundary conditions. The number of degrees of freedom of the KS equation grows as $N_{KS} \sim (Rk_m)^2$, where R is the transverse radius of the laser beam in the cavity. However we can write (using the value of a after the equations (1,2,3))

$$N_{KS} \sim \frac{\delta|E_0|^2}{C} Fr = \frac{Fr}{Fr_0}$$

as a function of the Fresnel number (Fr) of the laser cavity. For small $Fr(\approx Fr_0)$, such that $N_{KS} \approx 1$, the KS equation leads to low dimensional chaos. In this case, the spatial correlations, which do not decrease exponentially, are determined only by the crystal structure: essentially a long range order. Using the *Ansatz* of Ref.[19], the correlation decays algebraically as ($\Delta x = |\mathbf{x} - \mathbf{x}'|$, etc.):

$$C(\Delta x, \Delta t) \sim J_0(q_0 \Delta x) \times e^{-\frac{\Delta t}{\tau}}.$$

However, for a large Fresnel number, KS becomes turbulent and the correlation function is screened by a factor $e^{-\Delta x/\lambda_c}$, with $\lambda_c \sim k_m^{-1}$. The resulting state may be called optical (weak-)turbulence.

It is worth noting that there may be topological defects of the complex field A. (A defect is a point at which A vanishes and at which there is a dislocation of the periodic structure.) These defects differ from the ones predicted by Coullet *et al.* [7] in that these do not correspond to zeros in the electric field, but instead to zeros of A. This means that in the core of these defects $|E| = |E_0|$.

In conclusion, we reduce the complete Maxwell-Bloch equations (1,2,3) to coupled KS and CSH equations. The temporal signal $|E(\mathbf{x}, t)|^2$ at each point presents oscillations due to a Hopf bifurcation and a disordered signal resulting, from the phase field ϕ which is essentially chaotic and non correlated in time. Moreover, for large Fresnel numbers, we show the existence of phase turbulence in lasers, which seems to agree with the experimental measurements [5]. However, the temporal mean value of the intensity, $\langle|E(\mathbf{x}, t)|^2\rangle$, becomes regular in space and retains the symmetry of the crystal (A), see FIG. 1-a. Our analysis is simply based upon amplitude equations and subsequently may be applied to other systems wherever a Hopf bifurcation is found.

This physical picture has also been observed in recent experiments on the hydrodynamical instability [15, 16] in which the pattern is very irregular in space and time, but the average time pattern recovers its symmetry. It would be interesting to investigate such situations with our approach (KS+CSH eqn.).

Finally, it is a pleasure to thank N.B. Abraham, for his constant interest and usefull advices. We also acknowledge F.T. Arecchi, P. Coullet, M.C.

Martinoni, Y. Pomeau, A. Pumir, E. Tirapegui and J.R. Tredicce for stimulating discussions, and J. Clark for her help. These new instabilities in the MB equations were discovered thanks to the software *NxEigenvalues* created by P. Coullet to whom we thank for his help.

References

[1] P. Manneville, *"Dissipative structures and weak turbulence"*, Acad. Press. (1990); G. Nicolis and I. Prigogine, "Self-organization in nonequilibrium systems, from dissipative structures to order through fluctuation", Wiley (1977); M.C. Cross and P.C. Hohenberg, Rev. Mod. Phys. **65**, 851 (1993).

[2] W.W. Rigrod, Appl. Phys. Lett. **2**, 51 (1963); P. Goldsborough, Appl. Opt. **3**, 267 (1964).

[3] N.B. Abraham and W. Firth, J. Opt. Soc. Am. B **7**, 951 (1990).

[4] C. Green, G.B. Mindlin, E.J. D'Angelo, H.G. Solari and J.R. Tredicce, Phys. Rev. Lett. **65**, 3124 (1990). E.J. D'Angelo, E. Izaguirre, G.B. Mindlin, G. Huyet, L. Gil and J.R. Tredicce, Phys. Rev. Lett. **68**, 3702 (1992).

[5] G. Huyet, M.C. Martinoni, S. Rica and J.R. Tredicce, Submitted to Phys. Rev. Lett., November 1994.

[6] D. Dangoise, D. Hennequin, C. Lepers, E. Louvergneaux, and P. Glorieux, Phys. Rev. A **46**, 5955 (1992).

[7] P. Coullet, L. Gil, and F. Rocca, Opt. Commun. **73**, 403 (1989).

[8] G.L. Oppo, G. D'Alessandro, and W.J. Firth, Phys. Rev. A **44**, 4712, (1991).

[9] Q. Feng, J.V. Moloney, and A.C. Newell, Phys. Rev. Lett. **71**, 1705 (1993).

[10] F.T. Arecchi, Il Nuovo Cimento **107A**, 1111 (1994).

[11] Y. Kuramoto and T. Tsuzuki, Prog. Theor. Phys. **55**, 356 (1976); G.I. Shivasinsky, Acta Astronautica **4** 1177 (1977); see also Y. Kuramoto, *Chemical Oscillations, waves and turbulence*, Springer (1984). For lasers

the KS equation has been introduced by R. Lefever, L.A. Lugiato, W. Kaige, N.B. Abraham and P. Mandel, Phys. Lett. A **135**, 254 (1989).

[12] J. Swift and P.C. Hohenberg, Phys. Rev. A **15**, 319 (1977).

[13] P. Manneville, Phys. Lett. A **84**, 129 (1981); Y. Pomeau, A. Pumir and P. Pelce, J. Stat. Phys. **37**, 39 (1984) for 1D, and G.I. Shivasinsky, Ann. Rev. Fluid Mech. **15**, 179 (1983) for the 2D case.

[14] P. Chossat and M. Golubitsky, Physica **32 D**, 423 (1988).

[15] B.J. Gluckman, P. Marcq, J. Bridger, and J.P. Gollub, Phys. Rev. Lett. **71**, 2034 (1993).

[16] L. Ning, Y. Hu, R. Ecke, and G. Ahlers, Phys. Rev. Lett. **71**, 2216 (1993).

[17] As an example, for the numerical values of FIG. 2 we obtain $\mu \sim (\delta - \delta_c)$; $\Omega = 0.3304\kappa$; $v = 0.7674\kappa$; $\alpha = \kappa(0.0480 - i0.1596)$ & $aq_0^2/2 = 0.3650$.

[18] A.C. Newell and J.V. Moloney, *Nonlinear Optics*, Addison-Wesley, Redwood City, CA, (1992).

[19] A.C. Newell and Y. Pomeau, J. Phys. A: Math. Gen. **26** L429 (1993).

LONG WAVELENGTH OSCILLATORY INSTABILITY
IN BINARY FLUIDS

T. Clune[1], M. C. Depassier[2] & E. Knobloch[3]

(1) JILA, U. of Colorado, Boulder, CO 80309 U.S.A.

(2) Facultad de Física, U. Católica de Chile, Santiago 22, Chile

(3) Department of Physics, U. of California, Berkeley, CA 94720, U.S.A.

ABSTRACT. Binary fluid convection takes the form of oscillations for sufficiently negative separation ratios. When the confining plates are of poor thermal conductivity these oscillations have a long wavelength. Nonlinear planform equations governing the resulting instability are derived for both two and three dimensions. The technique uses a multiple scale analysis combined with reconstitution. The validity of the resulting equations is discussed and compared with asymptotically exact alternatives. The results are found to differ demonstrating the failure of the reconstitution procedure.

Introduction

Binary fluid convection exhibits both oscillatory and stationary instabilities depending on the boundary conditions and parameter values. The nonlinear development of the instabilities as well as the stability properties of the patterns that may arise has been studied both theoretically and experimentally in a number of different cases. We consider here a binary fluid mixture confined between poorly conducting plates in which convection arises as a long wavelength stationary or oscillatory instability depending on the values of the separation ratio S and the Lewis number τ. The appearance of a long wavelength instability enables one to derive the equations that govern the nonlinear evolution of the instability using the wavenumber of the instability as a small parameter. The resulting equations thus describe fully nonlinear solutions of the system, but take a much simpler form than the original equations. In the literature two approaches have been used to implement this idea. The first and asymptotically exact procedure, used by Pismen (1988), resulted in equations that were nonlocal in physical space. Pismen used these equations to demonstrate that at onset an oscillatory pattern called alternating rolls was stable. More recently, techniques based on the reconstitution of the asymptotic expansion were introduced (Cox and Leibovich 1994, Cox 1994) yielding evolution equations local in physical space. In the following we use the binary fluid problem to compare these two approaches and show that the reconstitution procedure yields incorrect stability results.

Basic Equations

We consider a binary fluid mixture confined between rigid (i.e., no-slip) poorly conducting

E. Tirapegui and W. Zeller (eds.), Instabilities and Nonequilibrium Structures V, 75–84.

plates. The equations for the dimensionless perturbations to the static state are, for infinite Prandtl number and two-dimensional motion,

$$\Delta^2 \psi + R(1+S)\theta_x + SR\phi_x = 0 \tag{1}$$

$$\theta_t + J(\psi, \theta) = \psi_x + \Delta\theta \tag{2}$$

$$\phi_t + J(\psi, \phi) = \tau \Delta \phi - \Delta\theta \tag{3}$$

(cf. Knobloch 1989). Here ψ is the stream function, θ and Σ denote the temperature and concentration perturbations, and $\phi \equiv \Sigma - \theta$. The quantity $J(a, b)$ denotes $a_x b_z - a_z b_x$. The parameters R, S and τ are the Rayleigh number, the separation ratio and the Lewis number, respectively. We assume that the fluid is bounded above and below by rigid, poorly conducting, surfaces in which case the boundary conditions are

$$\psi = D\psi = 0 \qquad \text{on } z = \pm 1 \tag{4}$$

$$D\theta \pm Bi\,\theta = 0, \ D\phi = 0, \qquad \text{on } z = \pm 1, \tag{5}$$

where $D \equiv \partial/\partial z$. In what follows we assume that the Biot number Bi is small and write $Bi = \epsilon^4 b$ where ϵ is a small parameter. Since the instability for small Biot number occurs with long wavelength we define new scaled variables $X = \epsilon x$, $T = \epsilon^2 t$ and $T' = \epsilon^4 t$. A consistent solution is found provided $\tilde{\psi} = \epsilon \psi$. The scaled equations become, after dropping the tilde from ψ,

$$D^4\psi + 2\epsilon^2 D^2\psi_{XX} + \epsilon^4\psi_{XXXX} = -R(1+S)\theta_X - RS\phi_X \tag{6}$$

$$D^2\theta + \epsilon^2\theta_{XX} = \epsilon^2\theta_T + \epsilon^4\theta_{T'} + \epsilon^2(\psi_X D\theta - \theta_X D\psi) - \epsilon^2\psi_X \tag{7}$$

$$\tau D^2\phi + \epsilon^2\tau\phi_{XX} = \epsilon^2\phi_T + \epsilon^4\phi_{T'} + \epsilon^2(\psi_X D\phi - \phi_X D\psi) + D^2\theta + \epsilon^2\theta_{XX} \tag{8}$$

subject to

$$\psi = D\psi = D\phi = 0 \qquad \text{on } z = \pm 1 \tag{9}$$

$$D\theta \pm \epsilon^4 b\,\theta = 0 \qquad \text{on } z = \pm 1. \tag{10}$$

The equations are then solved by an asymptotic expansion of the form

$$\psi = \psi_0 + \epsilon^2\psi_2 + \ldots$$

$$\theta = \theta_0 + \epsilon^2\theta_2 + \ldots$$

$$\phi = \phi_0 + \epsilon^2\phi_2 + \ldots$$

$$R = R_0 + \epsilon^2 R_2 + \ldots$$

Solution

The leading order solution is given by

$$\theta_0 = f(X, T, T'), \qquad \phi_0 = g(X, T, T') \qquad \psi_0 = -R_0 K_X P(z),$$

where f and g are arbitrary functions, P is a polynomial such that $P^{(iv)} = 1$, $P(\pm 1) = 0$, $P'(\pm 1) = 0$, and $K \equiv (1+S)f + Sg$. At the next order, the solvability conditions for θ and ϕ determine a coupled system of linear equations that relate f and g:

$$f_T - f_{XX} + \frac{R_0}{45} K_{XX} = 0 \tag{11}$$

$$g_T - \tau g_{XX} + f_{XX} = 0, \tag{12}$$

and the solution at this order is

$$\theta_2 = R_0 K_{XX} T_1(z) + R_0 K_X f_X T_2(z) + f_2(X, T)$$

$$\phi_2 = \frac{R_0}{\tau} K_{XX} T_1(z) + \frac{R_0}{\tau}(K_X f_X + K_X g_X) T_2(z) + g_2(X, T)$$

$$\psi_2 = R_0 K_{XXX} (2 P_1(z) - R_0(1 + S + S/\tau) P_2(z)) - R_2 K_X P(z) - R_0 K_{2X} P(z)$$

$$- R_0^2 [(1 + S + S/\tau)(K_X f_X)_X + (S/\tau)(K_X g_X)_X] P_3(z),$$

where f_2 and g_2 are arbitrary functions, $K_2 \equiv (1+S) f_2 + S g_2$, and the polynomials that appear satisfy $T_1'' = P - 1/45, T_1'(\pm 1) = 0, T_2'' = P', T_2'(\pm 1) = 0, P_1^{(iv)} = P'', P_1(\pm 1) = P_1'(\pm 1) = 0, P_2^{(iv)} = T_1, P_2(\pm 1) = P_2'(\pm 1) = 0, P_3^{(iv)} = T_2, P_3(\pm 1) = P_3'(\pm 1) = 0$. The linear theory that follows from equations (11) and (12) does not contain the correction due to small Biot number and predicts, incorrectly, that the most unstable wavenumber is $k = 0$. To calculate the correct critical wavenumber it is necessary to go to higher order.

At fourth order the solvability conditions for θ_4 and ϕ_4 give a set of two coupled equations for f_2 and g_2. These are,

$$f_{2T} - f_{2XX} + \frac{R_0}{45} K_{2XX} = -bf - f_{T'} + \frac{R_0}{2} K_{XXXX} [I_1 + 2N_1 - R_0(1 + S + S/\tau) N_2]$$

$$- \frac{R_0}{2} I_1 K_{XXT} - \frac{R_2}{45} K_{XX} - \frac{R_0^2}{2} J_2 (K_X^2 f_X)_X \tag{13}$$

and

$$g_{2T} - \tau g_{2XX} + f_{2XX} = bf - g_{T'} - \frac{R_0}{2\tau} I_1 K_{XXT} - \frac{R_0^2}{2\tau} J_2 (K_X^2 f_X + K_X^2 g_X)_X, \tag{14}$$

where $I_1 \equiv \int_{-1}^{1} T_1 dz = 31/7560$, $N_1 \equiv \int_{-1}^{1} P_1 dz = -4/945$, $N_2 \equiv \int_{-1}^{1} P_2 dz = 181/3742200$ and $J_2 \equiv \int_{-1}^{1} P' T_2 dz = -4/2835$.

Following Cox and Leibovich we obtain local evolution equations that include the correction to the critical wavenumber and Rayleigh number due to the small Biot number by reconstructing from the above results equations for $F \equiv f + \epsilon^2 f_2$, and $G \equiv g + \epsilon^2 g_2$. Going back to the original variables x and t, and recalling that $R = R_0 + \epsilon^2 R_2$ and $Bi = \epsilon^4 b$ we find that the small parameter ϵ cancels out and the evolution equations for F and G, correct to order ϵ^2, become

$$F_t + \left[\frac{R(1+S)}{45} - 1 \right] F_{xx} + \frac{RS}{45} G_{xx} = -BiF + \gamma F_{xxxx} + \delta G_{xxxx} - \frac{R^2}{2} J_2 (K_x^2 F_x)_x \tag{15}$$

$$G_t - \tau G_{xx} + F_{xx} = BiF - \frac{R\,I_1}{2\,\tau}\left[\alpha F_{xxxx} + \beta G_{xxxx}\right] - \frac{R^2}{2\,\tau}J_2\left[K_x^2 F_x + K_x^2 G_x\right]_x, \qquad (16)$$

where

$$\alpha = 1 - \frac{R}{45}(1+S)^2$$

$$\beta = \tau S - \frac{R}{45}S(1+S)$$

$$\gamma = \frac{R}{2}(1+S)\left[I_1 + 2N_1 - (1+S+\frac{S}{\tau})RN_2\right] - \frac{\alpha R}{2}I_1$$

$$\delta = \frac{R}{2}S\left[I_1 + 2N_1 - (1+S+\frac{S}{\tau})RN_2\right] - \frac{\beta R}{2}I_1,$$

and $\frac{\partial}{\partial t} \equiv \epsilon^2\frac{\partial}{\partial T} + \epsilon^4\frac{\partial}{\partial T'}$. Note, however, that although ϵ is formally absent from these equations, the equations are only valid in the limit $\epsilon \to 0+$. In this limit all the nonlinear terms, as well as the fourth derivative terms, provide *small* corrections to the linear problem (11,12). When this is not the case the equations no longer describe the correct dynamics.

Linear Theory

Consider now the linear theory that follows from equations (15) and (16). If we let $F = F_0 e^{\lambda t}e^{ikx}$, $G = G_0 e^{\lambda t}e^{ikx}$ we find the characteristic equation for the eigenvalues λ:

$$\lambda^2 + (a_1 + a_4)\lambda + (a_1a_4 - a_2a_3) = 0.$$

Oscillatory instability with $\lambda = i\omega$ is possible when $a_1 + a_4 = 0$, in which case $\omega^2 = a_1a_4 - a_2a_3 > 0$. The first condition implies

$$R^o = \frac{45(1+\tau)}{(1+S)} + \frac{45}{(1+S)}\left[\frac{Bi}{k^2} + \left(\frac{\beta R^o I_1}{2\tau} - \gamma\right)k^2\right].$$

Recalling that the Biot number is of order ϵ^4 and the horizontal scale is of order ϵ, an error of order ϵ^4 is made if the quantity accompanying k^2 is evaluated at $R_0^o = 45(1+\tau)/(1+S)$. Doing so we obtain

$$R^o = \frac{45(1+\tau)}{(1+S)} + \frac{45}{(1+S)}\left[\frac{Bi}{k^2} + \nu k^2\right], \qquad (17)$$

where

$$\nu = \frac{2}{231}\left(\frac{1+\tau}{1+S}\right)\left[17 + 12S - 5(\tau + S\tau + \frac{S}{\tau})\right]. \qquad (18)$$

The critical wavenumber calculated from $dR/dk^2 = 0$ is

$$k_c^4 = \frac{Bi}{\nu} \qquad (19)$$

and the critical Rayleigh number and frequency are then given by

$$R_c^o = \frac{45}{1+S}\left(1 + \tau + 2\sqrt{\nu Bi}\right), \qquad (20)$$

$$\omega_c^2 = -k_c^4\left(\tau^2 + \frac{S(1+\tau)}{1+S}\right), \qquad (21)$$

as obtained by Pismen (1988). From the above expressions we see that overstability is possible for $\nu > 0$ and for $\tau^2 + S(1+\tau)/(1+S) < 0$. This last condition is, for positive τ, more restrictive than $\nu > 0$. The only condition for the existence of the oscillatory instability is then,

$$\tau^2 + \frac{S(1+\tau)}{1+S} < 0. \tag{22}$$

In contrast steady convection arises if $a_1a_4 - a_2a_3 = 0$. The critical Rayleigh number for the onset of stationary convection is, to leading order,

$$R_c^s = \frac{45\tau}{S + \tau + \tau S} \tag{23}$$

which is possible whenever $S + \tau + S\tau > 0$. Note that $\omega_c^2 > 0$ implies that $R_c^s > R_c^o$, i.e., that the bifurcation to oscillations then precedes the steady state bifurcation. In Figure 1 we show the critical values of the Rayleigh number for the onset of steady and oscillatory convection as a function of S for $\tau = 0.01$.

Discussion

When the Prandtl number is infinite vertical vorticity cannot be excited. The above derivation then extends immediately to two horizontal directions:

$$F_t + \left[\frac{R(1+S)}{45} - 1\right]\triangle F + \frac{RS}{45}\triangle G = -BiF + \gamma\triangle^2 F + \delta\triangle^2 G - \frac{R^2}{2}J_2\nabla\cdot[(\nabla K)(\nabla K\cdot\nabla F)] \tag{24}$$

$$G_t + \triangle F - \tau\triangle G = BiF - \frac{R\,I_1}{2\,\tau}\left[\alpha\triangle^2 F + \beta\triangle^2 G\right] - \frac{R^2}{2\,\tau}J_2\nabla\cdot[\nabla K\,(\nabla K\cdot(\nabla F + \nabla G))]. \tag{25}$$

Pattern selection in the resulting equations can be studied for $|R - R_c^o| \ll R_c^o$. In the absence of system anisotropies it is usual to consider spatially periodic patterns with wavelength $2\pi/k_c$ in two independent directions. There are then two possibilities that are of fundamental interest: patterns on the square lattice and patterns on the hexagonal lattice. The basic theory has been worked out by Swift (1984) and was completed by Roberts et al (1986) and Silber and Knobloch (1991). The theory not only establishes the nature of the different types of solutions that bifurcate at $R = R_c^o$ but also their stability properties with respect to perturbations in the form of all the competing patterns. On the square lattice there are generically five distinct temporally periodic solutions that bifurcate simultaneously; an additional unstable periodic solution may also be present as can a quasiperiodic solution with two temporal frequencies or a solution in the form of an attracting heteroclinic cycle. The situation is yet richer on the hexagonal lattice, where as many as eleven temporally periodic solutions bifurcate simultaneously. In certain regions in parameter space solutions with multiple frequencies can be stable, as can chaotic solutions. The hexagonal lattice results have recently been applied to the closely related doubly diffusive problem (Renardy 1993) in which the concentration gradient is maintained externally rather than being set up in response to the applied temperature gradient. A more detailed application of both sets of results to convection in an imposed vertical magnetic field is also available (Clune and Knobloch 1994) and focuses on the more complex solutions allowed by the respective amplitude equations.

In the two examples mentioned above the horizontal wavelength of the instability is comparable to the layer depth resulting in a more involved reduction to the relevant amplitude equations. Equations (24,25) offer a significantly simpler system to which these techniques can be applied. The results for the square lattice and $Bi = 10^{-4}$ are shown in fig. 2 for $0 < \tau < 1.0$, $-1.0 < S < 0$. The region labelled TW indicates small amplitude stable travelling waves; as one crosses the boundary of this region the travelling waves become subcritical, and there are no stable solutions near the Hopf bifurcation. The region SS indicates the absence of a Hopf bifurcation. The stability boundaries shift only slightly as Bi varies from 10^{-12} to 10^{-2}. Identical results obtain on the hexagonal lattice. This result is both interesting and important since Pismen (1988), in analyzing the same problem (equations 1-5), concluded that among the patterns on the square lattice there is a stable *two*-dimensional oscillatory pattern, and that all one-dimensional patterns (such as TW) are unstable with respect to transverse perturbations. This pattern (called the antiphase pattern by Pismen) corresponds to the alternating roll pattern studied by Silber and Knobloch (1991). We have checked Pismen's result and find that it is correct. Consequently the difference in the prediction using the local equations and Pismen's nonlocal calculation must be traced to the method used to solve the problem. We now discuss the two techniques in turn.

The technique employed in deriving equations (24,25) involves a resummation (or reconstitution) of the first two equations obtained using an order by order asymptotic expansion. Procedures of this type ought to be suspect in spite of their common usage. The reason is simply that the very essence of the technique is taking the limit as $\epsilon \to 0$ order by order. Only because of such a limiting process is one allowed to conclude that the first order terms must balance, and so on. If one assumes (implicitly) that ϵ is finite (as we have done) the corresponding conclusion need not hold. Moreover one must now address the convergence of the expansion in order to truncate the resummed equations. This is in general hard, and in fact is unlikely to be true since "most" asymptotic expansions do not in fact converge. In the following we relate the discrepancy between the two sets of stability results to this point.

We begin by discussing a simple example, the derivation of the coupled complex Ginzburg-Landau equations for counterpropagating waves on a line. We start with our equations (1-5) in the symbolic form

$$(L_0 + \epsilon^2 L_2)\psi = \epsilon N(\psi, \psi), \tag{26}$$

where L_0 is the linear operator at R_c^o, $\epsilon^2 \equiv (R - R_c^o)/R_c^o$ and N denotes nonlinear terms. As usual we expand

$$\psi = \psi_0 + \epsilon\psi_1 + ... \tag{27}$$

The solution of the linear problem $L_0\psi_0 = 0$ is assumed to be of the form

$$\psi = \epsilon\{A_1(X, T, T')e^{i(\omega t + kx)} + B_1(X, T, T')e^{i(\omega t - kx)} + c.c.\} + ..., \tag{28}$$

where $X = \epsilon x$, $T = \epsilon t$ and $T' = \epsilon^2 t$ are the slow scales and A_1, B_1 describe the slowly varying amplitudes of left and right travelling waves, respectively. At leading order one gets the dispersion relation relation $\omega = \omega(k)$ specifying the wavenumber dependence of the frequency. At second order one obtains the equations

$$\frac{\partial A_1}{\partial T} - c_g\frac{\partial A_1}{\partial X} = 0, \quad \frac{\partial B_1}{\partial T} + c_g\frac{\partial B_1}{\partial X} = 0 \tag{29}$$

showing that $A_1 = A_1(\eta, T')$, $B_1 = B_1(\xi, T')$, where $\eta = X + c_g T$, $\xi = X - c_g T$. At next order one obtains equations of the form

$$\frac{\partial A_2}{\partial T} - c_g \frac{\partial A_2}{\partial X} = -\frac{\partial A_1}{\partial T'} + \mu A_1 + \gamma A_{1XX} + \alpha |A_1|^2 A_1 + \beta |B_1|^2 A_1 \tag{30}$$

$$\frac{\partial B_2}{\partial T} + c_g \frac{\partial B_2}{\partial X} = -\frac{\partial B_1}{\partial T'} + \bar{\mu} B_1 + \bar{\gamma} B_{1XX} + \bar{\alpha} |B_1|^2 B_1 + \bar{\beta} |A_1|^2 B_1. \tag{31}$$

The resulting expansion will be asymptotic for $(x, t) = O(\epsilon^{-2})$ if and only if A_2, B_2 remain of order one on these scales. Consequently the right hand sides of equations (30,31) must vanish when averaged over the corresponding comoving variables:

$$\frac{\partial A_1}{\partial T'} = \mu A_1 + \gamma A_{1\eta\eta} + \alpha |A_1|^2 A_1 + \beta \langle |B_1|^2 \rangle A_1 \tag{32}$$

$$\frac{\partial B_1}{\partial T'} = \bar{\mu} B_1 + \bar{\gamma} B_{1\xi\xi} + \bar{\alpha} |B_1|^2 B_1 + \bar{\beta} \langle |A_1|^2 \rangle B_1. \tag{33}$$

These equations are the required asymptotically exact evolution equations and were derived by Knobloch and De Luca (1990); their properties were discussed in detail by Knobloch (1992).

If instead we followed the resummation procedure employed above, we would add to equation (29) ϵ times equation (30) obtaining instead

$$\frac{\partial A_1}{\partial T} + \epsilon \frac{\partial A_2}{\partial T} + \epsilon \frac{\partial A_1}{\partial T'} - c_g \left(\frac{\partial A_1}{\partial X} + \epsilon \frac{\partial A_2}{\partial X}\right) = \epsilon \mu A_1 + \epsilon \gamma A_{1XX} + \epsilon \alpha |A_1|^2 A_1 + \epsilon \beta |B_1|^2 A_1 \tag{34}$$

with a similar equation for the right travelling waves. We next define the reconstituted amplitudes $A = \epsilon A_1 + \epsilon^2 A_2 + ...$, $B = \epsilon B_1 + \epsilon^2 B_2 + ...$ and suppose these to depend on a combined slow time t such that $(\partial/\partial t) = \epsilon(\partial/\partial T) + \epsilon^2(\partial/\partial T')$. It follows that ϵ once again disappears from the resummed equations and one obtains

$$\frac{\partial A}{\partial t} - c_g \frac{\partial A}{\partial X} = (R - R_c^o + i\Omega)A + \gamma A_{xx} + \alpha |A|^2 A + \beta |B|^2 A + ... \tag{35}$$

$$\frac{\partial B}{\partial t} + c_g \frac{\partial B}{\partial X} = (R - R_c^o - i\Omega)B + \bar{\gamma} B_{xx} + \bar{\alpha} |B|^2 B + \bar{\beta} |A|^2 B + ..., \tag{36}$$

where Ω denotes the shift in ω due to $R - R_c^o$. Note that these equations are of the form obtained above *except* for the absence of the averages. In other words, the main distinction between the asymptotically exact equations and the resummed ones is that the former are *nonlocal* while the latter are *local*. This is the same dichotomy as for the binary fluid problem treated above.

It is important to observe that in the above example the distinction between the two approaches is not merely academic: the stability of uniform wavetrains with respect to slow modulations are *different* in the two formulations (see Knobloch 1992 for a discussion of this point). We have noted a similar discrepancy between the two approaches in the example studied in this paper.

To understand the origin of this discrepancy recall that Pismen does not carry out any resummation of his asymptotic series. Instead he proceeds by explicitly solving the second order problem (11,12) for the slow evolution of the planform functions f, g and then employs these in equations (13,14) to identify those terms on the right hand side that resonate with the linear operator on the left. If one applies his procedure to the derivation of the coupled complex Ginzburg-Landau equations one again recovers the Ginzburg-Landau equations with the mean field coupling, as one would expect since both techniques are asymptotically exact. For the binary mixture convection Pismen's approach is more involved, mostly because equations (11,12) are somewhat more complicated than in the Ginzburg-Landau case. But they also ultimately lead to evolution equations that are nonlocal in space. To overcome this difficulty Pismen writes his equations in Fourier space in which his equations *are* local. Owing to the symmetries of the problem these Fourier space equations are precisely of the type studied by Silber and Knobloch (1991). Using the coupling coefficients computed by Pismen (which we checked) and the general stability criteria derived by Silber and Knobloch it is easy to verify Pismen's conclusion that alternating rolls are in fact stable near onset, and that the competing patterns are all unstable. More succinctly, Pismen takes the limit $\epsilon \to 0+$ *before* computing the stability of the patterns. Recall that these stability calculations are themselves carried out using a perturbation expansion in powers of a *different* ε, namely the amplitude of the pattern, with the eigenvalues of interest being of order ε^2. This ε bears in general no relation to the ϵ introduced by the smallness of the Biot number. In contrast the reconstitution procedure requires one to take ϵ finite, and calculate the $O(\varepsilon^2)$ eigenvalues. In effect it requires that one first takes the limit $\varepsilon \to 0+$ before taking the limit $\epsilon \to 0+$. Pismen's procedure takes these limits in the opposite order, and our results indicate that these two limiting procedures do not commute. A similar lack of commutativity is known to be responsible for the difference in the stability properties of the standing wave solutions of the local and nonlocal coupled complex Ginzburg-Landau equations (Knobloch 1992).

Conclusions

In this paper we have compared a resummation procedure used to derive a coupled set of local equations that describe the onset of oscillatory convection in a binary fluid confined between poorly conducting plates with the results of an asymptotically exact analysis due to Pismen (1988). The former are local but not asymptotically exact; the latter involves nonlocal but asymptotically exact equations. Given that the two techniques differ dramatically, and lead to dramatically different evolution equations, we have thought it worth while to present both techniques somewhat in parallel, in order to facilitate comparison between them. The outcome of this comparison is unambiguous and important. The resummed equations fail to give the correct stability properties of the patterns and the technique, however appealing, should be avoided, unless rigorously justified.

It should be noted that nonasymptotic equations have been used with success in appropriate circumstamces. Perhaps the best known is the case of the Takens-Bogdanov bifurcation (cf Knobloch and Proctor 1981, Guckenheimer and Knobloch 1983) which arises in equations (1-5) when $R_c^o = R_c^s$. Here, however, one can appeal to a "theorem" (Arnol'd 1977) that there exists a *finite* range of ϵ for which the resulting equations hold. In particular the perturbation expansion converges, and hence the amplitude equations can be found by a nonasymptotic procedure. Corresponding results are not, however, usually available for other problems and in particular for ones involving slow spatial scales. A notable exception is provided by the recent work of Collet and Eckmann (1990, 1992; see also van

Harten 1991 and Schneider 1994). Results of this type appear to suffice for the validity of the reconstitution procedure since they guarantee that the two limiting procedures involved in the stability assignments do in fact commute.

Acknowledgement

Most of this work was done during E.K.'s visit to the Universidad Católica de Chile in December 1993. Two of us (M.C.D. and E.K.) would like to thank Professor E. Tirapegui for his invitation to the Fifth International Workshop on Instabilities and Nonequilibrium Structures. Partial support from the California Space Institute is also acknowledged. The work of M.C.D. was partially supported by Fondecyt 193/0559.

References

[1] V.I. Arnol'd, Functional Anal. and Applics. 11, 85 (1977).

[2] T. Clune and E. Knobloch, Physica D, in press (1994).

[3] P. Collet and J.-P. Eckmann, Commun. Math. Phys. 132, 139–153 (1990).

[4] P. Collet and J.-P. Eckmann, Nonlinearity 5, 1265–1302 (1992).

[5] S. Cox, J. Eng. Math., in press (1994).

[6] S. Cox and S. Leibovich, preprint (1994).

[7] J. Guckenheimer and E. Knobloch, Geophys. Astrophys. Fluid Dyn. 23, 247–272 (1983).

[8] A. van Harten, J. Nonlinear Sci. 1, 397–422 (1991).

[9] E. Knobloch, Phys. Rev. A 40, 1549–1559 (1989).

[10] E. Knobloch, in *Pattern Formation in Complex Dissipative Systems*, S. Kai, ed., World Scientific, Singapore, pp. 263–274 (1992).

[11] E. Knobloch and J. De Luca, Nonlinearity 3, 975–980 (1990).

[12] E. Knobloch and M.R.E. Proctor, J. Fluid Mech. 108, 291–316 (1981).

[13] L. Pismen, Phys. Rev. A 38, 2564–2572 (1988).

[14] Y. Renardy, Phys. Fluids A 5, 1376–1389 (1993).

[15] M. Roberts, J.W. Swift and D. Wagner, Contemp. Math. 56, 283–318 (1986).

[16] G. Schneider, J. Nonlinear Sci. 4, 23–34 (1994).

[17] M. Silber and E. Knobloch, Nonlinearity 4, 1063–1106 (1991).

[18] J.W. Swift (1984), Ph.D. Thesis, University of California, Berkeley.

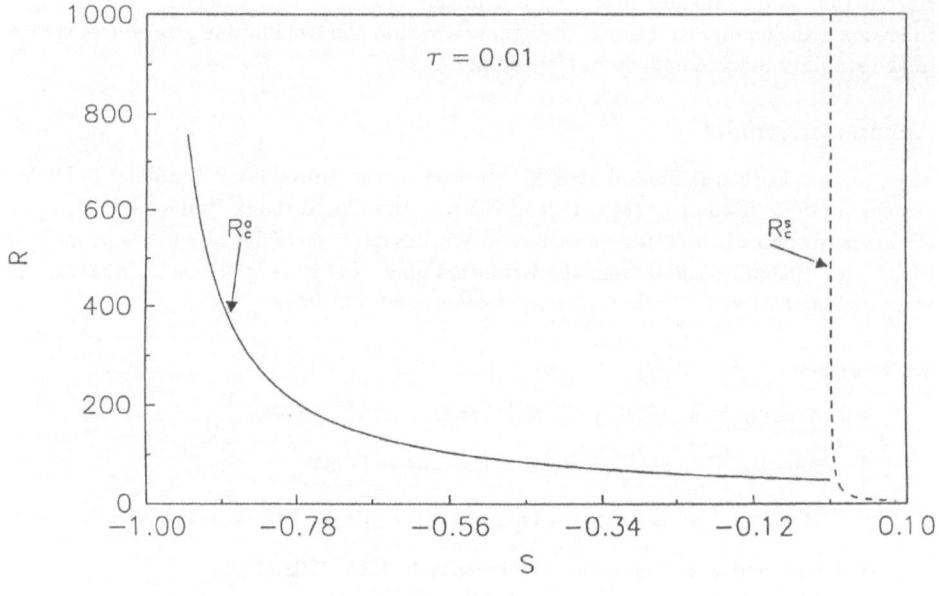

Figure 1

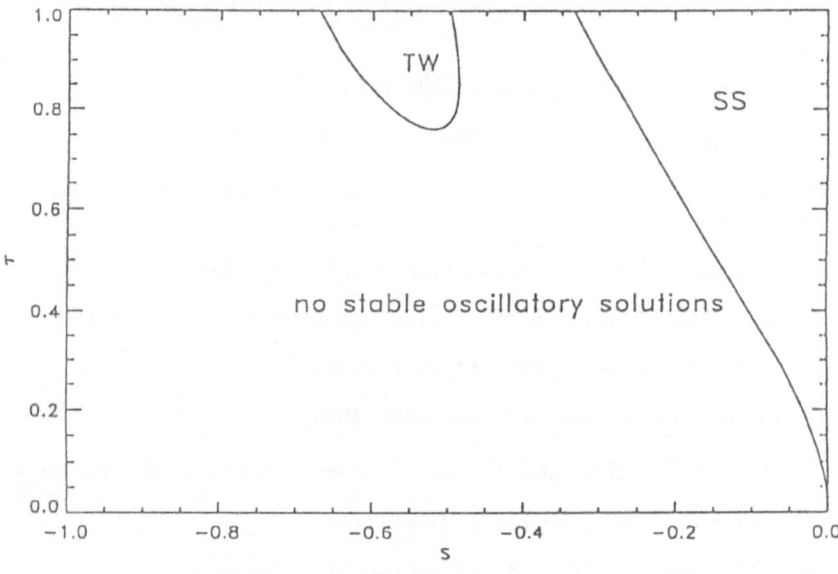

Figure 2

MULTIPLE FRONT PROPAGATION IN A POTENTIAL NON-GRADIENT SYSTEM

M. SAN MIGUEL, R. MONTAGNE[1], A. AMENGUAL, E. HERNÁNDEZ-GARCÍA.
Departament de Física
Universitat de les Illes Balears
E-07071 Palma de Mallorca, Spain

ABSTRACT. A classification of dynamical systems in terms of their variational properties is reviewed. Within this classification, front propagation is discussed in a non-gradient relaxational potential flow. The model is motivated by transient pattern phenomena in nematics. A front propagating into an unstable homogenous state leaves behind an unstable periodic pattern, which decays via a second front and a second periodic state. An interface between unstable periodic states is shown to be a source of propagating fronts in opposite directions.

1. Introduction

In this paper we will consider pattern formation in systems which approach, asymptotically in time, an homogeneous stable state. However, they exhibit long-lived pattern dynamics as transient states. Such patterns may originate in finite wavenumber fluctuations triggering the decay of an initially homogeneous stable state[1,2], or be a consequence of fronts propagating through the system[3]. We will focus here on patterns created by front propagation into unstable states. A physical motivation for this study is the long lived transient patterns observed in the Fréedericksz transition in nematic liquid crystals[4].

The type of systems discussed here are described by dynamical systems for which a Lyapunov functional, also called potential, exists. Time evolution proceeds then by minimizing such potential. However, we wish to emphasize that potential minimization can be attained along many different directions in the phase space of the system, depending on other ingredients of the dynamical system beyond the potential. In fact, the existence of rich transient dynamical behavior in many systems can be traced back to these other dynamical ingredients of the potential system. For pedagogical purposes we review in Sect. 2 a classification of dynamical systems in which the different role of relaxational vs. non-relaxational (non-dissipative) contributions to the dynamics is pointed out.

A general model for transient pattern dynamics is introduced in Sect. 3. The model is motivated by the physics of the Fréedericksz transition in nematics and justified from the nematodynamic equations[5]. Within the classification of Sect. 2, the model belongs to the class of potential and relaxational non-gradient flows. The non-gradient contributions originate in an approximation to non-relaxational dynamics associated with hydrodynamic coupling in the nematodynamic equations. In Sects. 4 and 5 we study front propagation in this model. A front connecting homogeneous stable and unstable states is seen to propagate leaving behind it a periodic spatial pattern which is linearly unstable. The decay of this pattern is via an intermediate periodic pattern of larger wavelength. This second pattern is still linearly unstable, but it is more stable in terms of the Lyapunov potential. It appears behind a second front which propagates into the original periodic

---

[1]on leave from Universidad de la República (Uruguay).

E. Tirapegui and W. Zeller (eds.), Instabilities and Nonequilibrium Structures V, 85–97.

state. The velocity of the two propagating fronts, as well as the wavenumber of the periodic states, is successfully determined from marginal stability arguments[6]. Finally, in Sect. 5 we consider the fronts which originate in the interface between two periodic unstable states of different periodicity. Such interface is shown to be a source of fronts propagating in opposite directions while the system evolves through states of greater stability in its search for the global potential minimum.

2. What is a potential system?

It is commonly stated that the rich variety of dynamical states that occur in non-equilibrium systems originates in the non-potential or non-variational character of the dynamical models which describe them. The main argument is that, in these cases, the study of the dynamics can not be reduced to the minimization of a potential which plays the role of the free energy of equilibrium systems. This general statement needs to be qualified, since it is well known that models used to study equilibrium critical dynamics[7] include mode-mode coupling terms such that its dynamical evolution is not simply given by the minimization of the free energy. In the following we review a classification of dynamical systems that, although well established in the context of stochastic dynamics[8,9] it is often overlooked in general discussions of deterministic spatio-temporal dynamics.

Non-potential dynamical systems are usually defined as those for which there is no Lyapunov potential giving the time evolution. Unfortunately, this definition is also applied to cases in which there is no *known* Lyapunov potential. To be more precise, let us consider dynamical systems of the form

$$\dot{\psi} = A[\psi] \tag{2.1}$$

where ψ represents the set of dynamical variables. For generality, we take them to be complex, and the notation ψ^* represents the complex conjugate of ψ. Here the dynamical variables will be spatially dependent fields: $\psi = \psi(\mathbf{x}, t)$. $A[\psi]$ is a functional of them. Let us now split A into two contributions:

$$A[\psi] = G[\psi] + N[\psi] , \tag{2.2}$$

where G, the *relaxational* part, will have the form

$$G[\psi] = -\Gamma \frac{\delta F[\psi]}{\delta \psi^*} , \tag{2.3}$$

with F a real and scalar functional of ψ. Γ is an arbitrary hermitic and positive-definite operator (possibly depending on ψ). In the case of real variables there is no need of taking the complex conjugate, and hermitic operators reduce to symmetric ones. The functional $N[\psi]$ in (2.2) is the remaining part of $A[\psi]$. The important point is that, if the splitting (2.2) can be done in such a way that the following orthogonality condition is satisfied (c.c. denotes the complex conjugate expression):

$$\int d\mathbf{x} \left(\frac{\delta F[\psi]}{\delta \psi^*} N[\psi(\mathbf{x})]^* + \text{c.c.} \right) = 0 , \tag{2.4}$$

then the terms in N neither increase nor decrease the value of F, which due to the terms in G becomes a decreasing function of time:

$$\frac{dF[\psi(\mathbf{x},t)]}{dt} \le 0 . \tag{2.5}$$

If F is bounded from below then it is a Lyapunov potential for the dynamics (2.1). Equation (2.4), with $N = A - G$ can be interpreted as an equation for the potential F associated to a given dynamical system (2.1). It has a Hamilton-Jacobi structure. Its solution is in general a difficult task, but several non-trivial results exist in the literature[8,10].

Once this notation has been set-up, we can call relaxational systems those such that there is a solution F of (2.4) such that $N = 0$, that is all the terms in A contribute to decrease F. Potential systems can be defined as those for which there is a nontrivial (i.e. a non-constant) solution F to (2.4). A more detailed classification is the following:

1.- **Relaxational gradient flows:** Those dynamical systems for which $N = 0$ and Γ is a constant. In this case the time evolution of the system follows the lines of steepest descent of F. A well known example is the Fisher-Kolmogorov equation, also known as model A of critical dynamics[7], or (real) Ginzburg-Landau equation for a real field $\psi(\mathbf{x}, t)$:

$$\dot\psi = r\nabla^2\psi + c\psi - b \mid \psi \mid^2 \psi , \tag{2.6}$$

where r, c, and b are real coefficients. This equation is of the form of Eq. (2.1)-(2.3) with $N = 0, \Gamma = 1$, and $F = F_{GL}[\psi]$, the Ginzburg-Landau free energy:

$$F_{GL}[\psi] = \int d\mathbf{x} \left(\frac{r}{2} \mid \nabla\psi \mid^2 - \frac{c}{2} \mid \psi \mid^2 + \frac{b}{4} \mid \psi \mid^4 \right) \tag{2.7}$$

2.- **Relaxational non-gradient flows:** Still $N = 0$, but Γ is not constant, so that the dynamics does not follow the lines of steepest descent of F. A well known example of this type is the Cahn-Hilliard equation of spinodal decomposition, or model B of critical dynamics[7]:

$$\dot\psi = (-\nabla^2) \left(-\frac{\delta F_{GL}[\psi]}{\delta\psi} \right) , \tag{2.8}$$

The symmetric and positive-definite operator $(-\nabla^2)$ has its origin in a conservation law for ψ.

3.- **Non-relaxational potential flows:** N does not vanish, but the potential F, solution of (2.4) exists and is non-trivial. Most models used in equilibrium critical dynamics[7] belong to this category. A simple example of this is

$$\dot\psi = -(1 + i)\frac{\delta F_{GL}[\psi]}{\delta\psi^*} , \tag{2.9}$$

where now ψ is a complex field. Notice that we can not interpret this equation as being of type 1, because $(1 + i)$ is not a hermitic operator, but still F_{GL} is a Lyapunov functional for the dynamics. Equation (2.9) is a special case of the Complex Ginzburg- Landau Equation

(CGLE), in which $A[\psi]$ is the sum of a relaxational gradient flow and a nonlinear Schrödinger-type equation $N[\psi] = -i\frac{\delta F_{GL}[\psi]}{\delta\psi^*}$. The general CGLE[11] is of the form (2.6) but ψ is complex and r, c, and b are arbitrary complex numbers. Calculations by Graham and coworkers indicate[12,13] that the CGLE, a paradigm of complex spatio-temporal dynamics, might be classified within this class of non-relaxational potential flows. The difficulty is that the explicit form of the potential to be obtained as a solution of (2.4), is only known in an uncontrolled small-gradient expansion.

4.- **Non-potential flows:** Those for which the only solutions F of (2.4) are the trivial ones (that is F = constant). Hamiltonian systems as for example the nonlinear Schrödinger equation are of this type.

In this paper we only deal with potential situations. The model we introduce in the next section belongs to the class of relaxational non-gradient dynamics. The non-gradient character has its origin in the non-relaxational and more complex dynamics of nematic liquid crystals in a magnetic field, from which our model is obtained after some approximations.

3. A general model for transient pattern dynamics

Transient pattern formation is well documented experimentally for different instabilities in nematic liquid crystals[14]. We discuss here a general model whose physical motivation is the magnetic Fréedericksz transition. In this transition the reorientation of the nematic director in response to a large enough applied magnetic field does not proceed homogeneously: a transient striped pattern with a characteristic wavelength emerges. At long times the pattern disappears leading to the homogeneously reoriented final equilibrium state. We consider a twist geometry in which the sample is contained between two plates separated a distance d and perpendicular to the z-axis. The nematic material is prepared with its director field $\vec{n}^o$ aligned with the x-axis and a magnetic field $\vec{H}$ is applied in the y-direction. When the magnetic field is switched-on at time $t = 0$ from an initial value smaller than a critical value to a final value above it a striped pattern appears in the x-y plane with domain walls parallel to the y-axis.

The dynamics of this system can be described in terms of the nematodynamic equations. We assume homogeneity in the y-direction and that the reorientation takes place in the x-y plane, so that the director field $\vec{n}(\vec{r})$ is written in terms of an angle ϕ as $n_x(x, z) = \cos\phi(x, z)$, $n_y(x, z) = \sin\phi(x, z)$. The director field is coupled to a velocity field $\vec{v}(x, z)$ which in the geometry described above we assume to be oriented along the y-direction. In a minimal coupling approximation the equations for ϕ and v_y become[15]:

$$d_t\phi(x, z) = -\frac{1}{\gamma_1}\frac{\delta F}{\delta\phi} + \frac{1}{2\rho}(1 + \lambda)\partial_x\frac{\delta F}{\delta v_y} , \qquad (3.1)$$

$$d_t v_y(x, z) = \frac{1}{2\rho}(1 + \lambda)\partial_x\frac{\delta F}{\delta\phi} + \frac{1}{\rho^2}(\nu_2\partial_z^2 + \nu_3\partial_x^2)\frac{\delta F}{\delta v_y} . \qquad (3.2)$$

ρ is the mean density, γ_1, λ , ν_2, and ν_3 are viscosity coefficients and F is a free energy:

$$F = \int d\vec{r}\{\frac{1}{2}\left[k_1(\vec{\nabla}\cdot\vec{n})^2 + k_2(\vec{n}\cdot\vec{\nabla}\times\vec{n})^2 + k_3(\vec{n}\times(\vec{\nabla}\times\vec{n}))^2\right] - \frac{1}{2}\chi_a(\vec{n}\cdot\vec{H})^2 + \frac{1}{2}\rho\vec{v}^2\} \quad (3.3)$$

The first three terms in (3.3) form the Oseen-Frank free energy associated with distortion of the director field, with k_1, k_2 and k_3 being splay, twist and bend elastic constants respectively. The next term is the magnetic contribution with χ_a being the anisotropic susceptibility, and the last term gives the hydrodynamic contribution.

This dynamical model falls within the general category of non-relaxational potential flows discussed above. The free energy F is a Lyapunov functional, but the dynamical model contains non-relaxational terms which give a vanishing contribution to the time derivative of F. The term proportional to $1/\gamma_1$ in (3.1) gives a gradient relaxational dynamics for ϕ, and the terms proportional to ν_2 and ν_3 in (3.2) give a non-gradient relaxational dynamics associated with viscosity for the velocity flow. On the other hand, the terms proportional to $(1 + \lambda)$ in (3.1) and (3.2), which give the coupling between ϕ and v_y, are of non-dissipative hydrodynamical origin and produce a non-relaxational dynamics for the whole system.

Dynamics can be described in terms of the amplitude $\psi(x)$ of the most unstable Fourier mode of $\phi(x, z)$ in the z-direction. A useful approximation in the limit of small inertia is to eliminate adiabatically the velocity field, which leads to a closed equation for $\psi(x, t)$. In appropriate dimensionless units such equation reads[16]

$$\partial_t \psi(x, t) = \Gamma(\partial_x)[\partial_x^2 \psi + c\psi - b\psi^3], \, b, c > 0 \qquad (3.4)$$

where $\Gamma(\partial_x)$ is a complicated kinetic coefficient which contains the remanent effect of the hydrodynamic coupling of director and velocity field after the adiabatic elimination of the latter. In the limit of long wavelengths that we consider in the following $\Gamma(\partial_x)$ becomes

$$\Gamma(\partial_x) \approx a - \partial_x^2, \quad a > 0 \qquad (3.5)$$

In the absence of hydrodynamic coupling Γ becomes a constant and (3.4) describes a gradient flow. In general (3.4) describes a non-gradient relaxational dynamics, with the ∂_x^2 term in (3.5) being the leading contribution from the non-relaxational dynamics in (3.1)-(3.2). By comparison with (2.8) we call (3.4) Modified Cahn-Hilliard Equation (MCH).

Eqs. (3.4) and (3.5) define a generic model to study transient pattern dynamics[1]. For this study it is first important to recall the existence of at least three kinds of bounded stationary solutions: (a) $\psi(x) = \pm\sqrt{c/b} \equiv \psi_\pm$, (b) $\psi(x) = 0$, and (c) a family of periodic solutions $\psi_q(x)$ of fundamental wavenumber q in the range $(0, \sqrt{c})$. Regarding stability with respect to small perturbations, both solutions included in type (a) are linearly stable, and all the solutions in (b) and (c) and linearly unstable. The dynamical evolution from the unstable solution (b) to the stable solution (a) proceeds via the formation of long-lived transient patterns which locally are close to solutions of type (c). The initial stages of pattern formation can be understood by the fact that the linear stability analysis of (3.4)-(3.5) around the solution $\psi = 0$ identifies a mode with finite wavenumber as the most unstable one. The instability becomes of zero wavenumber in the limit of gradient-flow dynamics in which Γ is a constant. What we show in the next section is that the periodic solutions (c) can also be realized by front propagation: A front connecting solutions (a) and (b) advances into (b) and leaves behind it a solution of type (c). This solution is shown to decay via a secondary front which separates states close to two different solutions of type (c).

4. Fronts propagating into periodic unstable states

The description of a uniform stable state advancing into a uniform unstable one is well known[17,18]. It is also known that front propagation can produce unstable periodic patterns, as explicitly demonstrated for the Extended Fisher-Kolmogorov model (EFK)[19], which is a relaxational gradient system. For the MCH model a new situation occurs, as shown in Fig. 1.

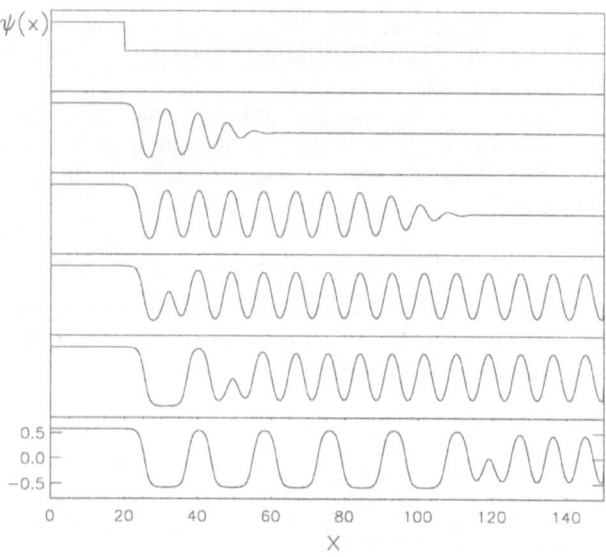

Figure 1 Graph of $\psi(x, t)$ for $t = 0, 20, 50, 125, 195$, and 455 for $a = 0.2$ and $b = 3$.

First, a front advances into the initial homogenous state leaving behind a periodic state. The novelty is that a second front appears and moves into the periodic state leaving behind a second periodic state with different periodicity. This was observed for different values of a. Figure 2 shows space-time plots of $\psi(x, t)$ obtained from a numerical solution of eqs. (3.4-3.5) for several values of a. The horizontal axis represents space, and time runs along the vertical axis. White corresponds to regions with ψ near the value of the stationary stable solution $1/\sqrt{b}$ and black to those with a value near $-1/\sqrt{b}$. The location of the fronts is clear and the measured slopes determine their velocities.

Our numerical results indicate that the mechanism of decay of the periodic state left behind the first front is similar to the decay of the initial $\psi = 0$ unstable state. That is, a front replaces the unstable state by some more stable state. Relative stability is here defined in terms of the Lyapunov functional of the model. This interpretation leads naturally to the study of fronts propagating into periodic unstable states as relevant for the understanding of the *second-front* phenomenon.

The first front, the one moving into the homogeneous phase, has a velocity well reproduced by marginal stability theory[19–21,6,18]. We will now show that the second front, understood as a front moving into an unstable periodic state, can also be described by a generalization of such theory[3]. There are several ways of formulating the marginal stability hypothesis for propagation into an homogeneous state. In all of them the dynamics of the front is analyzed in the leading edge, where the field is small enough so that the equation describing its evolution can be linearized. We use here the steepest descent (or saddle point) approach[19,22], since it is easily generalized to front propagation into a periodic state.

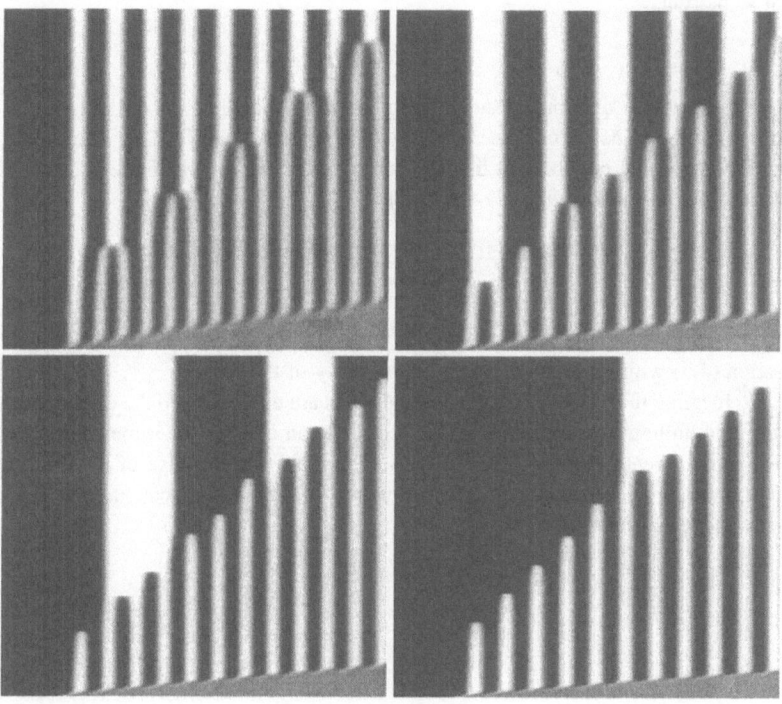

Figure 2 Space-time plots of $\psi(x, t)$ in gray levels, for $b = 3$ and $a = 0.0, 0.3, 0.6$ and 0.9. Space is represented along the horizontal axis (system size: 137.5) and time runs along the vertical one from 0 to 575.

We assume that the periodic pattern left by the first front is close to one of the stationary solutions with dominant wavenumber $q = q_i$, that will be denoted by ψ_{q_i}. We linearize around the periodic state $\psi(x, t) \equiv \psi_{q_i}(x) + \delta\psi(x, t)$ obtaining an equation for $\delta\psi$ of the form

$$\dot{\delta\psi} = [\mathcal{L} + \mathcal{U}_{q_p}] \delta\psi \ . \tag{4.1}$$

$\mathcal{L}$ is the linear operator giving the dispersion relation corresponding to linearization around the uniform solution $\psi = 0$, and $\mathcal{U}_{q_p}$ stands for the remaining part: a periodic operator of periodicity q_p related to q_i. Eq. (4.1) is a linear equation with periodic coefficients whose formal solution is given by Bloch (or Floquet) theory in terms of the eigenfunctions and eigenvalues of the linear operator

$\mathcal{L} + \mathcal{U}_{q_p}$. Given $\delta\psi_0(x) = \delta\psi(x, t = 0)$, the solution of (4.1) can be expressed in terms of the eigenfunctions $f_k(x)$ (of the Bloch form) and the eigenvalues $\epsilon(k)$ of the linear operator $\mathcal{L} + \mathcal{U}_{q_p}$

$$\delta\psi(x, t) = \int_{-\infty}^{+\infty} e^{\epsilon(k)t} f_k(x) \delta\hat{\psi}_0(k) dk \qquad (4.2)$$

where

$$\delta\hat{\psi}_0(k) = \int_L f_k^*(x) \delta\psi_0(x) dx , \qquad (4.3)$$

The integral is evaluated using the saddle point method in a referential frame $z = x - vt$ moving with the front velocity v,

$$\delta\psi(z, t \to \infty) \sim e^{h(q'_s)t + iq'_s z} \qquad (4.4)$$

where $h(q') = iq'v + \epsilon(k = q' - mq_p)$, and q'_s is the dominant saddle point of the function $h(q')$ extended to the complex plane. In our case[3], $mq_p \equiv -2q_i$, so that $k = q' + 2q_i$. The saddle point condition and the additional requirement that the perturbation $\delta\psi$ remains finite and bounded in time in the vicinity of the leading edge ($z \sim 0$) lead to

$$\begin{cases} \left.\frac{d\epsilon(k)}{dk}\right|_{k=q'_s - mq_p} = -iv \\ \\ \mathrm{Re}[h(q'_s)] = 0 \end{cases} \qquad (4.5)$$

From equation (4.5) with one can determine the velocity of the front $v = v_s$ and the complex number $q' = q'_s$ locating the saddle point. These equations are equivalent to the ones obtained for a front moving into an homogeneous state. The interpretation of q'_s is the same as in such case: Eq.(4.4) shows that the real part of q'_s gives the periodicity of $\delta\psi$ at the edge of the front, and its imaginary part characterize its steepness. The difference with the homogeneous case lies in the different eigenvalue spectrum $\epsilon(k)$.

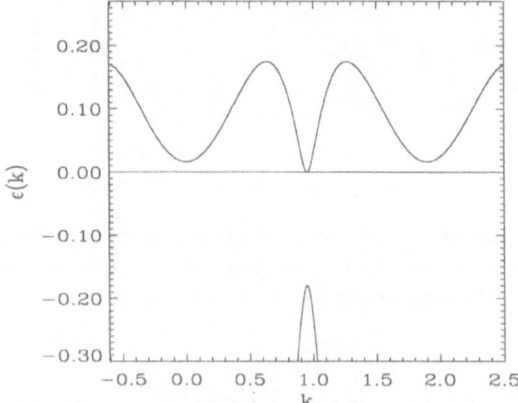

Figure 3 The spectrum $\epsilon(k)$ in the weak coupling approximation with $a = 0.02$, $c = 1.0$, $b = 3.0$ and an initial wavenumber $q_i = 0.95$

Figure 3 shows two branches of the spectrum $\epsilon(k)$ calculated using a *weak coupling approximation*[3] for an initial wavenumber $q_i = 0.95$. Since the upper branch is the positive one, it is the only to be used in (4.5).

The wavelength λ of the periodic pattern left behind by the moving front can be calculated following a standard prescription[19]: we assume that the oscillations created at the leading edge by the linear instability will become quenched by the nonlinearities, but their periodicity will not be modified. In the moving frame of speed v_s, linear theory predicts that the leading edge (4.4) oscillates at a frequency such that a number of nodes Φ is created in the unit of time, with $\Phi = \pi^{-1} \left(\text{Im} \left[\epsilon(k_s) \right] + v_s \text{Re} \left[q'_s \right] \right)$. Behind the front, where the pattern has a wavelength λ, the flux of nodes passing in the unit of time through a point fixed in the moving frame is $2v_s/\lambda$. From this we get

$$\lambda^{-1} = \frac{1}{2\pi} \left(\frac{\text{Im} \left[\epsilon(k_s) \right]}{v_s} + \text{Re} \left[q'_s \right] \right) , \qquad (4.6)$$

λ is determined by the two different wavenumbers k_s and q'_s . The hypothesis behind this formula are that no nodes are created nor destroyed far from the leading edge, and that every node that linear analysis predicts to be created has to be really created.

The velocities of the front propagating into the periodic state obtained from the theory and from numerical solution of the MCH equation are shown in Fig.4a. The mean periodicities left behind the second front are shown in Fig.4b together with those obtained from Eq. (4.6).

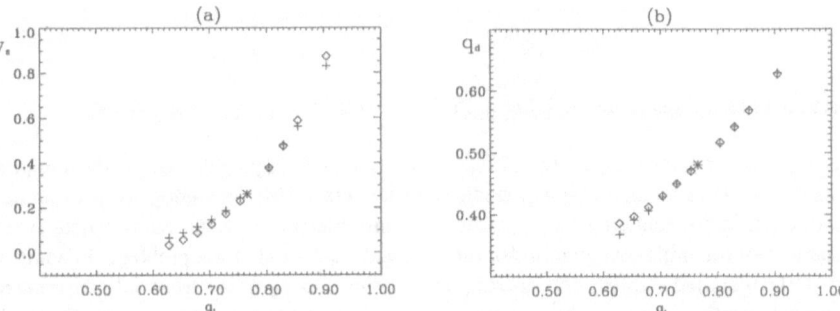

Figure 4 Velocity of the second front and wavenumber of the structure created when the front propagates into stationary states $\psi_{q_i}(x)$ with different dominant wavenumber ($a = 0$, $c = 1$). $+$: numerical simulations, $\diamond$: theory. The point represented by $*$ corresponds to the value of q_i obtained in the wake of the first front.

The initial conditions $\psi_0(x)$ for the numerical simulations were states in which a part of the system is in the ψ_+ state and the rest in the periodic unstable state $\psi_{q_i}(x)$ state. Simulations with several values of q_i are shown in Fig. 4. The $\psi_{q_i}(x)$ states, which are periodic functions containing the mode q_i and its harmonics, were obtained numerically by integrating the MCH equation with an approximation to the stationary solution as initial condition: $\psi_{q_i}(x) \approx [4(c - q_i^2)/3b]^{1/2} \sin(q_i x)$.

For $q_i > 0.9$ the pattern is already very unstable and numerically it decays by roll annihilation before the appearance of the front. On the other hand, for $q_i < 0.6$ the velocity of the front tends to be so small that the computer time needed to observe it becomes prohibitive. For intermediate q_i,

Fig. 4 shows good agreement between theoretical and numerical values for the velocity of the front and the wavenumber of the periodic pattern left behind (determined from the average wavelength of the pattern). Specially for $q_i \lesssim \sqrt{c}$, for which the weak coupling approximation is justified.

Once the theory for front propagation into periodic unstable states has been proved to be accurate, we used it to describe the second-front phenomenon of Figs. 1 and 2, that is, to predict the speed and the periodicity left behind a front that advances into the periodic state created by a first front, which is slightly different from the $\psi_{q_i}(x)$ used before. For small a the agreement is good[3] and becomes poorer for large a. The reason is that for large a the pattern created by the first front has a wavenumber quite different from $\sqrt{c}$, so that the weak coupling approximation used for calculating $\epsilon(k)$ is not accurate. It is interesting to note that for nematics such as PBG in solution[23], and for applied magnetic fields of about 8 kG, the parameters in the MCH model are $b \approx 3$, $c \approx 1$, and $a \approx 0.02$, which are in the range of validity of the theory. The theoretical prediction in such case is that the speed and periodicity behind the first front are of the order of 6 μm/s and 100 μm, respectively, whereas for the second front the values predicted are approximately 1μm/s and 170μm. Experiments to check these predictions would be welcome.

It is finally interesting to point out analogies and differences between the MCH model (a relaxational non-gradient flow) studied here and the EFK model[20] (a relaxational gradient flow). In the decay of an initial homegeneous unstable state a transient pattern with selected periodicity occurs in the MCH model, while the zero wavenumber is the most unstable mode in the EFK model. In front propagation into the unstable homogeneous state a pattern behind the front appears in both models. However, the second front described here possibly occurs in the EFK model only in situattions which are very difficult to observe numerically.

5. Interface between periodic unstable patterns: A source of propagating fronts

The results obtained in the previous section can be understood by saying that within the non-gradient potential flow (3.4)-(3.5), an unstable periodic state in contact with the homogeneously stable state decays through intermediate periodic and linearly unstable states. Such intermediate states are more stable than the initial one in terms of the Lyapunov potential of the problem. In addition the validity of the marginal stability criterion implies that the velocity of the front only depends on the initial periodic state and not on the state originally on the stable side of the front. To check the generality of these ideas we have considered the evolution of an interface between two periodic states of different wavenumbers $q_1 < q_2$. Both are linearly unstable, but from the argument above we expect the generation of a front moving into the less stable state (the one with larger wavenumber q_2) and leaving behind it a third state of wavenumber $q_3 < q_2$. If this process is still described by the marginal stability criterion the front velocity should be independent of q_1.

Numerical results corroborating such expectations are shown in Fig. 5. We note that the interface is far away from the boundaries of the system and, as the propagation of the front is analyzed also far enough from them, the boundaries of the system are meaningless in this discussion. We also note that two periodic solutions can be joined either through their maxima or through points of zero amplitude. In the first case, the initial condition will have a small jump since the steady amplitude depends on q_i. In the second case the change is smoother. However, when the matching is done with the homogeneous steady solution only the first case is meaningful. From Fig. 5 it is apparent that the way in which the matching is done does not affect the steady movement of the front: despite there is a slightly different delay for the first white region to disappear, the velocity of both fronts is the same and equal to 0.22. To check the independence of front velocity on q_1 we have also considered the situations with $q_1 = 0.785, 0.628$, and 0.419 joined to a region with $q_2 = 0.698$. In all cases we have found the same front velocity (0.21 ± 0.01) propagating into the region of q_2.

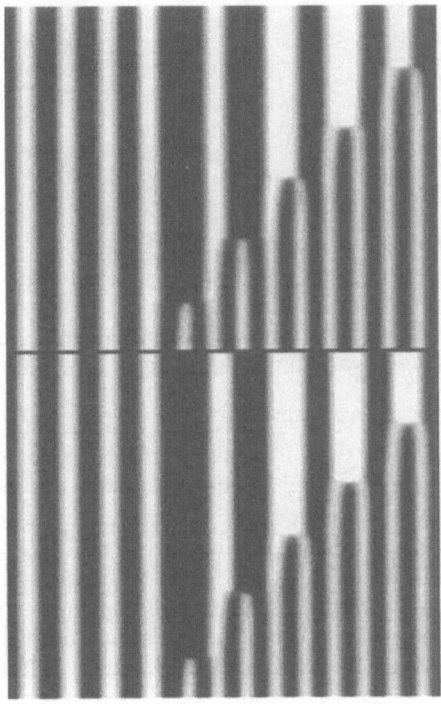

Figure 5 Space-time plots of $\psi(x, t)$ in gray levels (system size=130, the time for both plots run from 0 to 500). The initial condition correspond to an interface between two periodic stationary solutions with $q_1 = 0.503$ and $q_2 = 0.698$ connected through their maxima (top) and through a zero amplitude point (bottom)

The iteration of the mechanism that we have described of front propagation between two unstable periodic states naturally leads to a source of fronts propagating in opposite directions: if in the example in Fig. 5 it turns out that $q_1 > q_3$ we should expect a new front, now moving into the q_2-region. Such behavior of an interface as a source of fronts is seen in Fig. 6. We have considered here $q_1 = 0.698 < q_2 = 0.785$. A first front appears moving to the right with a velocity of 0.43. The pattern left behind by this front has a dominant wavenumber $q_3 < q_1 = 0.698$. The situation is then similar to the one found in Fig. 5 where we had a front propagating into a region with wavenumber 0.698, and, indeed, we find a front moving towards the left with the expected velocity of 0.21. This front leaves behind a new state with dominant wavenumber q_4 and a new front should

96

emerge in the interface between the q_2 and q_4 regions already created by front propagation.

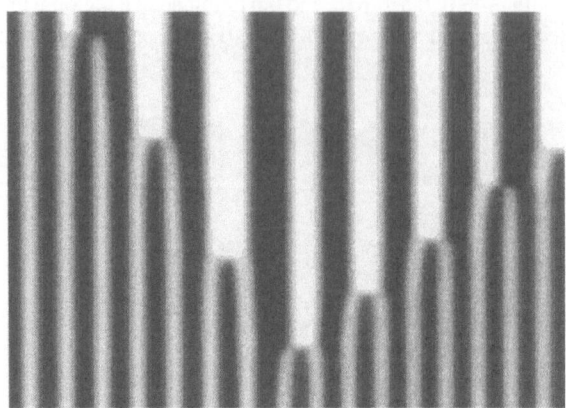

Figure 6 Space-time plots of $\psi(x, t)$ in gray levels. Two fronts moving in opposite directions emerge at the interface between the periodic patterns ($q_1 = 0.698$ and $q_2 = 0.785$; (system size=137.5 and time from 0 to 300).

We finally note that, in practice, the patterns left behind a front are not perfectly periodic. This makes the decay of periodic unstable states by a bulk mechanism more efficient and limits the number of successive fronts that are observed. In any case, and as a speculative comment, it is interesting to note some analogy between the source of fronts discussed here and the sources of traveling wave solutions found in non-relaxational flows such as the CGLE[11].

ACKNOWLEDGMENTS: RM and EHG acknowledge financial support from DGYCIT (Spain) Project PB92-0046. R.M. also acknowledges partial support from the Programa de Desarrollo de las Ciencias Básicas (PEDECIBA, Uruguay), the Consejo Nacional de Investigaciones Científicas Y Técnicas (CONICYT, Uruguay) and the Programa de Cooperación con Iberoamérica (ICI, Spain)

References.

1- A. Amengual, E. Hernández-García, and M. San Miguel, Phys. Rev. E **47**, 4151 (1993).

2- M. San Miguel, A. Amengual and E. Hernández-García, Phase Transitions **48**, 65 (1994).

3- R. Montagne, A. Amengual, E. Hernández-Gracía and M. San Miguel, Phys. Rev. E **50**, 377 (1994).

4- M. San Miguel and F. Sagués, in *Patterns, defects and materials instabilities*, edited by D. Walgraef and N. Ghoniem (Kluwer, Dordrecht, 1990), and references therein.

5- P. G. de Gennes and J. Prost, *The Physics of Liquid Crystals* (Clarendon, Oxford, 1993).

6- W. van Saarloos, Phys. Rev. A **37**, 211 (1988).

7- P. C. Hohenberg and B. I. Halperin, Rev. Mod. Phys. **49**, 535 (1978).

8- R. Graham, in *Instabilities and Nonequilibrium Structures*, edited by E. Tirapegui and D. Villarroel (Reidel, Dordrecht, 1987), p. 271.

9- R. Graham, in *Theory of continous Fokker-Plank systems*, Vol. 1 of *Noise in nonlinear dynamical systems*, edited by F. Moss and P. V. E. M. Clintock (Cambridge University, Cambridge, 1989), p. 225.

10- R. Graham and T. Tel, in *Instabilities and Nonequilibrium Structures III*, edited by E. Tirapegui and W. Zeller (Reidel, Dordrecht, 1991), p. 125.

11- W. van Saarloos and P. Hohenberg, Physica D **56**, 303 (1992).

12- O. Descalzi and R. Graham, Phys. Lett. A **170**, 84 (1992).

13- O. Descalzi and R. Graham, Z. Phys. B **93**, 509 (1994).

14- B. L. Winkler, H. Richter, I. Rehberg, W. Zimmermann, L. Kramer, A. Buka, Phys. Rev. A **43**, 1940 (1991).

15- M. San Miguel and F. Sagués, Phys. Rev. A **36**, 1883 (1987).

16- F. Sagués and M. San Miguel, Phys. Rev. A **39**, 6567 (1989).

17- A.N. Kolmogorov, I.G. Petrovskii, and N.S. Piskunov, Bull. Univ. Moscou, Ser. Int., Sec. A **1**, 1 (1937), translated in *Dynamics of curved fronts*. P. Pelcé ed. (Academic, San Diego, 1988).

18- W. van Saarloos, Phys. Rev. A **39**, 6367 (1989).

19- G. Dee and J. Langer, Phys. Rev. Lett. **50**, 383 (1983).

20- E. Ben-Jacob, H. Brand, G Dee, L. Kramer and J.S. Langer, Physica D **14**, 348 (1985).

21- W. van Saarloos, Phys. Rev. Lett. **58**, 2571 (1987).

22- J.A. Powel, A.C. Newell and C. K.R.T Jones, Phys. Rev. A **44**, 3636 (1991).

23- S. F. Srajer, G. and R. B. Meyer, Phys. Rev. A **39**, 4828 (1989).

PATTERN SELECTION AND DIFFUSIVE INSTABILITY

M.F' Hilali, S. Métens, G. Dewel and P. Borckmans
Centre for Non-Linear Phenomena and Complex Systems
Université Libre de Bruxelles C.P. 231
Boulevard du Triomphe B-1050 Bruxelles

Abstract. The stationary pattern selection problem in one and two dimensions is discussed on the generalized Swift-Hohenberg model, in the presence of a large quadratic coupling term.

1 Introduction

Self-organization resulting from symmetry breaking instabilities, in spatially extended dissipative systems driven away from equilibrium, has been the focus of great activity in fields as diverse as hydrodynamics, liquid crystals, chemistry,and non-linear optics [1, 2, 3, 4]. In chemistry the systems involve in general a great number of species and most of the time, the exact kinetic mechanisms are not completely elucidated. We adopt a pragmatic approach which consists to reduce the intrinsic complexities of the observed phenomena to a simple model. Here, the pattern formation and selection problem is presented on the generalized Swift-Hohenberg model [5], which can be derived from a reaction- diffusion type equations [6]. It reproduces qualitatively various experimental observations obtained in an open gel reactor for the CIMA reaction (Chlorite - Iodide - Malonic acid [7]). The G.S.H evolution equation takes the following form :

$$\frac{\partial u}{\partial t} = ru - (\nabla^2 + q_c^2)^2 u + vu^2 - gu^3 \tag{1}$$

where r is chosen as the bifurcation parameter. This model has been used to describe the Bénard-Marangoni and non-Boussinesq Rayleigh- Bénard problems. In these cases, the quadratic coupling term v is small since it is related for instance with the weak temperature dependence of the transport coefficients. In the context of chemical experiences in open gel reactors, this term depends on the rate of the kinetic reactions and the diffusion coefficients and is not necessarily small. We show that the bifurcation analysis of the G.S.H model, under these conditions, exhibits surprisingly rich nonlinear behaviors.

2 Bifurcation analysis of G.S.H

The model admits three homogeneous steady states $u = 0$ and $1/2g(v \pm \sqrt{(v^2 + 4g(r - q_c^4))})$ which are destabilized by pattern forming instabilities at points $O(r = 0)$, $A(r = r_{T1})$ and $C(r = r_{T2})$ in (fig.1a). In well mixed reactor conditions (CSTR), the CIMA reaction can exhibit bistability between a reduced and an oxidized uniform steady states which correspond here to $u_\pm$. We perform a theoretical

E. Tirapegui and W. Zeller (eds.), Instabilities and Nonequilibrium Structures V, 99–104.
© *1996 Kluwer Academic Publishers.*

study via a standard weakly non-linear analysis and compare with results obtained by numerical integration of equation (1). In the vicinity of each symmetry breaking instability points, the dynamics may be reduced to equations for the complex amplitudes $A(T)$ (where $T = \epsilon^2 t$, ϵ is the expansion parameter) time see [8]) of the pattern. We focus the discussion on the pattern selection between solutions of different symmetry and neglect the spatial modulations of the amplitude.

2.1 ONE DIMENSIONAL SYSTEMS

When v becomes sufficiently large, the standard asymptotic expansion must be perform up the fifth order terms in order to saturate the instability, the amplitude equation takes then the following form :

$$\frac{dA}{dT} = rA - g_3|A|^2 A - g_5|A|^4 A \quad g_3 < 0 \text{ and } g_5 > 0$$

The expression of the coefficients are given in [8]. The stripes structures emerge subcritically through an inverted pitchfork bifurcation, they are stable until $r = r_b$ at higher values of the control parameter it only remains the two uniform $u_\pm$ states. At $r = r_b$, the branches of periodic structures annihilate with the unstable branches emanating from subcritical bifurcations at $r = r_{T1}$ or r_{T2} on $u_\pm$. These results are in complete agreement with the numerical simulations of (1). The complete bifurcation diagram is presented in (fig.1b). We observe two regions of bistability between stripes pattern and uniform steady states. Localized structures such a front that interconnect one of the uniform state and the Turing pattern (fig.2a), or a droplet of Turing structures (fig.2b), with various numbers of wavelengths in the core, embedded in a uniform background, have been found numerically. The stabilizing mechanism is provided by the pinning of the front by the periodic pattern [9, 10, 8]. For the parameters considered, the pinning force appears to be large as the pinning band covers nearly the whole domain of bistability between the $u_\pm$ uniform states and the stripes. We observe that the droplets present spatially oscillating tails. This gives rise to an effective potential between localized states that is a periodic function of the distance between them. As in condensed matter physics this provides a efficient tool for the onset of frozen-in disorder even in gradient systems.[11]

2.2 TWO DIMENSIONAL SYSTEMS

Similarly we have also derived up to the fifth order the amplitude equations for the three resonant pairs of active modes characterizing hexagonal patterns:

$$\frac{dA_i}{dT} = rA_i + 2v_1 A_j^* A_k^* - g_3|A_i|^2 A_i - h_3(|A_j|^2 + |A_k|^2)A_i - 2v_2 A^* j A_k^*|A_i|^2 \quad (2)$$

$$- v_2 A_j^* A_k^*(|A_j|^2 + |A_k|^2) - v_2 A_i^2 A_j A_k - g_5|A_i|^4 A_i - h_{51}(|A_j|^4 + |A_k|^4)A_i$$

$$- h_{52}[|A_j|^2 + |A_k|^2]|A_i|^2 A_i - h_{53}(|A_j|^2|A_k|^2)A_i - wA_i^* A_j^{*2} A_k^{*2}$$

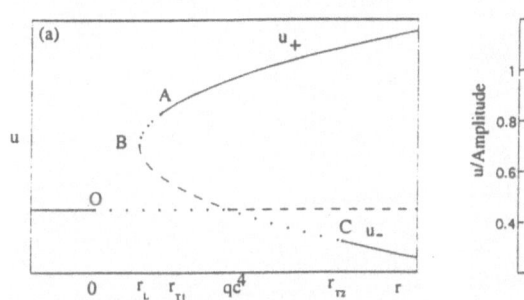
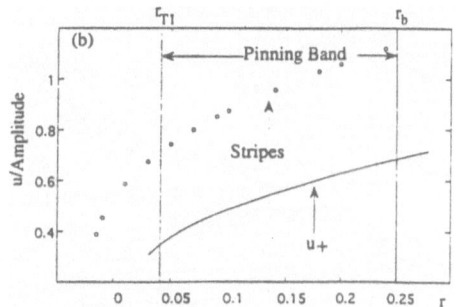

fig.1a: Bifurcation diagram exhibiting the stability of the uniform states for $q_c = 0.5$, $g = 1$, $v = 0.41$. The same values of the parameters are used throughout this paper. The plain, dashed and circled lines respectively correspond to linearly stable steady states, steady states already unstable to uniform perturbations and those unstable with respect to the space dependent perturbations as a result of symmetry breaking bifurcations at O, A and C.

fig1.b: Bifurcation diagram in one space dimension obtained from the numerical integration of Eq.(1). The amplitude u of the uniform $u_\pm$ state and the peak-to-peak amplitude of the striped structures are represented.

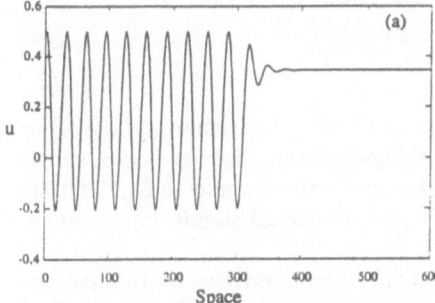
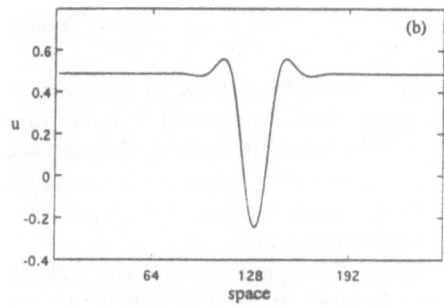

fig.2.a Pinned(immobile) 1D front separating a striped structure from the u_+ uniform steady state that coexist in space for $r = 0.04$

fig.2.b 1D localized structure between the uniform u_+ state and 1D stripes obtained numerically from Eq.(1) at $r = 0.1$

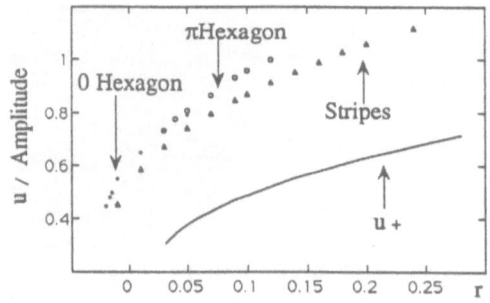

fig.3: Bifurcation diagram in two space dimensions obtained from the numerical integration of Eq.(1).

and two other equations of the same type for A_j and A_k The expression of the coefficients are given in [8]. If we express the complex amplitude in terms of a module and a phase as $A_i = R_i \exp(i\phi_i)$, then the sum of the phases, $\Phi = \sum_{i=1}^{3} \phi_i$, satisfies the following equation, where we put $R_1 = R_2 = R_3 = R$:

$$\frac{d\Phi}{dt} = \lambda_1 \sin(\Phi) + \lambda_2 \sin(2\Phi) \tag{3}$$

with $\lambda_1 = -6Rv + 9/2R^3v_2$ and $\lambda_2 = 3R^4w$

It follows from Eq.(3) that Φ monotonically relaxes to $\Phi = 0$ (triangular lattice) or $\Phi = \pi$ (honeycomb lattice). Linear stability analysis of the phase equation shows that a region of coexistence between both hexagonal patterns may exist. The 2D pattern selection problem is summarized in (fig.3), numerical simulations confirm the bifurcation analysis.

First H_0 appear subcritically followed by stripes. A subcritical tristability domain between H_0, stripes and the trivial state exist in the subcritical region. Then we observe the H_π hexagonal phase which first coexist with H_0 (fig.4). The coexistence of these two phases seem to have recently observed experimentally [12] in the non-Boussinesq Bénard convection in SF_6 near liquid-gas critical point [13]. Various 2D localized structures (fig.4) can be induced related to the multistability character of the system. their shape are also conditioned from the fact that the pinning is more important when the domain boundary is nearly perpendicular to the wave vectors associated to the patterns. Inhomogeneous spotty states have also been observed numerically. Note that when the bifurcation parameter r is

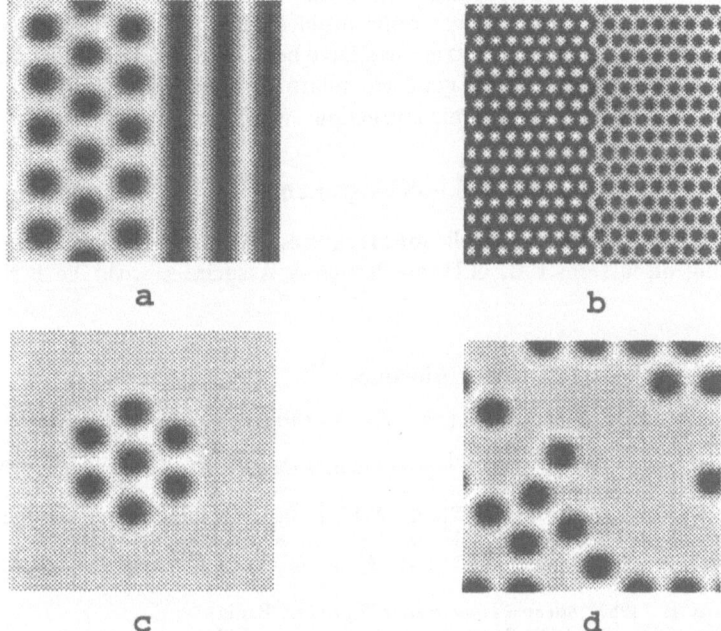

fig.4 Localized structures obtained numerically from Eq.(1) at r=0.1 except(b).

(a) Stable stationary front between H_π hexagons and stripes

(b) Stable stationary front between H_π and H_0 at $r = 0.035$

(c) H_π hexagonal droplet embedded in the uniform u_+ state

(d) "spotty" state

decreased the number of spots increases finally giving rise to the full hexagonal structure H_π.

3 Conclusions

This simple variational model exhibits a rich variety of patterns and localized states that have not been documented before, when the quadratic coupling term v is longer. The stability of the one and two dimensional localized structures must be understood in the framework of non-adiabatic effects [9]. Square patterns

104

and other kind of fronts structure have also been observed recently in this model
[14]. Taking into account the concentration profiles corresponding to the feeding
into the open gel reactor, ramped structures have been extensively studied on this
model [15]. Finally, it provides a good candidate to investigate the 3D Turing
structures and the effects of harmonic corrections.

4 Acknowledgements

We wish to thank Professor G.Nicolis for his interest in this work, and J.Lauzeral
for stimulating discussions.P.B. G.D. are Research Associates with the F.N.R.S.
(Belgium).

References

1. Cross, M. and Hohenberg, P.C. (1993) *Pattern Formation outside of Equilibrium*, Rev. Mod. Phys. **65**, 854-1112
2. Kai, S. (1992) *Physics of Pattern Formation in Complex Dissipative Systems*, (World Scientific, Singapore)
3. Borckmans P., Dewel G., De Wit A., Walgraef D. in *Chemical Waves and Patterns*, Kapral R. and Showalter K., eds., Kluwer, Amsterdam (to appear 1994)
4. Newell, A.C. and Moloney, J.V. (1992) *Nonlinear Optics*, (Addison-Wesley, Redwood City)
5. Haken, H. (1983) *Advanced Synergetics*, (Springer, Berlin)
6. Walgraef D., Dewel G., Borckmans P. (1982) *Nonequilibrium Phase Transitions and Chemical Instabilities*, Adv. in Chem.Phys. **49**, 311-355
7. Boissonade J., Dulos E. and De Kepper P. in *Chemical Waves and Patterns*, Kapral R. and Showalter K., eds., Kluwer, Amsterdam (to appear 1994)
8. Hilali M'F., Métens S., Dewel G., Borckmans P. *Pattern Selection in the Generalized Swift-Hohenberg model*, submitted to Phys.Rev.E
9. Pomeau, Y. (1986) *Front Motion, Metastability and Subcritical Bifurcations in Hydrodynamics*, Physica D **23**, 3-11
10. Jensen, O. Pannbacker, V.O., Dewel G., Borckmans P. (1993) *Subcritical Transitions to Turing Structures* , Phys.Lett.A **179**, 91-96
11. Binder, K. and Young, A.P. (1986) *Spin Glasses: Experimental Facts, Theoretical Concepts, and Open Questions*, Rev. Mod. Phys. **58**, 801-976
12. Assenheimer, M. and Steinberg, V. *Private communication,*
13. Assenheimer, M. and Steinberg, V. (1993) *Rayleigh-Bénard Convection Near the Gas-Liquid Critical Point*, Phys.Rev.Lett **70**, 3888-3891
14. Hilali M'F., Dewel G., Borckmans P. *in preparation*
15. Hilali M'F., Dewel G., Borckmans P. *preprint*

MONOTONIC FRONTS OF THE REACTION-DIFFUSION EQUATION IN ONE SPACE DIMENSION

R. D. Benguria and M. C. Depassier
Facultad de Física, U. Católica de Chile, Santiago 22, Chile

ABSTRACT. We study the speed selection mechanism for the nonlinear diffusion equation $u_t = u_{xx} + f(u)$ with $f(0) = f(1) = 0$ and $f > 0$ in $(0,1)$. It is known that sufficiently localized initial conditions evolve into a front which joins the state $u = 1$ to $u = 0$ which propagates with the lowest speed for which there is a monotonic front joining these two states. We give a lower bound on the speed which enables one to estimate a priori the point of transition from the linear to nonlinear regime. For polynomial $f's$ of odd degree we show that the exact point of transition may be calculated without knowledge of an exact solution $u(x - ct)$. We calculate the exact point of transition for a general quintic polynomial.

Introduction

The nonlinear diffusion equation $u_t = u_{xx} + f(u)$ models phenomena in diverse fields such as population growth, kinetics of phase transitions, chemical reactions and many others. Of special interest is the case when the function f is such that there exist two steady states, one stable and one unstable. We shall assume that the equation has been scaled so that the unstable state is $u_u = 0$ and the stable state is $u_s = 1$, and we consider functions f which are positive in $(0, 1)$. Then sufficiently localized initial conditions evolve into a travelling front which joins the two steady states[1]. The speed at which the front propagates, c^* is equal or greater than the linear marginal stability value $c_L = 2\sqrt{f'(0)}$. Aronson and Weinberger [1] have shown that any positive initial condition $u_0(x) < 1$ for all x, which decays exponentially or faster at infinity will evolve into a front propagating with speed c^*. This asymptotic speed is the lower speed for which equation (1) has a monotonic front joining the stable state $u = 1$ to the unstable state $u = 0$. Moreover,

$$2\sqrt{f'(0)} \leq c^* < 2\sqrt{\sup(f(u)/u)}. \tag{1}$$

For the special case of the Fisher–Kolmogorov equation $f(u) = u - u^3$, $f'(0) = 1$ and $\sup(f(u)/u) = 1$ so that $c^* = 2$. In general [2], for any concave $f(u)$, $\sup(f(u)/u) = f'(0)$, and $c^* = 2\sqrt{f'(0)}$. The value $c^* = 2$ is the value which had been derived by Kolmogorov, Petrovsky and Piskunov [3] using an heuristic argument (the linear marginal stability mechanism) which is equivalent to the conjecture that the asymptotic speed of the front is that for which a perturbation to the front is marginally stable in the frame moving with the

E. Tirapegui and W. Zeller (eds.), Instabilities and Nonequilibrium Structures V, 105–110.

front speed. In many cases the asymptotic speed of propagation is exactly the linear value $c_L = 2\sqrt{f'(0)}$ obtained by the linear marginal stability criteria [3, 4]. There are cases however when the front propagates at a speed greater than this value, case which is referred to as that in which a nonlinear speed selection mechanism [5, 6, 7] operates. Explicit expressions for this special nonlinear front or strong heteroclinic connection and its speed have been obtained for particular choices of f. All the known solutions correspond to functions f of the form $f(u) = \mu u + u^n - u^{2n-1}$ which, for μ positive but smaller than a critical value μ_c, are strongly heteroclinic [8].

We give a new lower bound on f which allows one to estimate the point of transition from the linear to nonlinear speed selection regime and show using as an example a general quintic polynomial $f = \mu x(1 - x)(1 + \alpha x + \beta x^2 + \gamma x^3)$ that the exact point of transition can be calculated without knowledge of the solution in the case of odd polynomials. The main ingredient in the results given here is the analysis of the fronts in phase space.

Equations

We consider the reaction diffusion equation

$$u_t = u_{xx} + f(u)$$

with $f(0) = 0$, $f(1) = 0$, $f'(0) > 0$ and $f > 0$ in $(0, 1)$. Given these conditions on f then there exist fronts that connect the unstable fixed point $u = 0$ to the stable fixed point $u = 1$. Travelling wave fronts $u(x - ct)$ satisfy the ordinary differential equation

$$u_{zz} + cu_z + f(u) = 0 \qquad \lim_{z \to -\infty} u = 1, \quad \lim_{z \to \infty} u = 0, \tag{2}$$

where $z = x - ct$ and we assume that c is positive. A front joining the stable fixed point 1 to the unstable point 0 is monotonic if in addition its derivative du/dz does not change sign. If we search for monotonic fronts it is convenient to consider the dependence of z as a function of u, or rather the dependence of $v(u) = -(dz/du)^{-1}$ as a function of u. For a monotonic solution of equation (1), $u(z)$ decreases monotonically as z goes from $-\infty$ to ∞, therefore the function $v(u)$ is well defined and is positive between 1 and 0 and vanishes at the fixed points. One readily finds that the equation for $v(u)$ is

$$v(u)\frac{dv}{du} - cv(u) + f(u) = 0, \tag{3}$$

with

$$v(0) = v(1) = 0, \qquad \text{with } v > 0. \tag{4}$$

Bound on the Selected Speed

As shown by Aronson and Weinberger, sufficiently small positive initial conditions evolve into a front which propagates at the minimum speed for which there is a monotonic front, that is, at the minimum speed for which equation (3) has a solution. Let g be any positive function in (0,1) such that $h = -dg/dq > 0$. Multiplying equation (3) by g/p and integrating with respect to q we find that

$$\int_0^1 \left(h\,p + \frac{f(q)}{p}\,g \right) dq = c^* \int_0^1 g\,dq \qquad (5)$$

where the first term is obtained after integration by parts. However since p, h, f, and g are positive, we have that for every fixed q

$$h\,p + \frac{f(q)\,g}{p} \geq 2\,\sqrt{f\,g\,h}$$

hence we obtain our main result,

$$c^* \geq 2\,\frac{\int_0^1 \sqrt{f\,g\,h}\,dq}{\int_0^1 g\,dq} \qquad (6)$$

where

$$g \geq 0 \quad \text{and} \quad h = -g' \geq 0 \quad \text{in} \quad (0,1). \qquad (7)$$

One possible choice for g is

$$g_1(q) = \int_q^1 f(x)\,dx$$

Then

$$c \geq \frac{4}{3}\frac{\left(\int_0^1 f(q)\,dq \right)^{3/2}}{\int_0^1 q\,f(q)\,dq}. \qquad (8)$$

This expression gives a simple estimate of the transition from the linear to the nonlinear regime for arbitrary f. Used in conjunction with the upper bound of Hadeler and Rothe[9] it allows to obtain a good estimate of the asymptotic speed of the front.

0.1 Exact Solution for $f = \mu x(1 - x)(1 + \alpha x + \beta x^2 + \gamma x^3)$

In this section we show that the transition from linear to nonlinear marginal stability can be formulated in a simple way starting from equation (3) and therefore it is not necessary that an exact solution $u(x - ct)$ for the front be known.

Since the endpoints 0 and 1 of equation (3) are singular we must determine the behavior near them analytically. If we consider functions f analytic around 0 and with $f'(0) > 0$, then near $u = 0$ we find

$$v(u) = a_1 u + a_{3/2} u^{3/2} + a_2 u^2 + a_{5/2} u^{5/2} + a_3 u^3 + \dots$$

where the first terms are given by

$$a_1^2 - ca_1 + f'(0) = 0 \qquad (9)$$

$$a_{3/2}\left(\frac{5}{2}a_1 - c \right) = 0 \qquad (10)$$

$$a_2(3a_1 - c) + \frac{1}{2}f''(0) = 0 \qquad (11)$$

$$a_{5/2}\left(c - \frac{7}{2}a_1\right) - \frac{7}{2}a_{3/2}a_2 = 0 \qquad (12)$$

and so on. That the leading term in the expansion of v near zero is linear in u is due to the fact that the front in the original coordinates $u(z)$ approaches the fixed points exponentially. Since v must be positive between 0 and 1, a_1 must be real and positive. The two roots for a_1 are given by $a_{1P} = (c + \sqrt{c^2 - 4f'(0)})/2$ and $a_{1M} = (c - \sqrt{c^2 - 4f'(0)})/2$. The minimum speed c for which there may be a monotonic front is the linear marginal speed value $c_L = 2\sqrt{f'(0)}$ value at which the roots coincide $a_{1P} = a_{1M} \equiv a_{1L}$. For speeds greater than this value $a_{1M} < a_{1L} < a_{1P}$. Strong heteroclinic solutions or special nonlinear front profiles are those associated with a_{1P}. From the expansion at the origin it follows that either $c = 5a_1/2$ or $a_{3/2} = 0$. In the first case we find that $c = 5\sqrt{f'(0)/6} \approx 2.041\sqrt{f'(0)}$ and $a_1 = \sqrt{2f'(0)/3} = a_{1M}$. As it is known, these solutions are not a preferred asymptotic state. Strong heteroclinic connections can be achieved only if $a_{3/2} = 0$, all half integer coefficients vanish then and $v(u) = a_1 u + a_2 u^2 + \ldots$.

Near $u = 1$, assuming $f'(1) < 0$,

$$v(1 - u) = b_1(1 - u) + b_2(1 - u)^2 + b_3(1 - u)^3 + \ldots$$

where b_1 is the positive solution of

$$b_1^2 + cb_1 + f'(1) = 0.$$

There is only one positive solution for b_1, the rest of the coefficients follow easily.

It is convenient to introduce a new parameter λ defined by $c = \lambda a_1$. It is not difficult to realize that whenever $1 < \lambda < 2$ then the solution for v is strongly heteroclinic, that is, associated with a_{1P} and when $\lambda > 2$ it becomes associated with a_{1M}; hence for $\lambda > 2$ the linear marginal speed is selected. If $c = \lambda a_1$ then

$$c = \lambda\sqrt{\frac{f'(0)}{\lambda - 1}}, \qquad a_1 = \sqrt{\frac{f'(0)}{\lambda - 1}}. \qquad (13)$$

At $\lambda = 2$, the speed attains its linear value $c_L = 2\sqrt{f'(0)}$. The problem then is to determine the value of λ. This transition value $\lambda = 2$ is not associated with any specific nonlinearity, it is valid for any f which satisfies the conditions given above.

Now we apply this observation to the calculation of the exact point of transition from linear to nonlinear marginal stability for a quintic polynomial. The most general form of a quintic polynomial that vanishes at 0 and 1 is

$$f(x) = \mu x(1 - x)(1 + \alpha x + \beta x^2 + \gamma x^3) \qquad (14)$$

where μ, α, β and γ are four arbitrary parameters whose only restriction is given by the requirement $f'(0) > 0$ and $f > 0$ in $(0, 1)$. On the other hand, the most general closed form solution for v given a quintic f is given by

$$v(u) = a_1 u(1 - u)(1 + bu) \qquad (15)$$

where $b > -1$. Introducing again the parameter λ given above, so that a_1 and c are given by equation (13), equation (15) is the exact solution of equation (3) with

$$f(u) = f'(0)u(1 - u)\left(1 + \frac{(2 + \lambda b - 3b)}{\lambda - 1}u + \frac{b(5 - 2b)}{\lambda - 1}u^2 + \frac{3b^2}{\lambda - 1}u^3\right). \qquad (16)$$

In the solution for v we have three adjustable parameters, $\lambda, b, f'(0)$ whereas in the most general form for f, four adjustable parameters exist. Hence, an exact solution for v can be found chosing three parameters of f arbitrarily and the fourth one in terms of them. Choosing μ, β and γ arbitrarily, we identify

$$f'(0) = \mu$$

$$\lambda = 1 + \frac{75\gamma}{(3\beta + 2\gamma)^2}$$

and

$$b = \frac{5\gamma}{3\beta + 2\gamma}$$

and the exact solution exists if

$$\alpha = \frac{(2 + \lambda b - 3b)}{\lambda - 1}$$

For any other value of α a closed form solution does not exist and we cannot determine the value of λ. The criterion for the solution to be strongly heteroclinic $1 < \lambda < 2$ is expressed now in terms of the free parameters β and γ.

Now we show that an explicit solution for the front in the original coordinates exists only if an additional condition on b, hence a relation between the free parameters β and γ is satisfied. One readily finds that $u(z)$ is the solution of

$$e^{-(b+1)a_1 z} = \frac{u^{1+b}}{(1 - u)(1 + bu)^b}. \tag{17}$$

Writing $b = n/p$ the equation for u is

$$e^{-(n+p)a_1 z} = \frac{u^{n+p}}{(1 - u)^p(1 + bu)^n}. \tag{18}$$

This can be inverted to obtain the explicit solution for $u(z)$ if $n + p = 2, 3, 4$. The detailed inversion of all the solvable cases is not instructive, here we give one example. Choose $n = 2, p = 1$, then $b = 2$, and the front is a solution of the cubic equation

$$u^3(1 + 4e^{-3a_1 z}) - 3ue^{-3a_1 z} - e^{-3a_1 z} = 0. \tag{19}$$

This cubic has two complex roots and a single real positive root which is the desired front, given by

$$u(z) = \frac{2^{\frac{1}{3}}}{\sqrt{4 + e^{3a_1 z}}\left(e^{\frac{3a_1 z}{2}} + \sqrt{4 + e^{3a_1 z}}\right)^{\frac{1}{3}}} + \frac{\left(e^{\frac{3a_1 z}{2}} + \sqrt{4 + e^{3a_1 z}}\right)^{\frac{1}{3}}}{2^{\frac{1}{3}}\sqrt{4 + e^{3a_1 z}}} \tag{20}$$

Again this is an exact front for f of the form given by equation (14). It corresponds to a strongly heteroclinic connection for $\lambda < 2$. If one chooses the case $n + p = 4$ the quartic equation that arises has a pair of complex conjugate solutions, a negative solution and a positive solution which is the desired front. For values of b which do not allow the obtention of the explicit form of the front $u(z)$ we still have the speed selection criteria in terms of the free parameters of the polynomial. For arbitrary values of α, β and γ, after the value of λ is determined, if α exceeds the value for which an exact solution can be found, it is not difficult to see that the nonlinear front is selected.

Closed form solutions $v(u)$ for polynomial $f's$ can be obtained only if f is an odd polynomial. In general, if f is a polynomial of degree $2k+1$ that vanishes at 0 and 1, there are $2k$ free parameters (restricted only by the requirement of positivity of f), whereas the corresponding closed form solution for v has $k+1$ parameters, which implies that a closed form for v, and an explicit expression for λ is possible if $k-1$ parameters of f are chosen adequately in terms of the $k+1$ remaining free parameters.

Conclusion

The asymptotic state of sufficiently small positive initial conditions of the reaction diffusion equation (1) is a monotonic front which propagates at the lowest possible speed. We have shown that formulation in phase space of this statement enables one to estimate the point of transition from linear to nonlinear marginal stability for arbitrary f. The exact point of transition can be calculated for odd polynomial $f's$ in a given parameter range without knowledge of the explicit solution.

References

[1] D. G. Aronson and H. F. Weinberger, Adv. Math. **30**, 33 (1978).

[2] P. Collet and J. P. Eckmann. *Instabilities and Fronts in Extended Systems* (Princeton University Press, Princeton, 1990).

[3] A. Kolmogorov, I. Petrovsky, and N. Piskunov, Bull. Univ. Moscow, Ser. Int. A **1**, 1, (1937).

[4] G. Dee and J. S. Langer, Phys. Rev. Lett. **50**, 383 (1983).

[5] E. Ben-Jacob, H. Brand, G. Dee, L. Kramer, and J. S. Langer, Physica D **14**, 348 (1985).

[6] W. van Saarloos, Phys. Rev. A **37**, 211 (1988).

[7] W. van Saarloos, Phys. Rev. A **39**, 6367 (1989).

[8] J. Powell and M. Tabor, J. Phys. A **25**, 3773 (1992).

[9] K. P. Hadeler and F. Rothe, J. Math. Biol. **2**, 251 (1975).

POISEUILLE FLUX OF HARD PARTICLES: THEORY AND SIMULATIONS *

Dino Risso †

Departamento de Física, Universidad del Bío Bío, Concepción, Chile

Patricio Cordero

Departamento de Física, Universidad de Chile, Santiago, Chile

The Poiseuille flux is perhaps the best known examples in hydrodynamics. It consists of a viscous fluid flux along a channel of constant section. It is a classical example to study by means of the Navier Stokes equations [1, 2, 3]. Usually the Poiseuille flux is understood to be driven by an externally imposed pressure gradient but it is trivially equivalent to apply a gravitational force mg over each particle [4]. For small velocities (small Reynolds or Mach number) the flux is known to be laminar and stationary and the velocity profile is parabolic. There is however a critical Reynolds number above which a turbulent (*unstable*) regime starts. In [5] the authors studied a small system of Lennard Jones particles observing that in spite of the size the system has a good hydrodynamics behavior.

In this article we summarize our analytic and simulational results regarding the Poiseuille flux of a two dimensional system of hard particles. As we shall explain below our study is deeply inside the stable regime. An interesting feature of our system of hard disks is that it presents important thermal and compressibility effects. Hence this numerical system

---

*Work partially financed by *FONDECYT* Research Grant 1931105

†Work partially financed by *DIPRODE* Research Grant 9417051

E. Tirapegui and W. Zeller (eds.), Instabilities and Nonequilibrium Structures V, 111–118.
© 1996 *Kluwer Academic Publishers.*

represents a challenge to the hydrodynamics theory since its transport coefficients depend on the position. Notation and basic features of our simulations are in the appendix. Our simulations have been carried out using an efficient algorithm [6].

In [2] it has been shown that for a two dimensional incompressible Poiseuille flux in an infinitely long channel the predicted critical Reynolds number is $Re_C = 5772$ corresponding to long wave excitations [7]. But when the aspect ratio λ =width/length is finite the value of Re_C increases since these long wave excitations cannot exist. In [7] they have studied the stability curve λ versus Re_C that stems from a linear stability analysis. In particular from their results it is seen that for the aspect ratio $\lambda = 2$ that we use $Re_C \approx 10^6$. In view of what has been said above we can state that the numerical experiments that we are summarizing (details in [8]) are deeply in the stable region.

Our system consists of a rectangular $L_X \times L_Z$ box containing N hard disks that move under the effect of gravity along the Z direction. We take periodic boundary conditions at the horizontal walls while the collision rule with the vertical walls is such that (a) the macroscopic velocity field at the walls is zero and (b) they simulate contact with a heat bath at an imposed temperature T_w. This last condition is important since it is the only way the system looses the energy permanently injected by gravity. The aspect ratio was chosen to be $\lambda = L_X/L_Z = 2$, the fraction of area occupied by the disks was kept fixed at $\rho_A = \pi/16$ while different values for N from 100 to 5000 were considered. The value of g was controlled through an adimensional parameter a which naturally arises in the hydrodynamic study of the Poiseuille flux. Assuming that the system is sufficiently near *local* equilibrium and that the velocity field is $\vec{u} = u_Z(x)\hat{z}$, the density and temperature fields are $\rho = \rho(x)$ and $T = T(x)$ and also that the Fourier law for the heat flux: $\vec{q} = -k\nabla T$ is satisfied. The parameter a is

$$
\begin{aligned}
a &= \sqrt{\frac{\bar{\nu}}{\bar{\kappa}} \frac{1}{c_v T_w}} \frac{g L_X^2}{\bar{\nu}} \\
&= \sqrt{\lambda \, Pr \, N} \, n \frac{Fr}{\bar{\eta}_0}
\end{aligned}
\tag{1}
$$

where $\bar{\nu}$ and $\bar{\kappa}$ are the mean kinematic viscosity and thermal diffusivity coefficients and $\bar{\eta}_0$ is defined in the appendix.

In fact, to analyze the stationary regime we have studied the mass, momentum and energy balance equations for a fluid moving under the effect of an external force $-\rho g \hat{z}$ of a 2D infinite channel parallel to the Z axis with walls at $x = \pm L_X/2$. It is easy to derive that

$$\frac{\partial p}{\partial x} = 0 \tag{2}$$

and making the changes of scale:

$$\begin{aligned} x &\rightarrow L_X x \\ u_Z &\rightarrow \frac{g L_X^2}{\bar{\nu}} u \\ T &\rightarrow T_w t \end{aligned} \tag{3}$$

the equations for $u(x)$ and $t(x)$ become [8]

$$\begin{aligned} u'' + \frac{1}{\eta} \eta' u' &= -\frac{\bar{\nu}}{\nu} \\ t'' + \frac{1}{k} k' t' &= -a^2 u'^2 \end{aligned}$$

where $\kappa = k/\rho c_p$ and the prime indicates derivative with respect to x.

With the boundary conditions $u(0) = u(1) = 0$ and $t(0) = t(1) = 1$ and assuming that the density is uniform these equations become

$$u'' + \frac{t' u'}{2t} = -\frac{1}{\sqrt{t}} \tag{4}$$

$$t'' + \frac{t'^2}{2t} = -a^2 u'^2 \tag{5}$$

In our simulations with up to 5000 particles the signal to noise ratio is satisfactory when gravity is rather large, that is, the parameter a has been set to range from about 10 to 50 and sometimes even larger.

We have checked that the equation of state for hard disks (Henderson's equation of state [9]) is well satisfied in spite that the system is far from equilibrium. The prediction that the pressure remains uniform is well corroborated by our observations.

A rather technical remark but quite important is the well known result from kinetic theory that the temperature at the walls presents a discontinuity [10]. The phenomenon has already been studied in molecular dynamics off equilibrium simulations [11]. The final result is that in spite that there is an externally imposed temperature T_w at the walls it is necessary to measure the effective temperature at the walls T_w^{eff} and use this value to correct both the Froude number Fr and the value of the parameter a. These corrected values are the ones that we have used to numerically integrate the Navier Stokes equations to compare theoretical and simulational results.

The Mach number of the system $M = u_{\text{max}}/c$ that we obtain from the hydrodynamics analysis has been compared with the observed value of it. In figure 1 it is seen that they are in excellent agreement in a wide range for a when $N \geq 900$.

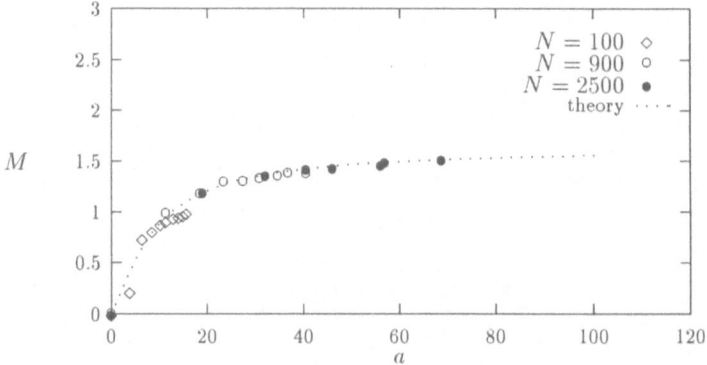

Figure 1: Mach number versus the adimensional parameter a that controls the intensity of the flux. The solid circles correspond to observations with $N = 2500$ particles, open circles to $N = 900$ and rhombus to $N = 100$. The light dotted line is our theoretical prediction. Only for $N = 100$ particles the deviations are significant.

From this figure it should be realized that the stationary behavior of the system is quite near that for $N \to \infty$ for values of N of 900 or larger. To have a better support for this

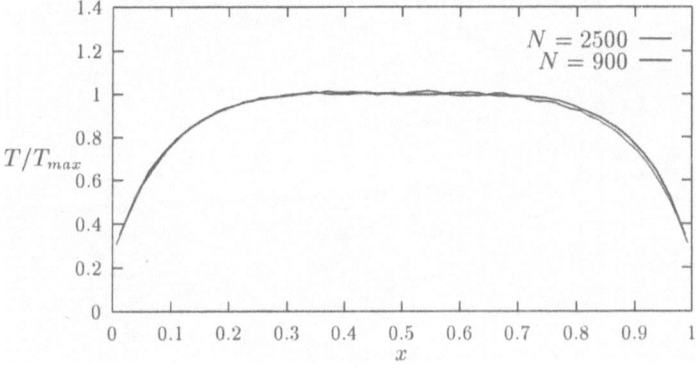

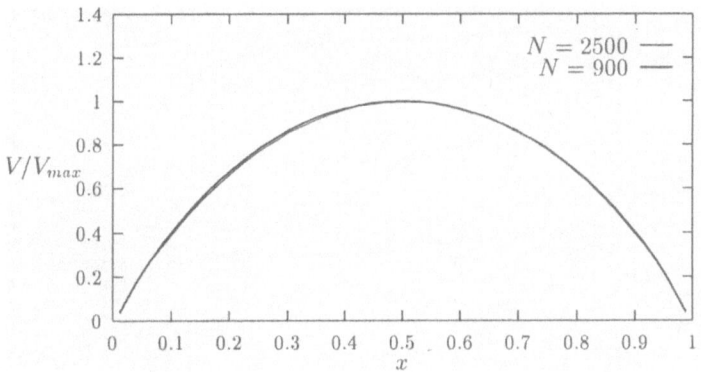

Figure 2: Comparison between the temperature (top) and velocity (bottom) profiles for the systems with $N = 2500$ and $N = 900$ both with $a = 40.5$. The coincidence between the profiles indicates size independence.

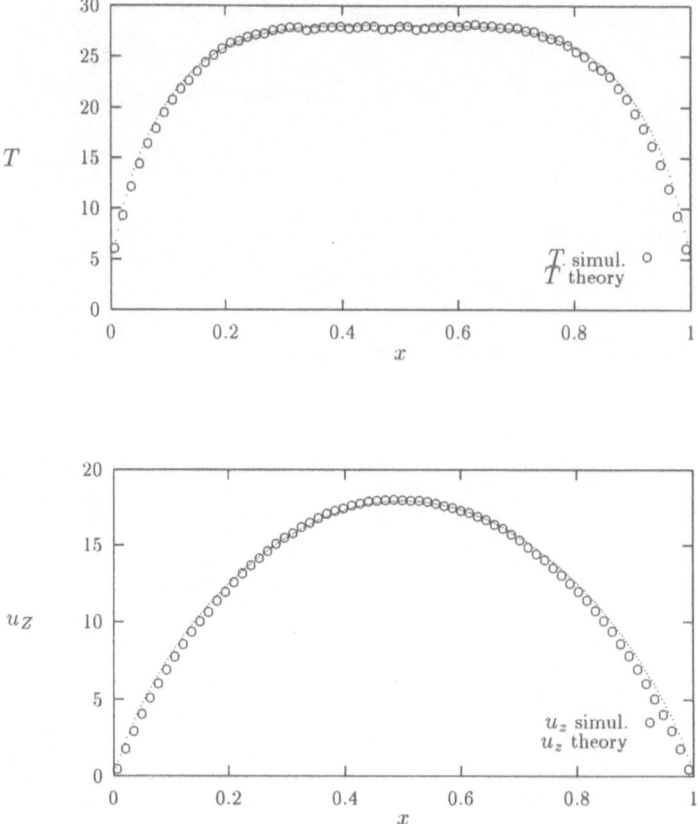

Figure 3: Comparison between the theoretical and observed temperature (top) and density (bottom) profiles against the transversal coordinate x for a system of $N = 2500$ particles and $a = 68.7$

statement we compare in figure 2 the temperature and velocity profiles for a system with $N = 2500$ particles, $N = 900$ and $a = 40.5$. We can say that the coincidence is almost perfect.

In figure 3 we compare the velocity and temperature profiles observed and predicted by our hydrodynamic analysis for a system of $N = 2500$ particles and $Fr = 8$, which implies $Fr_{\mathrm{eff}} = 1.87$ and $a = 68.7$. The agreement is very good except near the walls where the density variations are large. We remark that the predicted values for T_{max} and u_{max} differ by less than 2% from the predicted values.

There is no room in this summary to give any details about further analysis of our results but we would like to mention that: (1) the conductivity that comes from Enskog's theory is systematically observed to be 10-20% smaller than the observed one and (2) we have measured all the components of the stress tensor P_{ij}. The results show a slight but clear inhomogeneity between P_{xx} and P_{zz} and the equality $P_{xz} = P_{zx}$ has been checked within the precision of the measurements. Further the Newtonian behavior of the fluid is well verified at least not too near the walls.

APPENDIX

The basic parameters are: the width and height of the channel: L_X, L_Z, the number of particles N, their mass and diameter m, D, acceleration of gravity g and the temperature imposed at the lateral walls of the channel T_w.

From the above parameters it is convenient to define the aspect ratio $\lambda = L_X/L_Z$, the number density $n = N/L_X L_Z$, the fraction of area occupied by the disks $\rho_A = \frac{\pi D^2}{4} n$, the mass density $\rho = m\,n$.

Other derived quantities are more complex because they depend on quantities that can only approximately be calculated — from kinetic theory — from the basic parameters, namely the thermal conductivity k and the shear viscosity η. They are the kinematic viscosity $\nu = \eta/\rho$, the thermal diffusivity $\kappa = k/c_p\rho$, the Froude number $Fr = mg/k_B T_w$, the Prandtl number $Pr = \nu/\kappa$, the velocity of sound c.

For bare hard disks both k and η depend on the temperature through a factor $\sqrt{T}$,

118

namely [12]

$$\eta = \eta_0(n)\sqrt{T} \qquad k = k_0(n)\sqrt{T} \qquad (6)$$

In (1) η_0 is understood as η/T_w.

References

[1] L.D. Landau, L.M. Lifshitz (1959) *Fluid Mechanics*. Pergamon Press, Oxford (1959).

[2] S.A. Orzag, L.C. Kells, J. Fluid Mech. **96** 159 (1980).

[3] H.L. Swinney, J.P. Gollub, eds, (1985) *Hydrodynamic Instability and the Transition to Turbulence*, Springer Verlag, Berlin.

[4] M.Alaoui and A. Santos, *Phys. Fluids A* **4**, 1273 (1992).

[5] J. Koplik, J.R. Banavar and J.F. Willemsem, Phys. Rev. Letters **60** 1282 (1988).

[6] M. Marín, D. Risso, P. Cordero, *J. Comp. Phys.* **109** 306 (1993)

[7] L.P. Kadanoff, G.R. McNamara and G. Zanetti, Phys. Rev. A **40** 4527 (1989).

[8] D. Risso and P. Cordero, *in preparation*.

[9] D. Henderson, Mol. Phys. **30** 971 (1975).

[10] S. Chapman, T.G. Cowling, (1960) *The Mathematical Theory of Non-Uniform Gases*, University Press.

[11] A. Tenenbaum, G. Cicotti and R. Gallico, Phys. Rev. A **25** 2778 (1982).

[12] D. Gass, J. Chem. Phys. **54** 1898 (1971)

STATIONARY CONVECTION DUE TO RESISTIVITY AND VISCOSITY IN A CYLINDRICAL PLASMA WITH A FREE BOUNDARY

L. Gomberoff
Departamento de Física
Facultad de Ciencias
Universidad de Chile
Casilla 653
Santiago
Chile

ABSTRACT. Large scale stationary convection due to viscosity and resistivity in a current–carrying cylindrical plasma with a free boundary is investigated. By using the magneto-hydrodynamic (MHD) equations, it is shown that there are four states which are both marginal and stationary. These states are the plasma analog of stationary convection in ordinary hydrodynamics. Therefore, it is possible to define the critical Rayleigh number which characterizes the onset of steady convection. For Rayleigh numbers larger than the critical number, the whole nonlinear set of MHD equations possesses convective stationary solutions which bifurcate from the equilibrium solution.

1. Introduction

Stationary convection in plasmas has been the subject of several investigations.[1-7] A systematic study on large scale stationary convection in a current–carrying cylindrical plasma was started several years ago by Gomberoff and Maschke.[8] The studies were initiated by considering a plasma column limited by perfectly conducting walls. It was first shown that thermal conductivity and viscosity can trigger large scale stationary convection.[9,10] It was shown later that thermal conductivity and resistivity can also lead to steady convection.[11-14] Finally, it was shown that the combined effect of thermal conductivity, resistivity, and viscosity can also trigger stationary convection.[15,16] In all these cases, the convection cells are helically twisted tubes and the number of convection cells is equal to $2m$, where m is the azimuthal wave number.

The effect of Hall currents on the stationary convection of a plasma column including all the aforementioned effects has also been considered.[17,18]

More recently, with the purpose of studying the conditions under which stationary convection can take place in machines like $\theta - z$ pinches, where the obtained magnetic field and density profiles are similar to the model under consideration, a plasma column limited by a vacuum has been investigated. If steady convection can in fact be achieved, it can lead to stable configurations which, in the context of nonlinear theory, are usually

E. Tirapegui and W. Zeller (eds.), Instabilities and Nonequilibrium Structures V, 119–128.

thought to give rise to turbulence.[19-20] Moreover, stationary convection can also lead to anomalous transport effects. Thus, it has been recently shown that in a current–carrying cylindrical plasma with a free boundary both, thermal conductivity and resistivity, and thermal conductivity and viscosity, can lead to steady convection.[21,22]

We shall now show that when all three non ideal effects are present, namely, thermal conductivity, resistivity, and viscosity, there are four states which can bifurcate from the equilibrium solution and trigger large scale stationary convection.

The layout of the paper is as follows. In Section 2 the basic equations describing the system are solved. In Section 3 the boundary conditions are discussed. In section 4 the dispersion relation is derived and the existence of four states which, for finite adiabaticity coefficient and critical values of the pressure gradient, lead to the onset of steady convection in the plasma. In Section 5 the results are summarized and discussed.

2. The Basic Equations

The system consists of a uniform current–carrying plasma column surrounded by a vacuum. The length of the cylinder is L and the radius is a.

The equations describing the system are:

$$\rho(\frac{\partial}{\partial t} + v \cdot \nabla)\vec{v} = \frac{1}{c}\vec{j} \times \vec{B} - \vec{\nabla}p - \mu_\perp \vec{\nabla} \times (\vec{\nabla} \times \vec{v}), \tag{1}$$

$$\frac{\partial \rho}{\partial t} + \vec{\nabla} \cdot (\rho \vec{v}) = 0, \tag{2}$$

$$\frac{\partial p}{\partial t} + \vec{v} \cdot \vec{\nabla}p - \frac{2}{3}\kappa \nabla^2 p - \frac{2}{3}\eta |\vec{j}|^2 - S_0 = -\gamma p \vec{\nabla} \cdot \vec{v}, \tag{3}$$

$$\vec{E} + \frac{1}{c}\vec{v} \times \vec{B} = \eta \vec{j}, \tag{4}$$

$$\vec{\nabla} \cdot \vec{B} = 0, \tag{5}$$

$$\vec{\nabla} \times \vec{B} = \frac{4\pi}{c}\vec{j}, \tag{6}$$

$$\frac{1}{c}\frac{\partial \vec{B}}{\partial t} = -\vec{\nabla} \times \vec{E}, \tag{7}$$

where $\mu_\perp$ is the perpendicular viscosity,[23] κ is the thermal conductivity, S_0 is a constant heat source which maintains the equilibrium pressure profile, η is the resistivity, γ is the adiabaticity coefficient, and ρ is the density.

The equilibrium is characterized by a magnetic field given by

$$\vec{B}^p = (0, B_I \frac{r}{a}, B_0) \tag{8}$$

and

$$\vec{B}^v = (0, B_I \frac{a}{r}, B_0), \tag{9}$$

where p and v stand for plasma and vacuum respectively, and B_I and B_0 are constants.

The equilibrium velocity is zero, and the equilibrium pressure is given by:

$$p^{(0)} = p_0 - (B_I^2/4\pi)(r/a)^2, \tag{10}$$

where p_0 is a constant.

The rotational transform q is constant, and therefore, the magnetic field is shearless:

$$q = \frac{2\pi r B_z^{(0)}}{L B_\theta^{(0)}} = \frac{2\pi a B_0}{L B_I}. \tag{11}$$

Assuming the density to be nearly constant, $\rho \approx \rho_0$, the motion incompressible, and that all perturbed quantities behave like

$$f^{(1)}(r, \theta, z) = f^{(1)}(r) \exp(im\theta + ikz + \Omega t), \tag{12}$$

upon using Eq. (6), the linearized Eq. (1) reduces to

$$\rho_0 \Omega \vec{v}^{(1)} = -\vec{\nabla}(p^{(1)} + \frac{\vec{B}^{(0)} \cdot \vec{B}^{(1)}}{4\pi}) + \frac{iB_I}{4\pi a}(m - nq)\vec{B}^{(1)}$$
$$- \frac{2B_I}{4\pi a}(B_\theta^{(1)} \hat{e}_r - B_r^{(1)} \hat{e}_\theta) - \mu_\perp \vec{\nabla} \times (\vec{\nabla} \times \vec{v}^{(1)}). \tag{13}$$

On the other hand, using Eqs. (5–7), the linearized Eq.(4) yields:

$$\Omega \vec{B}^{(1)} = \frac{iB_I}{a}(m - nq)\vec{v}^{(1)} - \frac{\eta c^2}{4\pi} \vec{\nabla} \times (\vec{\nabla} \times \vec{B}^{(1)}). \tag{14}$$

In the ideal case, the solution of the linear magnetohydrodynamics satisfy the force free–field condition $\vec{\nabla} \times \vec{B} = \beta \vec{B}$. From Eq. (14) it then follows that $\vec{\nabla} \times \vec{v} = \beta \vec{v}$, with β a constant to be determined later.

Substituting Eq. (14) into Eq. (13) yields:

$$[\hat{\Omega}^2 + \hat{\Omega}\beta^2 a^2(\hat{\mu}_\perp + \hat{\eta}) + \hat{\mu}_\perp \hat{\eta}\beta^4 a^4 + (m - nq)^2]\vec{\xi} =$$
$$-a^2 \vec{\nabla}\hat{p}^{(1)} - 2i|m - nq|(\xi_\theta \hat{e}_r - \xi_r \hat{e}_\theta), \tag{15}$$

where

$$\hat{\Omega} = (4\pi a^2 \rho_0 / B_I^2)^{1/2}\Omega, \tag{16}$$

$$\vec{\xi} = \frac{1}{\Omega + \eta\beta^2 c^2/4\pi} \vec{v}^{(1)}, \tag{17}$$

$$\hat{\mu}_\perp = (4\pi/\rho_0 B_I^2 a^2)^{1/2}\mu_\perp, \quad \hat{\eta} = (\frac{\rho_0 c^4}{4\pi a^2 B_I^2})^{1/2}\eta, \tag{18}$$

and

$$\hat{p}^{(1)} = (4\pi/B_I^2)(p^{(1)} + (\vec{B} \cdot \vec{B})^{(1)}/8\pi), \tag{19}$$

is the total perturbed pressure.

In terms of the components of $\vec{\xi}$, Eq. (15) reduces to

$$\xi_r = (\frac{a^2}{\Lambda})(\frac{\partial \hat{p}^{(1)}}{\partial r} + \frac{m}{r}\sigma\hat{p}^{(1)}), \tag{20}$$

$$\xi_\theta = (\frac{ia^2}{\Lambda})(\sigma\frac{\partial \hat{p}^{(1)}}{\partial r} + \frac{m}{r}\hat{p}^{(1)}), \tag{21}$$

$$\xi_z = (ika^2/\Lambda)(1 - \sigma^2)\hat{p}^{(1)}, \tag{22}$$

where

$$\sigma = 2(m - nq)/\{(\hat{\Omega} + \hat{\mu}_\perp\beta^2 a^2)(\hat{\Omega} + \hat{\eta}\beta^2 a^2) + (m - nq)^2\}, \tag{23}$$

$$\Lambda = \{(\hat{\Omega} + \hat{\mu}_\perp\beta^2 a^2)(\hat{\Omega} + \hat{\eta}\beta^2 a^2) + (m - nq)^2\}(\sigma^2 - 1), \tag{24}$$

and $n = -kL/2\pi$.

Taking the divergence of $\vec{\xi}$ by using its components, Eqs. (20–22), the incompressibility condition yields:

$$\nabla^2\hat{p}^{(1)} + k^2\sigma^2\hat{p}^{(1)} = 0. \tag{25}$$

This is Bessel's equation whose regular solution at the axis of the cylinder, $r = 0$, is

$$\hat{p}^{(1)} = \alpha J_m(k(\sigma^2 - 1)^{1/2}r), \tag{26}$$

where α is a constant.

Noting that $\vec{B}^{(1)} \parallel \vec{v}^{(1)}$ (see Eq. 14), the constant β can be determined by taking the curl of Eq. (15).

The result is

$$\beta = k\sigma, \tag{27}$$

where σ is defined in Eq. (23).

Before analyzing the boundary conditions, let us calculate the perturbed magnetic field in the vacuum region.

Since

$$\vec{\nabla} \cdot \vec{B} = 0, \tag{28}$$

and in vacuum

$$\vec{\nabla} \times \vec{B} = 0, \tag{29}$$

it follows that

$$\vec{B}^{(1)} = \vec{\nabla}\psi, \tag{30}$$

and

$$\nabla^2\psi = 0. \tag{31}$$

Therefore,

$$\psi = AK_m(kr), \tag{32}$$

where $K_m(kr)$ are Bessel's functions of the second kind which go to zero at infinity, and A is a constant.

Thus, from Eq. (32) and (30), it follows that the perturbed magnetic field in vacuum is given by

$$B_r^{(1)} = AkK_m'(kr), \tag{33}$$

$$B_\theta^{(1)} = \frac{imK_m}{krK_m'}B_r^{(1)}, \tag{34}$$

$$B_z^{(1)} = \frac{iK_m}{K_m'}B_r^{(1)}. \tag{35}$$

where K_m' is the first derivative of K_m.

On the other hand, using Eqs. (14) and (17), the perturbed magnetic field in the plasma is given by

$$\vec{B}^{(1)} = \frac{iB_I}{a}(m - nq)\vec{\xi}. \tag{36}$$

3. Boundary Conditions

At the plasma–vacuum boundary, the total pressure in the plasma has to be equal to the magnetic pressure in vacuum. This condition implies,[10,13,14]

$$\tilde{p}^{(1)}(a) + \tilde{p}^{(0)}(a + \xi_r) = \frac{1}{8\pi}(\vec{B}^{(0)}(a + \xi_r) + \vec{B}^{(1)}(a)) \cdot$$
$$(\vec{B}^{(0)}(a + \xi_r) + \vec{B}^{(1)}(a))|^v, \tag{37}$$

where

$$\tilde{p} = p + \frac{\vec{B} \cdot \vec{B}}{8\pi}, \tag{38}$$

is the total pressure in the plasma.

From Eq.(37) it follows that

$$\tilde{p}^{(1)}(a) = \frac{1}{4\pi}\left(\vec{B}^{(0)} \cdot \frac{\partial \vec{B}^{(0)}}{\partial r}\bigg|_{r=a}^v \xi_r(a) - 4\pi\frac{\partial\tilde{p}^{(0)}}{\partial r}\xi_r(a) + \vec{B}^{(0)} \cdot \vec{B}^{(1)}|^v\right). \tag{39}$$

Using Eqs. (8–10), the last equation reduces to

$$\tilde{p}^{(1)}(a) = \frac{1}{4\pi}(\vec{B}^{(0)} \cdot \vec{B}^{(1)})|^v. \tag{40}$$

Since the tangential components of the electric field at the plasma boundary must be equal, i.e.,

$$E_{tg}^p = E_{tg}^v|_{r=a}, \tag{41}$$

from Eqs. (4) and (36), it follows that

$$B_r^{(1)}|^v = \frac{iB_I}{a}(m - nq)\xi_r. \tag{42}$$

Therefore, Eq. (39) reduces to:

$$\tilde{p}^{(1)}(a) = -\frac{B_I^2}{4\pi a^2}\frac{K_m}{kK_m'}(m - nq)^2\xi_r(a). \tag{43}$$

Noting that

$$\hat{p}^{(1)}(a) = \frac{4\pi}{B_I^2}\tilde{p}^{(1)}(a), \tag{44}$$

and using Eqs. (20) and (26), Eq. (43) reduces to:

$$\frac{\sigma}{\sigma^2 - 1}(\frac{ka(\sigma^2 - 1)^{1/2}J_m'}{J_m} + m\sigma) = -\frac{2kaK_m'}{(m - nq)K_m}, \tag{45}$$

where J_m' is the first derivative of J_m.

Eq. (45) is the boundary condition to be satisfied at the plasma–vacuum boundary.

On the other hand, when the plasma is surrounded by conducting walls, the boundary conditions are $v_r(r = a) = v_\theta(r = a) = 0$.[9,16] From Eqs. (21), (22) and (26) it follows that $\xi_r(r = a) = \xi_\theta(r = a) = 0$ when

$$1 - \sigma = \frac{x_0 J_{m-1}(x_0)}{m J_m(x_0)}, \tag{46}$$

and

$$\frac{\sigma - 1}{\sigma} = \frac{x_0 J_{m-1}(x_0)}{m J_m(x_0)} \tag{47}$$

respectively.

In general, the last equations are satisfied for some $\sigma = \sigma_0$. For such value, $ka(\sigma_0^2 - 1)^{1/2} = x_0$.

Before solving for σ_0, we shall discuss the dispersion relation.

4. The Dispersion Relation

By setting $\sigma = \sigma_0$ in Eq. (23), it follows that the growth rate is given by:[11]

$$\hat{\Omega} = -\frac{1}{2}(\hat{\mu}_\perp + \hat{\eta})\beta_0^2 a^2 + \{\frac{1}{4}(\hat{\mu}_\perp + \hat{\eta})^2\beta_0^4 a^4 - \hat{\mu}_\perp\hat{\eta}\beta_0^4 a^4 + \frac{2}{\sigma_0}|m - nq| - (m - nq)^2\}^{1/2}. \tag{48}$$

Note that the dispersion relation reduces to the well known ideal MHD dispersion relation when $\mu = \eta = 0$.[24-27]

We are interested in solutions which are both stationary, $Im\Omega = 0$, and marginal, $Re\Omega = 0$. From Eq. (48) it follows that these conditions are satisfied when:

$$m - nq = \frac{1}{\sigma_0}[1 \pm (1 - \sigma_0^2\lambda^2)^{1/2}], \tag{49}$$

where

$$\lambda^2 = \hat{\mu}_\perp\hat{\eta}a^4k^4\sigma_0^4. \tag{50}$$

Thus, the states characterized by:

$$\alpha_1 = (m - nq) = \frac{1}{\sigma_0}[1 + (1 - \sigma_0^2\lambda^2)^{1/2}] \tag{51}$$

and

$$\alpha_2 = (m - nq) = \frac{1}{\sigma_0}[1 - (1 - \sigma_0^2\lambda^2)^{1/2}], \tag{52}$$

are both marginal and stationary.

Throughout we have assumed the motion to be incompressible, $\vec{\nabla} \cdot \vec{v} = 0$, which implies that γ in Eq. (3) is equal to infinity. However, we need the heat conduction equation to be fulfilled everywhere in the plasma for finite adiabaticity coefficient.

In Ref. (16) it is shown that the states $\pm\alpha_1$ and $\pm\alpha_2$ satisfy the full heat conduction for finite γ when

$$(\eta/\kappa) = 8\pi/3, \tag{53}$$

and

$$\mu_\perp\kappa = -\frac{3}{ak^4}\frac{dp^{(0)}}{dr}, \tag{54}$$

respectively.

In both cases the heat conduction Eq. (3), is satisfied for finite adiabaticity coefficient, $B_\theta \gg B_z$, and $ka \gg 1$.[16]

Thus, we have shown the existence of four states which satisfy the complete set of nonideal Eqs. (1-6), and which are both stationary and marginal.

It has been shown that states satisfying these properties are the analog in a plasma of the well known stationary convection in ordinary hydrodynamics.[28] Therefore, it is possible to define the plasma analog of the critical Rayleigh number, R_{crit}, which characterizes the onset of steady convection.[28] For $R > R_{crit}$ the complete nonlinear Eqs. (1-6) possess stationary convective solutions which bifurcate from the equilibrium solutions.

Finally, we shall determine the σ-values consistent with the boundary condition, Eq. (45). Replacing in Eq. (45) $m - nq = \alpha_{1,2}$ and solving for σ_0, yields $\sigma_0 = \pm 1$. These values of σ_0 are consistent with Eqs. (46) and (47). In fact, since

$$(\sigma_0^2 - 1)k^2a^2 = x_0 \tag{55}$$

it follows that

$$\sigma_0^2 = 1 + \frac{x_0}{k^2 a^2}, \tag{56}$$

which, for any finite x_0 and sufficiently large ka, gives $\sigma_0^2 = 1$. Clearly, $\sigma_0 = \pm 1$ are solutions of Eqs. (46) and (47).

5. Summary

We have studied the stability of a current–carrying cylindrical plasma surrounded by a vacuum. Previously, we have studied the problem of stationary convection of a plasma column limited by conducting walls.[8-17] However, it is important to know the conditions for convection in machines like $\theta - z$ pinches, because it can lead to stable configurations which, within the nonlinear theory, are thought to give rise to turbulence. Moreover, stationary convection can lead to anomalous transport effects.

We have included thermal conductivity, resistivity, and viscosity, and we have shown the existence of four states which are both marginal, $(Re\Omega = 0)$, and stationary, $(Im\Omega = 0)$. These states satisfy the boundary condition at the plasma–vacuum interface, and are the plasma analog of the well known Rayleigh–Bénard convection problem. The states satisfy the full set of Eqs. (1–6). The heat conduction equation, Eq. (3), is satisfied by the states $\pm\alpha_1$ and $\pm\alpha_2$ when the condition given by Eq. (49) is fulfilled.

Under these conditions it is possible to define the Rayleigh number which, for the states $\pm\alpha_2$, is given by

$$R = \frac{1}{r_c} \frac{dp^{(0)}}{dr}\bigg|_{r=a/2} (\mu_\perp \kappa)^{-1}, \tag{57}$$

where

$$\frac{1}{r_c} = \frac{1}{a}\left(\frac{B_I}{B_0}\right)^2. \tag{58}$$

is the curvature of the helical field lines at $r = a$.

The condition for the existence of a marginal solution, Eq. (49), yields the critical Rayleigh number:

$$R_{crit} = k^6 a^6 / 6m^2, \tag{59}$$

and similarly for the other modes, $\pm\alpha_1$.

The condition $R = R_{crit}$ characterizes the onset of steady convection in the plasma. For $R > R_{crit}$ the complete nonlinear Eqs. (1–6), possess stationary convective solutions which bifurcate from the equilibrium solutions.

Following the same procedure used in Reference [16], it is possible to show that the convective cells are hellicaly twisted tubes and the number of convective cells is equal to $2m$. It is also possible to show that on the center line of each tube, $v_r^{(1)} = 0$ and $|v_\theta^{(1)}| \gg v_z^{(1)}$ provided that $B_\theta/B_z \gg 1$,[16] justifying the assumption that the parallel viscosity has no effect on the purturbed motion.

It is also interesting to note that as far as convection is concerned, resistivity and viscosity seem to play equivalent roles with one acting as the inverse of the other. Resistivity controls convection at the edge of the unstable spectrum whereas viscosity controls the central region of the spectrum.[16]

Acknowledgements

This work has been supported in part by Fondo Nacional de Ciencia y Tecnología (FONDE-CYT) grant N^{0} 1940360.

References

1. Simon, A. (1968) 'Convection in a weakly ionized plasma in a non–uniform magnetic field', Phys. Fluids 11, 1186–1191.

2. Kadomtsev, B. and Pogutse, O. (1970) 'Turbulence in toroidal plasmas' in M. Leontovich (ed.), Reviews of Plasma Physics, Plenum Press, New York, Vol 5, pp. 349–498.

3. Okuda,H. and Dawson, J. M. (1973) 'Theory and numerical simulations in plasma diffusion across a magnetic filed', Phys. Fluids, 16, 408–426.

4. Roberts, H. and Taylor, J. B. (1965) ' Gravitational resistive instability of an incompressible plasma in a sheared magnetic field', Phys. Fluids 8, 315–322.

5. Wobig, H. (1972) 'Convection of a plasma in a gravitational field', Plasma Phys. 14, 403–416.

6. Maschke, E. K. and Paris, R. B. (1973) in Proceedings of the Sixth European Conference on Controlled Fusion and Plasma Physics, Moscow, Vol 1, p. 205.

7. Dagazian, R. Y. and Paris, R. B. (1977) 'Stationary convection-like modes in a plasma slab with magnetic shear', Phys. Fluids 20, 917–927.

8. Gomberoff, L. and Mashke, E. K. (1981) 'Non–Ideal Effects on the Stability of a Current–Carrying Plasma', in E. Tirapegui(ed.), Field Theory, Quantization, Reidel, New York, pp. 123–145.

9. Gomberoff, L. and Hernández, M. (1983) 'Stationary convection in a cylindrical plasma', Phys. Rev., A27, 1244–1246.

10. Gomberoff, L. and Hernández, M. (1984) 'Large-scale stationary convection in a cylindrical current-currying plasma', Phys. Fluids 27, 392–398.

11. Gomberoff, L. (1984) 'Resistive convection in a cylindrical plasma', J. Plasma Phys. 31, 29–37.

12. Gomberoff, L. (1985) 'Resistive convection in a cylindrical plasma. Part 2', J. Plasma Phys. 34, 299–303.

13. Gomberoff, L. (1989) 'Resistive convection in, a current-carrying cylindrical plasma: An exact model', Phys. Fluids B 1, 10, 2126–2128.

14. Gomberoff, L. (1991) 'An exact model of resistive convection in a cylindricla plasma', in E. Tirapegui and W. Zellers (eds.), Instabilities and Nonequilibrium Structures III, Klubers Academic Publishers, The Netherlands, pp. 307–316.

15. Gomberoff, L. (1983) 'Resistive and viscous convection in a cylindrical plasma', Phys. Rev. A 28, 3125–3127.

16. Gomberoff., L. and Palma, G. (1984) 'Stationary convection due to resistivity, viscosity, and thermal conductivity in a cylindrical plasma', Phys. Fluids 27, 2022–2027.

17. Gomberoff, K. and Gomberoff, L. (1992) 'Effect of Hall currents on the steady convection of a current–carrying cylindrical plasma ', Phys. Fluids B 4, 3024–3030.

18. dos Santos, C. A. M. and Galvao, R. M. O. (1992) 'Influence of the Hall effect on convection in plasmas, Phys. Fluids, B 4, 4187–4189.

19. Shaper, U. (1983) 'Stabilizing and distabilizing influence of the Hall effect an a Z pinch with a step-like colume current profile', J. Plasma Phys. 30, 169–178.

20. Shaper, U. (1983) 'On the influence of the Hall effect on the spectrum of the ideal magnetohydrodinamic cylindrilcal pinch', J. Plasma Phys. 29, 1–19.

21. Gomberoff, L. and Gomberoff, K (1992) 'Convection in a cylindrical plasma with a free boundary', Phys. Fluids B 4, 1428–1431.

22. Gomberoff, L. (1994) 'Stationary convection in a cylindrical plasma', Phys. Rev. A (submitted for pubilcation).

23. Braginskii, S. I. (1965) Reviews in plasma Physics, Leontovich (consultant Bureau, New York, 1965), Vol. 1, p. 205.

24. Shafranov, V. D. (1958) 'The instability of a plasma column with a distributed current', in Plasma Physics and the Problem of Controlled Thermonuclear Research, Pergamon Press, New York, Vol. 2, pp. 71–80.

25. Friedberg, J. P. (1970) 'Magnetohydrodynamic Stability of a Diffuse Screw Pinch', Phys. Fluids 13, 1812–1818.

26. Goedbloed, J. P. and Hagebeuk, H. J. L. (1972) 'Growth Rates of Instabilities od a Diffuse Linear Pinch', Phys. Fluids 15, 1090–1101.

27. Coppins, M., Bond, D. J., and Haines, M. G. (1984) 'A study of the stability of a Z pinch under fusion conditions using the Hall fluid model', Phys. Fluids, 27, 2886–2889.

28. Chandrasekhar, S. (1961) Hydrodynamics and Hydromagnetic Stability, Oxford University Press, London.

GINZBURG–LANDAU EQUATIONS FOR OSCILLATORY CONVECTION IN OLDROYD-B FLUIDS

J. MARTINEZ-MARDONES,[1] R. TIEMANN[2] and W. ZELLER[1]
[1] Instituto de Física, Universidad Católica de Valparaíso (UCV)
Casilla 4059, Valparaíso, Chile.
[2] Facultad de Ciencias, Universidad de Playa Ancha (UPLACED)
Casilla 34-V, Valparaíso, Chile.

ABSTRACT. Coupled Ginzburgh-Landau amplitude equation supports the weakly nonlinear analysis of oscillatory convection from Oldroyd-B fluids in a thin horizontal fluid layer heated from below. Preliminary comprehension of nonlinear bifurcation of stationary structures and travelling and standing wave patterns has been achieved.

INTRODUCTION

Rayleigh-Bénard convection constitutes an important piece of thought on theoretical and experimental instability work [1-3]. Recently the interest has been extended to both binary mixtures [4-6] and viscoelastic fluids [7-10]. The rheology contributes with constitutive equations which, in general, relate stress with rate of strain tensors [11,12]. Focusing on polymeric solutions, there are several models to describe specific fluid characteristics [13-15]. Our previous work was centered on linear and weakly nonlinear Stuart-Landau analysis on Maxwell and Oldroyd fluid convection [16-17]. The present study introduces advective, diffusive, dispersive and coupling effects in a Ginzburg-Landau amplitude equation form [18,20] for the Oldroyd-B fluid model in order to acquire insight into nonlinear standing patterns and travelling and stationary wave structures formation.

GOVERNING EQUATIONS OF VISCOELASTIC CONVECTION

A shallow viscoelastic Oldroyd-B fluid layer of depth d heated from below remains at rest until a critical temperature gradient is reached. As usual, the outcome of the viscoelastic Rayleigh-Bénard problem, under Boussinesq approximation, can be described by mass, Cauchy and energy transport equations

$$\nabla \cdot \mathbf{v} = 0 \tag{1}$$

$$\rho_o [\partial_t \mathbf{v} + (\mathbf{v} \cdot \nabla)\mathbf{v}] = -\nabla p + \nabla \cdot \tau + \rho_o [1 - \alpha(T - T_o)]\mathbf{g} \tag{2}$$

$$\partial_t T + (\mathbf{v} \cdot \nabla)T = \kappa^2 \nabla T, \tag{3}$$

where $\mathbf{v}$ is the velocity field, p the pressure, τ the extra-stress tensor, T the temperature, ρ_o the reference density, T_o the temperature on the bottom, α

E. Tirapegui and W. Zeller (eds.), Instabilities and Nonequilibrium Structures V, 129–135.
© 1996 Kluwer Academic Publishers.

the thermal expansion coefficient, κ the thermal diffusivity and $\mathbf{g}$ the gravity. In general, based on rheological considerations, the stress tensor, $\tau = \tau^P + \tau^s$, may be decomposed into the polymeric contribution τ^P and the stress tensor $\tau^s = 2\eta_s \mathbf{D}(\mathbf{v})$ of the Newtonian solvent. The stress tensor τ^P of a polymeric solution has to be determined by a constitutive equation for the viscoelastic fluid of the form

$$[1+\lambda_1 D_t]\tau^P + f[\tau^P, \mathbf{D}(\mathbf{v})] = 2\eta_p \mathbf{D}(\mathbf{v}). \tag{4}$$

The convected contravariant derivative D_t of a tensor $\mathbf{S}$ is defined as

$$D_t \mathbf{S} = \partial_t \mathbf{S} + (\mathbf{v} \cdot \nabla)\mathbf{S} - (\nabla \mathbf{v})^T \cdot \mathbf{S} - \mathbf{S} \cdot \nabla \mathbf{v}. \tag{5}$$

The viscosity of the fluid, $\eta_o = \eta_s + \eta_p$, is given by the addition of the viscosities of the solvent and the polymer, η_s and η_p, respectively. We have also introduced the relaxation time λ_1 and the rate of deformation $\mathbf{D}(\mathbf{v}) = 1/2 [\nabla \mathbf{v} + (\nabla \mathbf{v})^T]$. The viscoelastic fluids, containing flexible unentangled and entangled polymer molecules –namely, diluted solutions and concentrated solutions and melts, respectively–, may be represented by means of the appropriate function $f[\tau^P, \mathbf{D}(\mathbf{v})]$, such as that corresponding to the Maxwell, Oldroyd, Johnson and Segelman, Phan-Thien and Tanner, Giesekus, and Larson models [21].

The Maxwellian model assumes that τ ($\equiv\tau^P$) is equivalent to the polymeric contribution alone. We selected the Oldroyd-B model for our present work, which contains only one single additional constant, namely the retardation time λ_2 ($0 \leq \lambda_2 < \lambda_1$), and which represents the properties of a low concentrated mixture fairly well. Oldroyd-B constitutive equation is given by the use of

$$f[\tau^P, \mathbf{D}(\mathbf{v})] = 0. \tag{6}$$

As a matter of fact, the function f has to take into account the rheological properties of the fluid. For example, based on molecular arguments, Giesekus introduced into the upper convected Maxwell equation the simplest term quadratic in stress in order to bound the extensional viscosity at all extension rates, and proposed

$$f[\tau^P, \mathbf{D}(\mathbf{v})] = \frac{\alpha\lambda}{\eta_p} \tau^P \cdot \tau^P, \tag{7}$$

where a single "mobility factor" α must be obtained from the experimental data which has to be considered in addition to the usual three parameters: the relaxation time λ, and the solvent and polymer contributions to the zero-shear-rate viscosity, η_s and η_p. The Larson model, given by

$$f[\tau^P, \mathbf{D}(\mathbf{v})] = \frac{2\alpha\lambda}{3\eta_p}\mathbf{D}(\mathbf{v}):\tau^P[\tau^P+(\eta_p/\lambda)\mathbf{I}], \tag{8}$$

where $\mathbf{I}$ is the unit tensor $\mathbf{I}$, takes into account important characteristics of melts and concentrated solutions with some basis on the molecular behavior of polymers. From a continuum-mechanical approach, Oldroyd produces the following general eight-constant constitutive equation

$$\tau + \lambda_1 D_t \tau + \mu_0 \, tr(\tau)D - \mu_1(\tau \cdot D + D \cdot \tau) + \nu_1(\tau:D)I =$$

$$2\eta_0[D + \lambda_2 D_t D - 2\mu_2 D \cdot D + \nu_2(D:D)I]. \qquad (9)$$

Obviously, the general Oldroyd equation, with eight experimental constants, is limited and extremely difficult to handle in practical calculations. Therefore, several simplified versions of this model, in which special values are assigned to some constant, have been used in the literature extensively.

ADIMENSIONAL DISTURBANCE EQUATIONS AND LINEAR ANALYSIS

We assume that the Oldroyd-B viscoelastic fluid is initially at rest, corresponding to a pure conductive reference state and subject to mechanically free and perfectly thermal conductive constraints. As usual, the physical system is perturbed, and its evolution is governed by an adimensional disturbance equation of the compact form

$$\partial_t Lu = Mu + N(u,u), \qquad (10)$$

The perturbation fields matrices L and M are linear, and N takes into account the nonlinearities. The perturbation vector $u(x,z,t) = [\psi, \Theta, \tau_{xz}, S, U]^T$ represents the stream function Ψ, the temperature Θ, the in-plane shear tensor τ_{xz}, the normal stress difference $S = \tau_{xx} - \tau_{zz}$ and the nonequilibrium stress trace $U = tr(\tau)$. We also need to introduce the following nondimensional parameters: the Rayleigh number $R = \rho_0 g\alpha |\Delta T| d^3/\eta_0 \kappa$, the Prandtl number $P = \eta_0/\rho_0 \kappa$, the stress relaxation time $\Gamma = \lambda_1 \kappa/d^2$, and the ratio between retardation and relaxation times $\Lambda = \lambda_2/\lambda_1 = \eta_s/\eta_0$.

Furthermore, according to the constraints for the system mentioned above, we select free ($\psi = \partial_{zz}^2 \psi = 0$) and perfectly conductive ($\Theta = 0$) boundary conditions, both on the bottom and at the upper surface ($z = 0, 1$).

As long as the perturbations are small enough, we assume that the nonlinear term $N(u,u)$ in (10) may be considered negligible. From linear stability analysis we obtain the marginal curves when the growth rate vanishes. In Fig. 1 are presented the marginal stationary (dot) and oscillatory curves (dashed) for diluted ($P=100$, $\Lambda=0.75$, $\Gamma=100$) and concentrated ($P=1000$, $\Lambda=0.0$, $\Gamma=0.2$) polymeric fluids. It is well known that the full threshold problem, either for mechanically free and thermal conductive or rigid and isolated, boundary conditions has been solved for many kinds of Rayleigh-Bénard convection [7-10].

WEAKLY NONLINEAR ANALYSIS

The weakly nonlinear study of spacio-temporal behavior of the patterns near threshold is here based on nonlinear Ginzburg-Landau equations [21-23]. These are obtained from a perturbation expansion of underlying governing equations for small amplitudes of the growing linear modes that characterize the convecting rolls. The usual constant amplitudes of the perturbation oscillatory modes used in the linear analysis are now slow variables in space and time, X and T, respectively. The solutions can be expressed by the perturbation vector as

$$u(x,z,t) = U(z)[A(X,T)e^{i(k_c x+\omega_c t)} + B(X,T)e^{-i(k_c x-\omega_c t)}] + c.c. \qquad (11)$$

where c.c. refers to the complex conjugate of the first term. Since the amplitudes $A(X,T)$ and $B(X,T)$ depend on the slow variables, this multiscale approach, which takes into account the gauge of the temporal and spatial coordinates and differentiation operators, leads to a broad Ginzburg-Landau problem.

The disturbances give rise to stationary structures (SS) for $\omega_c=0$ (making B=0), pure left and right travelling waves (TW), if either A=0 or B=0, and stationary waves (SW), if A=B. Therefore, according to Eq. (11), the nonlinear spatial-temporal behavior may be described by the following coupled Ginzburg-Landau amplitude equations

$$(\partial_T + v\partial_X)A = \mu A + (\alpha_1 + i\alpha_2)\partial_X^2 A - (\beta_1 + i\beta_2)|A|^2 A - (\gamma + i\delta)|B|$$

$$(11)$$

$$(\partial_T - v\partial_X)B = \mu B + (\alpha_1 + i\alpha_2)\partial_X^2 B - (\beta_1 + i\beta_2)|B|^2 B - (\gamma + i\delta)|A|^2$$

where the sets of linear and nonlinear coefficients are $(v, \mu, \alpha_1, \alpha_2)$ and $(\beta_1, \beta_2, \gamma, \delta)$, respectively. The linear coefficients may be obtained by the use of the results of the linear stability analysis, where v represents the group velocity, μ measures the deviation to the instability threshold, α_1 deals with diffusion effects, α_2 is related to dispersion effects, β_2 and δ are nonlinear renormalization frequences, and both β_1 and γ correspond to non-linear saturation parameters.

RESULTS AND DISCUSSION

In order to obtain information about nonlinear pattern structures, our present concern deals mainly with two nonlinear coefficients: i) γ, which rules the competition between travelling and standing waves corresponding to nontrivial homogeneous solutions, and ii) β_1, which determines the type of bifurcation. The nonlinear renormalization frequencies β_2 and δ are physically meaningless. In contrast, the relevant roles are played by the coefficients β_1 and γ, which provide information of the pattern structure of the physical system. As it has just been said, β_1 determines the type of bifurcation: subcritical ($\beta_1<0$) or supercritical ($\beta_1>0$). On the other hand, the value of γ provides information about the stability of both the travelling waves ($\gamma>\beta_1$) and the stationary waves $(-\beta_1 < \gamma < +\beta_1)$.

The nonlinear coefficients $\beta_1 = \beta_{TW}$ and γ for the travelling waves were calculated following the well-known standard methodology, and they are presented in Figs. 2a) and 2b) as a function of the nondimensional relaxation time Γ for both a concentrated Maxwellian fluid (P=1000.0, Λ=0.0) and a diluted polymeric fluid (P=100.0, Λ=0.75). The results show that coefficient β_1 is always positive and, therefore, we may expect only supercritical bifurcation. Furthermore, since γ appears to be negative for both fluids, the travelling waves are unstable.

In Fig. 2 c) the analysis for standing waves shows that the corresponding coefficient $\beta_1 = \beta_{SW}$ is given by the addition of γ and the previous coefficient β_{TW} evaluated for the travelling waves. The value of $\beta_{SW} = \gamma + \beta_{TW}$ is always negative for diluted polymers. Nevertheless, it is positive for high concentrated fluids when Γ is lower than 64, range in which there are supercritical solutions for standing waves. A tricritical bifurcation is found at this point, and from this point on, for higher values of Γ, the coefficient becomes negative, thus subcritical bifurcations appear and it becomes necessary to go further into a fifth-order nonlinear analysis of the amplitude equation in order to achieve insight into the evolution of the physical system. Moreover, the stability of the possible standing wave patterns on the supercritical region cannot be proved without performing the analysis of phase dynamics.

ACKNOWLEDGMENT

The authors would like to thank Professor Dr. Daniel Walgraef (Brussels) for his useful remarks and discussion. This work has been supported by FONDECYT-CHILE and DGIPG-UCV.

REFERENCES

1. Chandrasekhar, S. (1981) 'Hydrodynamics and Hydromagnetic Stability', Dover, New York.
2. Swinney, H.L. and Gollub, P., eds. (1981) 'Hydrodynamical Instabilities and the Transition to Turbulence', Springer-Verlag, Heidelberg.
3. Gershuni, G.Z. and Zhukhovitskii, E.M. (1976) 'Convetive Stability of Incompressible Fluids', Keter, Jerusalem.
4. Zielinska, B.J.A. and Brand, H.R. (1987) 'Exact solution of the linear stability problem for the onset of convection in binary mixtures' Phys. Rev. A 35, 4349.
5. Müller, H.M., and Lücke, M. (1988) 'Competion between roll and square convection patterns in binary mixtures', Phys. Rev. A 38, 2965.
6. Schöpf, W. and Zimmermann, W. (1993) 'Convection in binary fluids: Amplitude equations, codimension-2 bifurcation, and thermal fluctuation', Phys. Rev. E 47, 1739-1764.
7. Sokolov, M. and Tanner, R.I. (1972) 'Convective stabillity of a general viscoelastic fluid heated from below', Phys. Fluids 15, 534-539.
8. Vest, C.M. and Arpaci, A. (1969) 'Overstability of a viscoelastic fluid layer', J. Flui Mech. 36, 613-623.

134

9. Zielinska, B.J.A., Demay, Y (1987 'Couette-Taylor instability in visco-elastic fluids', Phys. Rev. A 38, 897-903.

10. Martínez-Mardones, J. and Pérez-García, C. (1990) J. Phys.: Condens. Matter 2, 1281.

11. Oldroyd, J.G. (1950) 'On formulation of rheological equations of state', Proc. R. Soc. A 200, 523-541.

12. Bird, R.B., Curtiss, C.F., Armstrong, R.C. and Hassager, O. (1987) 'Dynamics of Polymeric Liquids'. Vol.1, John Wiley and Sons, New York.

13. Joseph, D.D. (1990) 'Fluid Dynamics of Viscoelastic Liquids'. Springer-Verlag, Heidelberg.

14. Larson, R. (1988) 'Constitutive Equations for Polymeric Melts and Solutions', Butterworths, Boston.

15. Jackson, K.P., Walters, K., and Williams, R.W. (1987) 'A rheometrical study of Boger fluids', J. Non-Newt. Fluid Mech. 14, 173-188.

16. Martínez-Mardones, J. and Pérez-García, C. (1992) 'Bifurcation analysis and amplitude equations for viscoelastic convective fluids', Il Nouvo Cimento 14, 961-975.

17. Martínez-Mardones, J., Tiemann, R., Zeller, W., and Pérez-García, P. (1993) 'Amplitude equations for viscoelastic convective fluids', in E. Tirapegui and W. Zeller (eds.), Instabilities and Nonequilibrium Structures IV, Kluwer Academic Publisher, Dordrecht, pp. 317-324.

18. Newell, A.C. (1974) 'Envelope equations', Lect. Appl. Math. 15, 157-163.

19. Coullet, P., Fauve, S. and Tirapegui, E. (1985) 'Large scale instability on nonlinear standing waves', J. Physique Lett. 46, 787-791.

20. Cross, MC. and Hohenberg, P.C. (1993) 'Pattern formation outside of equilibrium', Rev. Mod. Phys. 65, 851-1112.

21. Larson, R.G. (1992) 'Instabilities in viscoelastic flows', Rheol. Acta 31, 213-263.

22. Lega, J. (1989) 'Defauts Topologiques Associes a la Brisure de L'Invariance de Traslation dans le Temps', Thèsis de Doctorat, Université de Nice.

23. Coullet, P., Elphick, C., Gil, L., and Lega, J. (1987) 'Topological Defects of Wave Patterns', Phys. Rev. Lett. 59, 884.

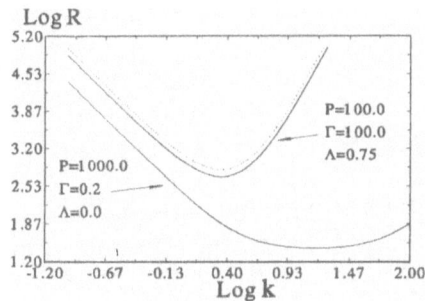

Fig.1: Marginal stability curve (dashed) and oscilatory curves of overstable states for concentrated (P=1000.0 ; Λ=0.0) and diluted (P=100.0 ; Λ=0.75) polymeric fluid.

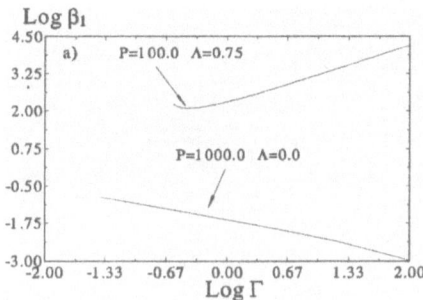

Fig.2: The most relevant nonlinear coefficients of Ginzburg-Landau amplitude equations for diluted (P=100.0 ; Λ=0.75) and concentrated (P=1000.0 ; Λ=0.0) polymers: a) $\beta = \beta_{TW}$, b) γ , and c) $\beta_{SW} = \beta + \gamma$ as a function of the nondimensional relaxation time Γ.

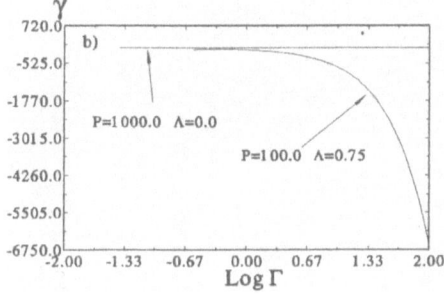

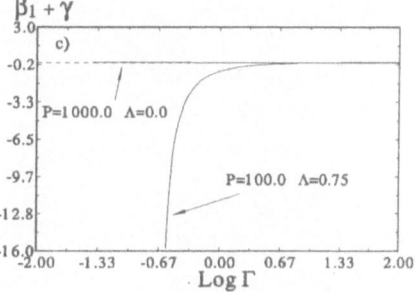

LASER MODEL WITH PERIODIC EXTERNAL INJECTED SIGNAL AND NOISE: SMALL NET GAIN LIMIT

R. C. BUCETA
Departamento de Física, Facultad de Ciencias Exactas y Naturales,
Universidad Nacional de Mar del Plata.
Funes 3350, (7600) Mar del Plata, Argentina.

M. S. TORRE and H. F. RANEA-SANDOVAL*
Instituto de Física "Arroyo Seco", Facultad de Ciencias Exactas,
Universidad Nacional del Centro de la Provincia de Buenos Aires.
Pinto 399 (7000) Tandil, Argentina.

ABSTRACT. A model for a laser with periodically modulated injected signal in the limit of gain approximately equal to the losses, is presented. By means of the normal-form theory we caracterize its dynamical behaviors arround of a Hopf bifurcation point, taking into account fluctuations in intensity and frequency of the injection also. We determinate the spectral contributions when the Hopf frequency is or it is not coincident with the modulation frequency of the injected signal.

I.   Deterministic Model

A class A laser with a periodically modulated injected signal can be modelled by a single differential equation for the complex electric field, through an adiabatic elimination of the polarization and the population difference variables in the Maxwell-Bloch [1] equations. Thence, the equation in our case is [2, 3]

$$\dot{E} = (i + g(E))\, E + F_b + \xi\, (1 + a \cos \omega t)\,, \tag{1}$$

where $g(E) = -\kappa + \gamma(1 + |E|^2)^{-1}$ is the net saturated gain in terms of the adimensional gain γ and loss κ parameters ($\gamma \geq \kappa$). The equation is normalized to the detuning $\Delta\nu$, that is, the difference between the injected signal frequency and the laser frequency. F_b is the excitation needed to bring the system to its Hopf Bifurcation Point (BP). If there is not periodic modulation ($a = 0$), the system is driven by an external signal

*Work supported by Grant No. PID-BID 0149 granted by Consejo Nacional de Investigaciones Científicas y Técnicas de la Republica Argentina.

E. Tirapegui and W. Zeller (eds.), Instabilities and Nonequilibrium Structures V, 137–143.
© 1996 Kluwer Academic Publishers.

whose intensity is $F = F_b + \xi$, being ξ a parameter which measures the system's offset from the BP. The system as it is modelled, is capable of reaching limit-cycle solutions for $\xi < 0$, or stable, fixed-point solutions for $\xi > 0$ (locking region). The deterministic and stochastic dinamics near to the BP has been analyzed by normal-form techniques [2, 3, 4].

We present an analysis in the region $\gamma \simeq \kappa$. Thus, the Eq. (1) reduces to

$$\dot{E} = (i + \delta - \gamma|E|^2) E + F_b + \xi (1 + a \cos \omega t) , \qquad (2)$$

with $\delta = \gamma - \kappa$, the net gain. For fixed gain and losses the system presents a Hopf bifurcation if

$$F_b^2 = \frac{\delta}{2\gamma} (1 + \frac{\delta^2}{4}) . \qquad (3)$$

At the BP the electric field modulus and phase are

$$|E_b|^2 = \frac{\delta}{2\gamma} , \qquad \varphi = - \cot^{-1} (\frac{\delta}{2}) , \qquad (4)$$

and the Hopf frequency is given by $\Omega^2 = 1 - \delta^2/8$.

By means of a perturbative expansion in Eq. (1) around E_b (i.e. with $E = E_b + v$) it is possible to obtain a differential equation for the perturbative variable $v \in \mathbb{C}$. Defining the vector $\mathbf{u} = v \ \epsilon^1 + \bar{v} \ \epsilon^2$ this equation be come (here $\{\epsilon^1, \epsilon^2\}$ is the canonical base of $\mathbb{C}^2$)

$$\dot{\mathbf{u}} = L_b \mathbf{u} + \mathbf{N}_b(\mathbf{u}) + \xi \ \mathbf{D}(t) , \qquad (5)$$

where the subindex b indicates that the coefficients are evaluated at the BP. Explicitly, the linear part is

$$L_b = \begin{pmatrix} i & -\gamma E_b^2 \\ -\gamma \overline{E_b}^2 & -i \end{pmatrix} , \qquad (6)$$

and the periodic driving vector is

$$\mathbf{D}(t) = (1 + a \cos \omega t)(\epsilon^1 + \epsilon^2) . \qquad (7)$$

The nonlinear term can be written in a condensed form (we use the convention that repeated indexes indicate sum), as

$$\mathbf{N}_b(\mathbf{u}) = \sum_{r>1} A_j^{j_1 \cdots j_r} u_{j_1} \cdots u_{j_r} \epsilon^j , \qquad (8)$$

with $j_l = 1, 2$ (for all $1 \leq l \leq r$) and $u_1 = \bar{u}_2 = v$. The nonlinear, non-null, second order coefficients are

$$2 A_1^{11} = 2 \overline{A_2^{22}} = A_2^{12} = \overline{A_1^{12}} = -\sqrt{2\delta\gamma} \ e^{-i\varphi} ,$$

and the non-null, third order coefficients are

$$A_1^{112} = \overline{A_2^{122}} = -\gamma$$

which are the only ones we give here, for the purpose of the analysis that follows. As it will soon become evident, this are the coefficients which give an explicit expression for the nonlinear change of variables, and the Normal Form Equation (NFE) for the critical variables.

We now determine the critical modes of the system, and to this end we diagonalize the linear operator (6) to get the eigenvalues and the critical eigenvectors. Let $\{\chi^1, \chi^2\}$ be the eigenvector base in $\mathbb{C}^2$ which diagonalize L_b, i.e. $L_b \, \chi^j = \Lambda_j \, \chi^j$. The corresponding eigenvalues are $\Lambda_{1,2} = \pm i\Omega$. The case $\Omega = 1$ ($\delta = 0$) is excluded from this analysis since E_b and F_b are both null and the differential equation (2) is the NFE for the critical variables. The eigenvectors χ^j in the canonical base ϵ^k are $\chi^j = \chi_k^j \, \epsilon^k$, where χ_k^j are the coefficients relative to the change of base, given by $\chi_1^1 = \overline{\chi_2^2} = e^{i2\varphi}$, $\chi_2^1 = \overline{\chi_1^2} = (2i/\delta)(1-\Omega)$. For the inverse transformation, $\epsilon^l = \epsilon_m^l \, \chi^m$, the coefficients are $\epsilon_1^1 = \overline{\epsilon_2^2} = \Delta^{-1} \, \chi_2^2$, $\epsilon_2^1 = \overline{\epsilon_1^2} = \Delta^{-1} \, \chi_1^2$, with $\Delta \simeq (1 - \delta^2/4)$ for $\delta \simeq 0$.

II.   Perturbative Theory in the Periodic Terms

We determine the nonlinear change of variables which allows the electric field to be expressed as a power series of the critical variables w_j, in order to obtain the NFE which will describe the complete dynamics of the system. This change of variables is (from here on, crossed indexes imply summation)

$$\mathbf{u}(\mathbf{w}, t) = \sum_{r,k=0}^{+\infty} \mathbf{u}^{[k,r]}(\mathbf{w}, t) , \qquad (9)$$

$$\mathbf{u}^{[k,r]}(\mathbf{w}, t) = U_j^{j_1 \cdots j_r, (k)}(t) \, w_{j_1} \cdots w_{j_r} \, \chi^j ,$$

The supra indexes k, r inside square brackets mean that the functions are of order k in ξ and of order r in $\{w_j\}$, and $\mathbf{w} = w_j \, \chi^j$ is the critical vector. The coefficients of the vectors $\mathbf{u}^{[0,r]} \equiv \mathbf{u}^{[r]}$ (with $\mathbf{u}^{[0]} \equiv 0$) are time-independent, while those of vectors $\mathbf{u}^{[k,r]}$ ($k > 1$) are time-dependent.

The NFE for the critical modes is

$$\dot{w}_j = \sum_{r,k=0}^{+\infty} F_j^{[k,r]}(\mathbf{w}, t) , \qquad (10)$$

where the functions $F_j^{[0,r]} \equiv F_j^{[r]}$ ($F_j^{[0]} \equiv 0$) are time-independent, while the functions $F_j^{[k,r]}$ ($k > 1$) are time-dependent.

The evaluation of the vector $\mathbf{u}^{[r]}$ and the function $F_j^{[r]}$, in powers of $\{w_j\}$, up to the third order, can be done by the same procedure as in Ref. [5]. At order zero in ξ, for small net gain δ, the non-null coefficients for $\mathbf{u}^{[2]}$ (the supraindex (0) is omitted) are

$$U_1^{11} = \overline{U_2^{22}} \simeq (i/2)\sqrt{2\gamma\delta}\; e^{i\varphi} ,$$
$$U_1^{12} = \overline{U_2^{12}} \simeq -i\sqrt{2\gamma\delta}\; e^{-i\varphi} , \tag{11}$$
$$U_1^{22} = \overline{U_2^{11}} \simeq -(i/6)\delta\sqrt{2\gamma\delta}\; e^{-i3\varphi} ;$$

and the non-null coefficients for $\mathbf{u}^{[3]}$ are

$$U_1^{111} = \overline{U_2^{222}} \simeq (3/8)\,\gamma\delta\; e^{i2\varphi} ,$$
$$U_1^{122} = \overline{U_2^{112}} \simeq (13/8)\,\gamma\delta\; e^{-i2\varphi} , \tag{12}$$
$$U_1^{222} = \overline{U_2^{111}} \simeq (11/96)\,\gamma\delta^2\; e^{-i4\varphi} .$$

Such coefficients will enable explicit evaluations at frequencies 2Ω, 3Ω, in the cross-correlation function for the electric field, at order zero in ξ. The nonlinear change of variables up to third order in the critical variables and order zero in ξ is

$$u_j = (w_i + w_{i_1}\, w_{i_2}\, U_i^{i_1 i_2} + w_{i_1}\, w_{i_2}\, w_{i_3}\, U_i^{i_1 i_2 i_3})\, \chi_j^i . \tag{13}$$

This procedure can be extended to higher orders. For odd orders, according to Refs. [5, 6], a minimal form for the NFE coefficients must be chosen in (10). For even orders, the coefficients ought to be zero for the resonances to be eliminated. Following Refs. [2, 3], the coefficients of the NFE are determined and it is possible for (10) to be written at third order in the critical variables and zeroth order in ξ explicitly, as

$$\dot{w}_j = i\Omega\, w_j + (\alpha + i\beta)\, |w_j|^2\, w_j , \tag{14}$$

where $j = 1, 2$. Note that $w_1 = \overline{w_2}$ so a single differential equation is sufficient to describe the complete dynamics of the system. The cubic coefficient are $\alpha = -\gamma(1 + \mathcal{O}(\delta^2))$, $\beta = -2\,\gamma\,\delta\,(1 + \mathcal{O}(\delta^2))$. Near the BP, the system must be described introducing the unfolding parameter μ, then the NFE to third order in the critical variables is

$$\dot{w}_1 = (\mu + i\Omega)\, w_1 + (-\gamma - i2\gamma\delta)\, |w_1|^2\, w_1 . \tag{15}$$

Note that in the region $\delta \simeq 0$, the real part of the cubic coefficient is always negative, so a third order theory is enough to describe limit-cycle solutions, i.e. with $\mu > 0$. An analysis of the regions of validity of the NFE in the space of parameters (κ, γ), is given in Ref. [3].

In order to characterize the dynamics of the critical variables in the vicinity of the BP, a perturbative theory of at least first order in ξ, must to be developed. To first order in ξ and order zero in w_j's we can consider two different situations: the modulation frequency of the external driving is coincident with the Hopf frequency, or it is not. The first case is what we call resonant case, the latter is called nonresonant. The change of variable vector and the NFE term can be written as

$$\mathbf{u}^{[1,0]}(\mathbf{w}, t) = \xi\, U_j^{(1)}(t)\, \chi^j , \qquad\qquad F_j^{[1,0]}(\mathbf{w}, t) = \xi\, F_j^{(1)}(t) .$$

The coeficients for the nonresonant case are

$$F_j^{(1)} = 0 , \qquad\qquad U_j^{(1)} = \xi \sum_{k=0,\pm 1} \frac{D_{j,k}}{ik\omega - \Lambda_j} e^{ik\omega t} ,$$

where $D_{j,k}$ is the Fourier coefficient of the j-th component of the driving term. For the resonant case, these result are

$$F_1^{(1)}(t) = \overline{F_2^{(1)}}(t) = \xi D_{1,1} e^{i\Omega t} ,$$
$$U_1^{(1)}(t) = \overline{U_2^{(1)}}(t) = i\xi \,/(2/\Omega)\,(2D_{1,0} + D_{1,-1} e^{-i\Omega t}) ,$$

which are the only possible solutions allowed for a T-periodic NFE ($T = 2\pi/\Omega$). Other nonperiodic solutions without normal-form term would drive the system to a nonstationary regime. From these expressions, we conclude, for the case of resonant injection, that there are no new spectral contributions due to the external modulation, either with the system at the limit-cycle region or in the phase-locking region; in fact, there is only an enhancement to the already present spectral components. On the contrary, in the nonresonant case there is a new peak at ω, either in the locking regime or in the limit cycle one. In the latter regime, can be identified new combination tones extending the perturbative analysis to higher orders [3].

The unfolding parameter μ, and the first correction ϖ to the Hopf frequency can be determined at first order in both the critical variables, and ξ in the small net gain limit. According to the procedure of Ref. [3] these are

$$\mu \simeq -\frac{5}{4}\xi \,\sqrt{2\delta\gamma}\sin\varphi , \qquad\qquad \varpi \simeq \frac{3}{20}\delta\mu . \qquad (16)$$

Hence, it is possible to obtain a T-periodic NFE, which account for the resonant and for the nonresonant injection cases, which is

$$\dot{w}_1 = (\mu + i(\Omega + \varpi))\, w_1 + (\alpha + i\beta)\, |w_1|^2\, w_1 + \xi D_{1,1}\, e^{i\Omega t}\delta_{\omega,\Omega} . \qquad (17)$$

Note that the NFE (17) in the resonant case ($\omega = \Omega$) is not invariant for a global-phase change. In fact, a symmetry breaking occurs which can be overcome by performing the known perturbative process. In this way, a new nonlinear change of variables is obtained, and therefore a new NFE. In the rotating-wave approximation it is possible to make the change of variables $w_1 = z\, e^{i(\Omega t+\psi)}$, $\psi = \arg(\xi\, D_{1,1})$ in Eq. (17). The equation for the new critical variables (up to third order) is

$$\dot{z} = (\mu + i\varpi)\, z + (\alpha + i\beta)\, |z|^2\, z + G ,$$

where $G = |\xi\, D_{1,1}|$. This is a NFE with a constant driving term and thus its analysis is similar to the one of previous works [2], for the case of a laser with constant injection. Through the usual procedure, i.e., by linearizing around stationary solutions and then by diagonalization of the linear part, it is possible to find the critical eigenvalues. Hence, the new critical modes can be determined. The corresponding NFE near the PB, which is global phase invariant, can be obtained by repeating the above described procedure.

III.   Stochastic Model

We shall consider fluctuations both in the amplitude of the injected signal and in the detuning between the laser and the external source for the injected signal. The temporal mean values of this parameters are, however, well defined. The Stochastic Differential Equation (SDE) is

$$\dot{E} = (i + g(E))\, E + F_b + \xi\,(1 + a\cos\omega t) + \zeta_A + i\zeta_M E \,, \tag{18}$$

where the ζ_A and ζ_M are time dependent random functions, statistically independent, both with zero mean values and non-null correlation functions. The additive noise ζ_A accounts for fluctuations in the intensity of the injected signal, and the multiplicative noise ζ_M, the detuning variations.

In the white- noise limit, the correlation functions are

$$\langle \zeta_J(t)\zeta_J(t')\rangle = \mathcal{D}_J\,\delta(\tau - \tau') \,, \tag{19}$$

for $J = A, M$; $\mathcal{D}_A = \Delta_c^2 T\epsilon_A/\Delta\nu^2$ and $\mathcal{D}_M = \epsilon_M/\Delta\nu^2$, being ϵ_A y ϵ_M the fluctuations in the intensity and in both frequency lasers respectively. The parameter Δ_c is the free spectral range of the injected laser cavity, and T the transmission coefficient of the coupling mirror.

Near the BP ($E = E_b + v$), the SDE (18) is

$$\dot{\mathbf{u}} \; = L_b\mathbf{u} + \mathbf{N}_b(\mathbf{u}) + \xi\,\mathbf{D}(t) + \mathbf{Z}(\mathbf{u},t) \,, \tag{20}$$
$$\mathbf{Z}(\mathbf{u},t) = (P_j(t) + P_j^k(t)\,u_k)\,\boldsymbol{\epsilon}^j \,,$$

where $P_1 = \overline{P_2} = \zeta_A + i\,E_b\,\zeta_M = p_R + i\,p_I$ and the non-null elements of $P_j^k(t)$ are $P_1^1 = \overline{P_2^2} = i\,\zeta_M$. The correlation functions of the real functions p_R and p_I are given by

$$
\begin{aligned}
\langle p_R(t)\,p_R(0)\rangle &= (\mathcal{D}_A + \mathcal{D}_M\,|E_b|^2\,\sin^2\varphi)\,\delta(t)\,,\\
\langle p_I(t)\,p_I(0)\rangle &= \mathcal{D}_M\,|E_b|^2\,\cos^2\varphi\,\delta(t)\,,\\
\langle p_R(t)\,p_I(0)\rangle &= \langle p_R(0)\,p_I(t)\rangle = -(1/2)\,|E_b|^2\,\mathcal{D}_M\,\sin(2\varphi)\,\delta(t)\,.
\end{aligned}
\tag{21}
$$

The noise terms, expressed in the critical base $\{\chi^i\}$ to first order in the $\{w_j\}$'s, are

$$\mathbf{Z}(\mathbf{w},t) = (Q_i(t) + Q_i^h(t)\,w_h)\,\chi^i \,, \tag{22}$$

being $Q_i = P_j\,\epsilon_i^j$, and $Q_i^h = \chi_i^j\,P_j^k\,\epsilon_k^h$. The stochastic NFE for the resonant and nonresonant cases is

$$
\begin{aligned}
\dot{w}_1 \; = \; & (\mu + i(\Omega + \varpi))\,w_1 + (\alpha + i\beta)\,|w_1|^2\,w_1 \\
& + \xi D_{1,1}\,e^{i\Omega t}\delta_{\omega,\Omega} + Q_1(t) + Q_1^1(t)\,w_1 + Q_1^2(t)\,w_2 \,.
\end{aligned}
$$

It is noticeable that this equation is not T-periodic due to the noise terms, in fact, Q_1 and $Q_1^2 w_2$ break the global phase invariance. In Eq. (23) the additive noise terms can be considered separately because the multiplicative noise terms are of higher order in the w_j's. The term considered contains information on the amplitude as well as on the detuning fluctuations.

IV.   Conclusions

We have investigated a model for a periodically modulated, injected signal laser in the presence of noise when the gain is near to the losses. By means of the normal-form theory, we have been able to completely characterize its dynamical behavior, giving an explicit NFE for both cases of resonant and nonresonant injection, which allows for elimination of the Hopf resonances up to all perturbations orders, and to finally find an autonomous NFE. A characterization of the unfolding parameter, which allows the system to get to the BP was also given, as well as the system's Hopf frequency. The simplicity of the coefficients for the change of variables, and the qualitatively similar results of the more general case, allows us for a complete description of more complex situations in this limit and to extend the results for a higher gain situation.

References

[1] A. E. Siegman, *Laser* (University Science Books, Mill Valley, CA, 1986), and references therein.

[2] R. C. Buceta, M. S. Torre and H. F. Ranea-Sandoval, Phys. Rev. A **48**, 3336, (1993).

[3] M. S. Torre, H. F. Ranea-Sandoval and R. C. Buceta, Phys. Rev. A **50**, 3427 (1994).

[4] H. Zeghlache and V. Zehnlé, Phys. Rev. A **46**, 6015 (1992); Phys. Rev. A **46**, 6028 (1992).

[5] P. Coullet, C. Elphick and E. Tirapegui, Phys. Lett. **111A**, 277, (1985).

[6] J. Guckenheimer and P. Holmes, *Nonlinear Oscillations, Dynamical Systems, and Bifurcations of Vector Fields*, edited by F. John, J. E. Marsden and L. Sirovich, Applied Mathematical Sciences **42**, (Springer-Verlag, Berlin, 1983).

FLOW TRANSITIONS IN RAYLEIGH–BÉNARD CONVECTION: INFLUENCE OF HEATING RATE

R.HERNANDEZ

Departamento de Ingeniería Mecánica
Universidad de Chile, Casilla 2777, Santiago, Chile

ABSTRACT. Temporal evolution of the 3–D Rayleigh–Bénard convection originated from different heating rates was obtained. The full 3–D governing equations for a Boussinesq fluid of Prandtl number $P = 0.71$ were solved inside a rectangular box of aspect ratios $\Gamma_z = 1.25, \Gamma_y = 2$ heated from below in the case of adiabatic and perfect conducting side walls. A final steady flow pattern, which consists of two counter rotating rolls with axes parallel to the shorter dimension of the box, is always found. However, the heating rate drives a flow transition for a wide range of Rayleigh numbers when perfect conducting side walls are considered. The flow transition consists of an abrupt change in the rotation sense of two counter rotating rolls and can be attributed to a thermal boundary condition competition. This transition agrees qualitatively with recent experimental results.

1 Introduction

The onset of the convective motion and the further behavior of the flow structures formed at the onset, have been widely studied in the Rayleigh–Bénard phenomenon (Chandrasekhar 1961). Different values for the critical Rayleigh number R_c can be found when the finite box problem is considered, due to an increase of the viscous influence of side walls (Stork & Müller 1972,1975).

The number of rolls becomes one of the most characteristic properties of RB convection. The number of rolls and the dimensionless wavelength λ which corresponds to the ratio of the width of one roll to the depth d of the fluid layer, are inversely related. The theoretical computations of Schlüter, Lortz & Busse (1965) predict a decrease of the wavelength (increase of number of rolls) with increasing supercritical Rayleigh number. Experimentally, however, the opposite result was found by Koschmieder (1969), and Krishnamurti (1970). Furthermore, Mukutmoni & Yang (1991) have shown through numerical computations that an increase of R beyond the critical value R_c affects significantly the steady flow pattern through the mechanism of loss of rolls.

When the heating in RB convection is developed through a steady and fast increase of the temperature difference between the upper and lower boundaries, the wavelength decreases (Koschmieder 1969). This fact shows that heating could be responsible of the change of pattern at a given supercritical Rayleigh number. In addition, Krishnamurti (1968), has shown that the slope η of a linear temperature variation can induce a different flow pattern depending on the value of η. At very low values of η the flow pattern was identified as two dimensional rolls and at high values of η hexagonal cells were found.

E. Tirapegui and W. Zeller (eds.), Instabilities and Nonequilibrium Structures V, 145–153.
© 1996 *Kluwer Academic Publishers.*

Moreover, the sign of η was the controlling parameter of the flow sense at the center of the hexagonal cells. Recently, the experiments of Arroyo & Savirón (1991) have shown that the velocity of the heating process leads to two different steady flow patterns in RB convection, which may be interpreted as two branches of a bifurcation.

In summary, the sensitivity of the convective pattern to the rate at which the heating process is performed must be explored accurately explaining the physical mechanism responsible of that transition. In that way, the full 3–D nonlinear equations must be solved retaining time derivatives which suggests, due to their complexity, that a detailed numerical study is an adequate tool to compute the time dynamics of the convective regime originated from different heating rates at increasing supercritical values of the Rayleigh number.

2 Formulation

2.1 Problem description

Consider a rectangular box of height $l_x = d$, horizontal dimensions l_y, l_z, of aspect ratios $\Gamma_y = l_y/d = 2$, $\Gamma_z = l_z/d = 1.25$ filled with a Boussinesq fluid of Prandtl number $P = 0.71$ (figure 1). The top and bottom walls are kept at temperatures T_c and T_h ($T_h > T_c$) respectively and the side walls are perfect conductors.

The dimensionless governing equations in Cartesian coordinates for a Boussinesq fluid in three dimensions, neglecting radiation, are:

$$\nabla \cdot \vec{v} \;=\; 0 \tag{1}$$

$$\frac{\partial \vec{v}}{\partial t} + (\vec{v} \cdot \nabla)\,\vec{v} \;=\; -\nabla p + R\,P\,\theta \hat{y} + P\nabla^2 \vec{v} \tag{2}$$

$$\frac{\partial \theta}{\partial t} + (\vec{v} \cdot \nabla)\theta \;=\; \nabla^2 \theta \tag{3}$$

R is the time dependent Rayleigh number which can be written as $R(t) = g\beta\Delta T(t)d^3/\nu\alpha$ and the Prandtl number is $P = \nu/\alpha$, where α is the thermal diffusivity, ν the kinematic viscosity, $\Delta T(t) = T_h - T_c$ is the temperature difference between the top and bottom walls (which is a function of time), g the gravitational acceleration and β the thermal expansion coefficient.

The time, velocity, pressure and temperature are scaled using d^2/α, α/d, $\rho(\alpha/d)^2$ and $(T_h - T_c)$, respectively, as reference quantities where ρ is the fluid density. The Cartesian coordinates are scaled with the height d of the cavity. The boundary conditions for the velocity and temperature in nondimensional form, $\vec{v} = (u, v, w)$, θ are the following:

$$y = 0, \Gamma_y; 0 \le x \le \Gamma_x; 0 \le z \le \Gamma_z; \vec{v} = 0; \theta = ltp$$

$$z = 0, \Gamma_z; 0 \le x \le \Gamma_x; 0 \le y \le \Gamma_y; \vec{v} = 0; \theta = ltp$$

$$x = 0, 1\,; 0 \le y \le \Gamma_y; 0 \le z \le \Gamma_z; \vec{v} = 0; \theta = 1, 0$$

At the vertical walls a linear temperature profile ltp between the temperature values of the hot and cold walls is imposed to emulate conduction (ElSherbiny, Hollands & Raithby 1985).

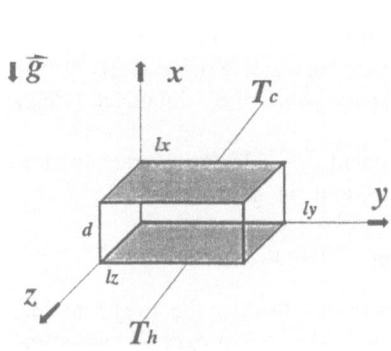

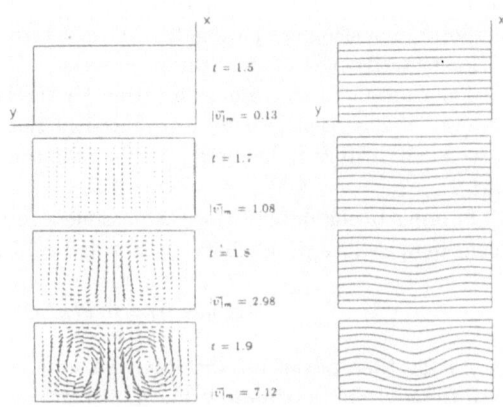

Figure 1: Physical configuration

Figure 2: Temporal evolution of velocity field (left) and isotherms (right) at the box middle plane at $R_s = 5 \times 10^3$ with $\tau = 1.4$

Heat transfer across the fluid can be represented by an overall Nusselt number $< Nu >$ which is computed, at each wall, averaging the corresponding local Nusselt number Nu. The non dimensional local Nusselt number is defined at each wall as the normal spatial derivative of the temperature θ,

$$Nu = \frac{\partial \theta}{\partial n}$$

where n is the normal coordinate to a given wall.

The numerical computation of the non dimensional heat flux has been performed integrating the local Nusselt number, at each wall of area Σ, using the following definition,

$$q = \int_\Sigma Nu \, d\sigma$$

In RB convection experiments the heating process is normally a quasi–steady operation where the time involved to reach a determined supercritical Rayleigh number is many times greater than the thermal diffusive time scale or vertical relaxation time $\tau_r \equiv d^2/\alpha$ of the system.

Usually in the experiments, the fluid is first thermostated at a given temperature say T_0, next the upper plate is stabilized at T_0 to proceed to raise the lower plate temperature slowly, until a predetermined R_s is reached. In this way the final flow pattern is reached through a successive number of intermediate steady flow configurations where the system is in thermal and mechanical equilibrium. Therefore, the temporal dependence of $R(t)$ is attributed to the temporal variation of the temperature difference $\Delta T(t)$.

In this work, the heating operation is a monotonic function of time which is introduced into the governing equations by means of a time dependent Rayleigh number $R(t)$ as follows:

$$R(t) = (R_s/\tau) \cdot t$$

where R_s is the prefixed supercritical value of R, t the dimensionless time and τ is a temporal dimensionless parameter which determines the slope and therefore the duration of the unsteady part of the heating process.

The choice of a time linear function for the heating process was made for convenience, however, it has been already introduced in the theoretical works of Krishnamurti (1968, part 1) and Swift & Hohenberg (1989) and in the experimental work of Krishnamurti (1968, part 2).

In order to simulate as close as possible a real experiment, the calculation starts with a linear temperature profile between hot and cold wall (ltp) and the fluid at rest.

$$\vec{v}(x,y,z,0) = 0; \; \theta(x,y,z,0) = ltp; \; R(0) = 0$$

As the time advances, the Rayleigh number is increased following the $R(t)$ function until the prefixed maximum value R_s is reached when $t = \tau$, then the programed variation of $R(t)$ ends (unsteady part of the heating) and the time evolution of the system continues with $R(t) = R_s$. Note that the heating process never ends because the forcing parameter $R(t)$ is always greater than zero.

The governing equations were solved in primitive variables. In the discretization of the physical domain a three–dimensional, uniform and staggered grid of $17 \times 21 \times 21$ points was used ($\Delta x = \Delta z = 0.0625$, $\Delta y = 0.1$) with a control volume formulation including a power law scheme which has been previously used (Hernández & Frederick, in press) to treat the convective–diffusive terms in the discrete formulation. The discrete equations were solved by an iterative tri–diagonal solver using a time step Δt of 0.01, which represent a 1 % of the vertical relaxation time τ_r. Tests of grid independent results suggest that the choice of this grid to perform the calculations represent a good compromise between accuracy and computational effort.

3 Results and Discussion

The heating rate which is controlled, at a fixed supercritical R_s, by the unsteady heating time τ produces an abrupt change in the steady flow pattern found in the box when different values of τ are considered. At a given supercritical Rayleigh number, the steady state flow pattern is formed by two counter rotating rolls with axes parallel to the shorter side of the box (l_z). This arrangement has been already predicted by many authors (Arroyo & Savirón 1991, Mukutmoni & Yang 1991, Stork & Müller 1972, Davis 1967).

At a fixed R_s, different slopes or heating rates which are a function of τ, will determine a very different flow structure. For example, when the steady state solution is reached at $R_s = 5 \times 10^3$ with $\tau = 1.4$ (figure 2), the convective motion found between the two counter rotating rolls (central region) is oriented downward and upward near the side walls (system state s_d). On the contrary, when $\tau = 1.5$ (not shown), the fluid rises in the central region and falls near the side walls (system state s_u). This overall behavior agrees very well with the experimental results of Arroyo & Savirón (1991). Figure 2 presents the time evolution of the flow pattern and temperature field at $z = 0.5l_z$ for s_d state of the system with $\tau = 1.4$ and $R_s = 5 \times 10^3$. It is shown x-y planes at $z = 0.5l_z$. The time sequence ranges from pure conduction to steady convection regime ($|\vec{v}|_m$ is the maximum velocity projection). The convective motion has opposite senses depending on the value of the τ parameter. When the heating routine is made with $\tau = 1.4$ a descending flow begins to appear at the central zone of the box, i.e., cold fluid begins to fall. This cold flow begins to increase with time due to the forcing parameter $R(t)$. At the same time, an ascending flow of hot fluid starts near the vertical walls by virtue of mass conservation and buoyancy forces, distorting the isotherms field at those regions. The same happens when $\tau = 1.5$ but in opposite sense. Hot fluid begins to ascends at the central region of the box originating a descending flow of cold fluid near the vertical walls which distorts the isotherms field in the opposite direction. This system behavior, which is clearly controlled by the parameter τ, seems to be a phase transition and it is possible, at least within time step accuracy, to find the closest values, τ^+ and τ^-, of the critical parameter τ_c at a given R_s.

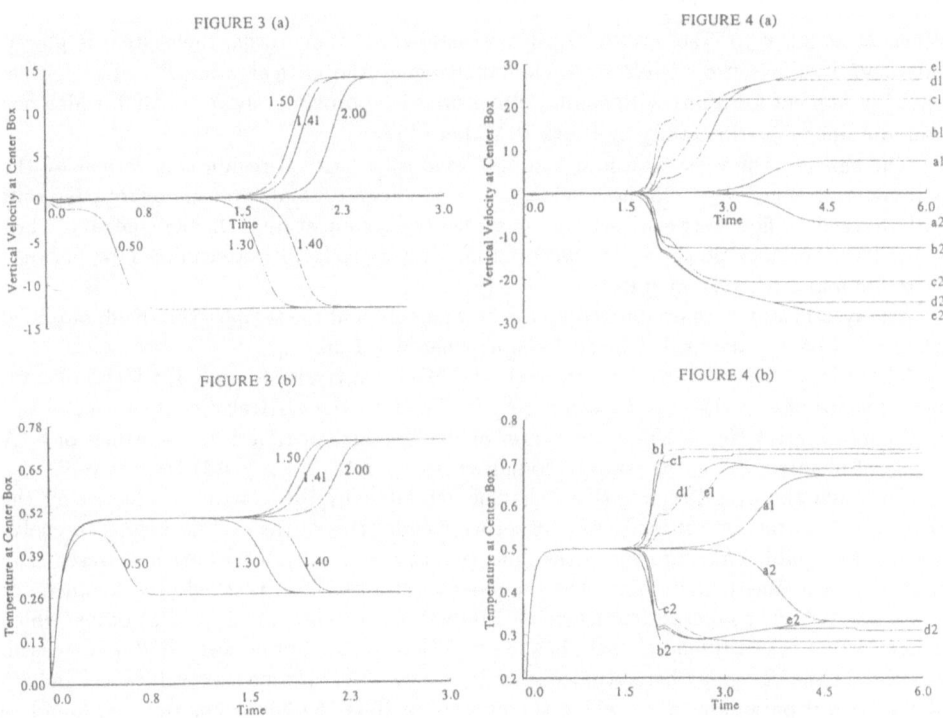

Figure 3: Sensitivity analysis as a function of parameter τ at the center of the box for (a) vertical velocity and (b) Temperature θ versus time t. The different values of τ are printed over the curves.

Figure 4: Vertical velocity u (a) and temperature θ (b) at the center of the box as a function of time t when $\tau = \tau^-, \tau^+$ for different Rayleigh numbers.

Table 1: Absolute values for maximum velocities and local Nusselt numbers at both limits of τ_c.

R_s	$[\tau^-, \tau^+]$	u_m	v_m	w_m	Nu_m
3600	$[1.05, 1.06]$	$[7.60, 7.60]$	$[4.07, 4.08]$	$[2.07, 2.07]$	$[2.88, 5.64]$
5000	$[1.40, 1.41]$	$[14.22, 14.25]$	$[8.91, 8.83]$	$[4.49, 4.51]$	$[5.43, 10.37]$
9000	$[2.55, 2.56]$	$[22.42, 22.42]$	$[17.33, 17.35]$	$[7.41, 7.41]$	$[9.47, 15.99]$
13000	$[3.68, 3.69]$	$[29.23, 29.26]$	$[22.76, 22.72]$	$[9.96, 10.11]$	$[12.79, 20.77]$
16000	$[4.53, 4.54]$	$[33.77, 33.80]$	$[25.94, 25.96]$	$[12.06, 12.06]$	$[14.90, 23.83]$

3.1 Flow Pattern; dependence on R_s

When $R_s = 3.6 \times 10^3$ the level of convection is low, in other words, the system is placed only slightly above the critical state characterized by the critical value R_c. The critical Rayleigh number for a box with similar dimensions is approximately $R_c \sim 3500$, which has been obtained experimentally by Stork & Müller (1972).

The change of flow pattern found at this weak supercritical region is governed by the parameter τ. When $\tau \leq 1.05$ the u velocity becomes negative and two counter rotating rolls with descending flow between them are observed (s_d system state). On the contrary, when $\tau \geq 1.06$ u becomes positive and two counter rotating rolls with ascending flow between them are found (s_u system state).

The system states, characterized by the flow pattern and the temperature field, obtained when $\tau \leq 1.05$ are identical. The same holds when $\tau \geq 1.06$.

This critical behavior can be seen in the sensitivity analysis of figure 3, which indicates the influence of τ in the case $R_s = 5 \times 10^3$. It is plotted the vertical velocity u (fig. 3 a)) and temperature θ (fig. 3 b)) at the center of the box against time t for six values of τ. A family of identical s_d and s_u states is found when $\tau \leq 1.40$ and $\tau \geq 1.41$ respectively.

Although the overall convective behavior exhibited by the system is selected by the heating rate, after the unsteady heating process ends ($t = \tau$) the system begins to evolve towards an equilibrium state where the flow pattern and heat transfer does not present any variation with time t, indicating that the steady state has been reached (See for instance figure 3). At higher supercritical Rayleigh numbers R_s the flow transition also occur, which indicates that it is a robust change, in the sense that it does not go away if R is increased, at least, in the R_s range here studied. Table 1 shows that the nearest values τ^- and τ^+ of the critical parameter τ_c at which the system exhibits the flow transition are found to increase with higher values of R_s.

The overall flow pattern configuration of all values of R_s is similar. However, the convective motion attributed to the velocities and heat transfer imposed by higher R_s values is clearly superior to low R_s. The convective increase of the system due to an increase of the Rayleigh number can be appreciated in Table 1, which shows the maximum velocities and local Nusselt numbers at each R_s when s_d and s_u states arise for the five cases here studied.

Owing to the strong increase of vertical velocities, the transversal velocities (v, w) begin to increase to preserve mass conservation on the whole physical domain. For instance, the maximum transversal velocity w_m of case $R_s = 3.6 \times 10^3$ represent approximately a 16 % of the corresponding w_m of the case $R_s = 1.6 \times 10^4$ (See Table 1). Figure 4 summarizes the abrupt change in the sense of rotation of the roll pattern at each R_s tested in this work as a function of the parameter τ. It is plotted the vertical velocity (figure 4 a)) and temperature (figure 4 b)) at the center of the box. The values of the parameter τ required to obtain the two senses of rotation at a given R_s are different. The subscripts 1,2 indicates τ^- and τ^+ respectively. ($R_s = 3.6 \times 10^3$ (a1,a2), $R_s = 5 \times 10^3$ (b1,b2), $R_s = 9 \times 10^3$ (c1,c2), $R_s = 1.3 \times 10^4$ (d1,d2), $R_s = 1.6 \times 10^4$ (e1,e2)).

For instance, in cases where $\tau \geq \tau^+$, the increase of vertical velocity with the Rayleigh number R_s is notorious. In fact, it is caused by the overall increase of buoyancy forces which are proportional to $R(t)$, a response of inertial terms to thermal source term in momentum equations. The behavior of the central temperature is not obvious, because there is not a net increase. The center temperature of cases $R_s = 3.6 \times 10^3$ and $R_s = 1.6 \times 10^4$ is almost the same, which implies the existence of a maximum value between them.

This fact is instructive because suggests that, as the R_s grows, the heat diffusion in transversal directions becomes important (right hand side of energy equation) which is a consequence of the presence of conducting side walls. The same occur when $\tau \leq \tau^-$ cases are inspected. There is a negative increase of vertical velocity at the center of the box with R_s but the center temperature does not decrease continuously. As R_s grows there is an increased distortion of isotherms at the side walls regions caused by an ascending or descending flow depending on the τ value. When τ^+ cases are considered, as the supercritical Rayleigh number R_s is increased, a portion of cold fluid coming from the top wall begins to be redirected towards the center box adopting a diagonal path. This trajectory will cool the center box, therefore decreasing the local temperature. On the contrary, when τ^- cases are considered, a portion of hot fluid coming from the bottom wall, which moves diagonally towards the box center, will raise the local temperature. Therefore, the temperature behavior at the center of the box is a result of the diagonal fluid paths.

The diagonal paths are caused by the thermal characteristics of side walls (ltp profile). When a descending or ascending fluid flow is imposed near those walls, the isotherms distortion cause a transversal heat flux which can be positive (incoming) or negative (outgoing). A positive heat flux will heat the descending cold fluid raising its local temperature and therefore the buoyancy forces. The increase of buoyancy make the fluid turn horizontally before to reach the bottom wall. On the contrary, if hot fluid ascends, a negative heat flux appears cooling it and therefore diminishing its buoyancy forces, then the fluid will turn horizontally before to reach top wall.

From the heat transfer analysis, at a determined R_s value, the absolute values of the heat fluxes at side walls are almost the same when s_u and s_d states are considered, and the top and bottom heat fluxes are inverted, i.e., the top heat flux value when $\tau = \tau^-$ coincides with the bottom heat flux when $\tau = \tau^+$ and vice versa. This fact with the help of the maximum velocity values sketched in Table 1 indicates that both s_u and s_d states are symmetrical under a transformation $x \to l_x - x$ and $u \to -u$.

This singular flow transition depends on the thermal boundary conditions imposed at vertical walls. When adiabatic instead of perfect conductor side walls are considered and the calculation starts with the same initial conditions, the flow pattern found does not present any change when τ is varied, i.e., the flow transition disappears, at least in the R_s range here considered.

The adiabatic flow pattern found (not shown) always corresponds to two counter rotating rolls with axes parallel to the shorter side of the box. Hot fluid raises at the center of the box and cold fluid falls near the side walls, which is similar to the perfect conductor case at s_u state, however, the boundary condition at side walls for the adiabatic case forbids the heat transfer between the fluid and the environment, then the incoming heat flux at the bottom wall is convected entirely towards the upper plate.

The comparison between both physical configurations allows to suspect that the mechanism responsible of the flow transition, at a given R_s, is the competition between the side wall and horizontal boundary conditions. The buoyancy forces generated by the ltp condition imposed at side walls will always drive a rising flow near these walls, independently of the Rayleigh number value ($R_s > 0$), when combined with adiabatic horizontal walls (Yang 1988). However, when the box is heated from below and the side walls are considered adiabatic, the convective motion starts when the Rayleigh number reaches a critical value R_c. Therefore, when both effects are superimposed, i.e., a box heated from below with ltp side walls, a rising flow near vertical walls accompanied by a downward flow at the box center (s_d state) would be observed for $R(t) < R_c$. This argument can be confirmed in figures 3 a) and 4 a) where a weak negative vertical velocity, during a short time interval, is observed at the center of the box in the perfect conductor case, indicating a very weak primary s_d state. However this state is then neutralized by virtue of the increase of buoyancy of the central flow, and therefore vanishes giving rise a pure conductive regime. This regime will be dominant until a critical Rayleigh number is reached, then the convective pattern starts either as an s_d or s_u state selected by τ^- or τ^+ respectively. This phenomenon is present in all cases independently of the final supercritical Rayleigh number. Thus the physical mechanism driving the transition can be regarded as the competition of buoyancy forces imposed near side walls against those imposed at the central region of the box when the critical state is reached. For $\tau < \tau_c$ convection starts near side walls and for $\tau > \tau_c$ the fluid motion will start at the central region giving rise to an s_d and an s_u state respectively. The computed values τ^- and τ^+ which determine s_d and s_u states at every R_s are, as it was mentioned before, the nearest values of the critical parameter τ_c allowed by the time step accuracy. In principle, a best approximation to τ_c could be possible with finer time steps, however, the restriction imposed by the machine precision and especially the excesive computing time forbide the numerical search of τ_c.

Finally, one further remark must be mentioned. The overall behavior exhibited by τ^-, τ^+ and therefore by τ_c in front of the corresponding supercritical R_s of Table 1, shows that as R_s approaches to the critical Rayleigh number (Stork & Müller 1972) the bounding limits of τ_c decrease and for R_s close to R_c ($R_s = 3.6 \times 10^3$) τ^- and τ^+ are found close to unity. In other words, for the lowest R_s value, the critical parameter is close to the vertical relaxation time of the system $\tau_c \sim \tau_r$. This fact suggests a tendency of the critical parameter, i.e. $\tau_c \to \tau_r$ when $R_s \to R_c$, but it is not possible here to assure it because lower R_s values were not considered due to an increased time step resolution to detect the flow transition would be required.

Part of this work was supported by FONDECYT Grant 1265-91 and CEC contract CI1*CT91-0947.

4 References

Arroyo, M.P. & Savirón, J.M. 1991 Rayleigh–Bénard convection in a small box: spatial features and thermal dependence of the velocity field, *J. Fluid Mech.*, **235**, 325–348.

Chandrasekhar, S. 1981 *Hydrodynamic and Hydromagnetic Stability, Chapter II*, Dover Publications, Inc.

Davis, S.H. 1967 Convection in a box: linear theory, *J. Fluid Mech.*, **30**, part 3, 465–478.

ElSherbiny, S.M., Hollands, K.G.T. & Raithby, G.D. 1985 Effect of Thermal Boundary Conditions on Natural Convection in Vertical and Inclined Air Layers, *ASME J. Heat Transfer*, **104**, 515–520.

Hernández, R. & Frederick, R.L. Spatial and thermal features of Rayleigh–Bénard convection, *Int. J. Heat Mass Transfer, in press.*

Koschmieder, E.L. 1969 On the wavelength of convective motions, *J. Fluid Mech.*, **35**, part 3, 527–530.

Krishnamurti, R. 1968 Finite amplitude convection with changing mean temperature. Part 1. Theory, *J. Fluid Mech.*, **33**, part 3, 445–455.

Krishnamurti, R. 1968 Finite amplitude convection with changing mean temperature. Part 2. An experimental test of the theory, *J. Fluid Mech.*, **33**, part 3, 457–463.

Krishnamurti, R. 1970 On the transition to turbulent convection. Part 1. The transition from two- to three–dimensional flow, *J. Fluid Mech.*, **42**, part 2, 295–307.

Mukutmoni, D. & Yang, K.T. 1992 Wave number selection for Rayleigh–Bénard convection in a small aspect ratio box, *Int. J. Heat Mass Transfer*, **35**, N°9, 2145–2159.

Schlüter, A., Lortz, D. & Busse, F. 1965 On the stability of steady finite amplitude convection, *J. Fluid Mech.*, **23**, 129–144.

Stork, K. & Müller, U. 1972 Convection in boxes: experiments, *J. Fluid Mech.*, **54**, part 4, 599–611.

Stork, K. & Müller, U. 1975 Convection in boxes: an experimental investigation in vertical cylinders and annuli, *J. Fluid Mech.*, **71**, part 2, 231–240.

Swift, J.B. & Hohenberg, P.C. 1989 Rayleigh–Bénard convection with time–dependent boundary conditions, *Phys. Rev. A*, **39**, N°8, 4132–4136.

Yang, K.T. 1988 Transitions and bifurcations in laminar buoyant flows in confined enclosures, *ASME J. Heat Transfer*, **110**, 1191–1204.

PART III
STOCHASTIC EFFECTS

NOISE-INDUCED PHASE TRANSITIONS

JUAN M.R. PARRONDO
Dep. Física Aplicada I. Universidad Complutense
28040-Madrid, Spain.
CHRISTIAN VAN DEN BROECK
L.U.C. B-3590 Diepenbeek, Belgium.

ABSTRACT. Models of macroscopic systems with an infinite number of microscopic degrees of freedom perturbed by external noise can undergo sharp phase transitions which are however modified or even induced by the external noise. These non-equilibrium phase transitions exhibit the features characteristic of equilibrium phase transitions such as symmetry breaking and non ergodicity, which are absent in the transitions studied in zero-dimensional systems.

1. Introduction

One of the main goals of Statistical Physics is to understand the role of fluctuations in macroscopic phenomena. Once the internal fluctuations of a system —those arising from its large number of degrees of freedom— were characterized through the fluctuation-dissipation theorem for systems close to equilibrium, a lot of efforts, mainly in the late 70's and the 80's, were devoted to study external noise with special emphasis on its effects on bifurcations or changes in the stability of steady states.

So far this problem has been studied mainly in systems with one or at most two degrees of freedom [1]. If one perturbs such a system with an external noise, then the system becomes stochastic and its state is given by a distribution probability. These *zero-dimensional* models have been successfully applied to systems that either are intrinsically zero-dimensional or that have macroscopic fluctuations.

Nevertheless, examples were found of transitions provoked by external fluctuations in which the system preserves its macroscopic non-fluctuating character. As an example, we cite the case of a photosensitive chemical reaction illuminated by a random source of light [1, 2]. The random source is realized by passing a light beam across a box containing small balls dancing in a turbulent airstream. In spite of the presence of external fluctuations, the chemical state of the system turns out to be equally sharply defined as the state in absence of these fluctuations. For instance, in bistability regions no spontaneous transition from one equilibrium state to the other was observed, indicating that in such regions the system is no longer ergodic. This result was considered to be "a surprise" by Horsthemke and Lefever as it does not agree with the behavior of zero-dimensional models. They pointed out that an appropriate explanation would require the inclusion of spatial dependence in the model [1].

More recently, a similar situation was encountered in numerical experiments with the Swift-Hohenberg equation perturbed by external multiplicative noise [3]: the noise shifts the location of the threshold above which patterns appear in the system, but the macroscopic

E. Tirapegui and W. Zeller (eds.), Instabilities and Nonequilibrium Structures V, 157–166.

state is still well defined. Moreover, a linear analysis of the equation yielded to a quite accurate estimation of the new threshold. This is a further surprise from the point of view given by the experience with zero-dimensional models since in these models the linear analysis often locates the critical point even in a qualitatively wrong way.

All this evidence indicates that the influence of external noise on spatially extended systems or, more generally, systems with an infinite number of degrees of freedom, is of a completely different nature from that on zero-dimensional systems. In addition, the behavior of spatially extended systems perturbed by external noise presents several features akin to those found in equilibrium systems: fluctuations decreasing with the volume and consequently well defined macroscopic states in the thermodynamic limit, sharp phase transitions, loss of ergodicity and spontaneous symmetry breaking when two or more phases coexist. These features do not occur in noise-induced transitions in zero-dimensional systems and some questions have been raised about the physical importance of the noise induced effects in these systems.

In this paper we present models of systems with a large number of degrees of freedom perturbed by external local noise. These models meet all the points of contention and discussion which were addressed above: it turns out that the fluctuations of the macroscopic states decrease with the size of the system and vanish in the thermodynamic limit, that it is possible to define a sharp phase transition and, moreover, that the theory explains the validity of the linear analysis in certain regimes. In section 2 we introduce the models and their mean-field version which can be solved analytically. Different effects of the noise are explored in section 3. We would like here to place special emphasis on the model of section 3.3 where a pure noise-induced phase transition occur, i.e., a phase transition in a system for which there is no transition at all in absence of external fluctuations. Finally a brief summary and our main conclusions are presented in section 4.

2. Spatially extended systems perturbed by external noise

The type of systems we consider are characterized by a scalar field x_i defined on a discrete lattice. The evolution equation for the field x_i at a given site i of the lattice is:

$$\dot{x}_i = -\frac{\partial U(x_i)}{\partial x_i} + \frac{D}{c}\sum_n (x_n - x_i), \qquad (1)$$

where the sum runs over sites n nearest neighbors of site i and c is the number of such nearest neighbors or the lattice *coordination number*. Eq. (1) is a standard field equation with a local potential $U(x)$ and a diffusive term (the discretization of the Laplacian) that tends to homogenize the field. One can consider a finite temperature dynamics at temperature T by including Gaussian additive noises $\xi_a^{(i)}$ white in space and time:

$$\langle \xi_a^{(i)}(t)\xi_a^{(j)}(t')\rangle = 2k_BT\delta_{ij}\delta(t-t').$$

In such a case the stationary distribution for (1) is given by the equilibrium Gibbs state

$$\rho_{st}(\{x_i\}) = \frac{1}{Z}e^{-H(\{x_i\})/k_BT} \qquad (2)$$

Z being a normalization constant —the partition function— and with the hamiltonian given by

$$H(\{x_i\}) = \sum_i U(x_i) + \frac{D}{2c} \sum_{(i,j)} (x_i - x_j)^2 \tag{3}$$

where the second sum runs over all possible nearest neighbors pairs (i,j). The macroscopic state of such a system is given by the quantity

$$\mu = \frac{1}{N} \sum_{i=1}^{N} x_i \tag{4}$$

where the sum runs over all the N sites of the lattice. In absence of long range correlations, the central limit theorem implies that the fluctuations of μ decrease as $1/\sqrt{N}$.

For certain forms of the local potential $U(x)$ and certain values of the temperature T it is possible to have coexistence of phases, that is, in the thermodynamic limit the stationary distribution (2) is no longer unique: one has different stationary states which are macroscopically different (i.e. they have different values of μ). The system is no longer ergodic: depending on the initial conditions and the thermal (microscopic) fluctuations it goes to one or other of these states. Finally, these states do not posses every symmetry present in the hamiltonian $H(\{x_i\})$. These are the key ideas of phase transitions and critical phenomena. Unfortunately, in most cases it is extremely difficult to carry out an exact analysis that confirms the existence of a phase transition. Nevertheless, a mean-field analysis, such as the Weiss theory of ferromagnetism, provides a very simple (albeit not completely reliable) technical tool to investigate the possibility of a phase transition. Moreover, this approach is exact in certain limits, such as the one in which every site is coupled to every other site.

We now move away from equilibrium theory and consider a multiplicative noise perturbing Eq. (1) in addition to the additive noise modeling finite temperature:

$$\dot{x}_i = f(x_i) + \frac{D}{c} \sum_n (x_n - x_i) + \xi_m^{(i)} g(x_i) + \xi_a^{(i)}. \tag{5}$$

The external noises $\xi_m^{(i)}$ are assumed to be Gaussian processes, white in space and time, and independent of the thermal noise, i.e.,

$$\langle \xi_{a,m}^{(i)}(t) \xi_{a,m}^{(j)}(t') \rangle = \sigma_{a,m}^2 \delta_{ij} \delta(t - t'), \tag{6}$$

and the terms of Eq. (5) with multiplicative noise will be interpreted following the *Stratonovich interpretation* [1].

In view of the numerical and chemical experiments discussed in the introduction, we expect that these noises do not induce macroscopic fluctuations, even though they can qualitatively modify the macroscopic behavior of the system. In order to study phase transitions in the system it will be necessary to solve Eq. (5). In the thermal equilibrium case or *potential* model, $\sigma_m^2 = 0$, we have at least a formal expression of the stationary solution for $\{x_i\}$, namely, the Gibbs state given by Eq. (2). This is not the case when we consider multiplicative noises and the usual techniques from Equilibrium Statistical

Mechanics cannot be applied. Any calculation or approximation must be based on the N-degree-of-freedom Langevin equation (5). The simplest one is a mean-field approximation *a la Weiss* where every site is coupled to the average value of the field. This approximation is actually exact for a model where every site is coupled to every other site of the lattice. This model can be solved in full detail and provides a first example of an exactly solvable system perturbed by multiplicative noise and with an infinite number of degrees of freedom.

2.1. MEAN-FIELD MODELS

In the mean-field model, every site is coupled to all the N other sites, according to the following evolution equation:

$$\dot{x}_i = f(x_i) + \frac{D}{N} \sum_{j=1}^{N}(x_j - x_i) + \xi_m^{(i)} g(x_i) + \xi_a^{(i)}. \tag{7}$$

Consider now the thermodynamic limit $N \to \infty$: we expect that μ, as defined by Eq. (4), converges to the mean field value $\langle x_i \rangle$, so that the field in all the sites obeys the same stochastic differential equation:

$$\dot{x} = f(x) + D(\mu(t) - x) + \xi_m g(x) + \xi_a. \tag{8}$$

In the stationary regime, $\mu(t)$ becomes constant. Therefore the stationary probability density for $x(t)$ can be easily solved in terms of μ (recall that we are using the Stratonovich interpretation):

$$P_{st}(x) = \frac{1}{Z(\mu)} \exp\left[\int^x dy \, \frac{f(y) + D(\mu - y) - \frac{\sigma_m^2}{2} g(y)g'(y)}{\frac{\sigma_a^2}{2} + \frac{\sigma_m^2}{2} g(y)^2} \right] \tag{9}$$

$Z(\mu)$ being a normalization constant. The value of μ is obtained from the following self-consistency equation

$$\mu = \phi(\mu) \tag{10}$$

where the function $\phi(\mu)$ is given by

$$\phi(\mu) = \frac{1}{Z(\mu)} \int_{-\infty}^{\infty} dx \, x \exp\left[\int^x dy \, \frac{f(y) + D(\mu - y) - \frac{\sigma_m^2}{2} g(y)g'(y)}{\frac{\sigma_a^2}{2} + \frac{\sigma_m^2}{2} g(y)^2} \right]. \tag{11}$$

Exactly as in the Weiss theory of paramagnetism, this nonlinear self-consistency equation may have more than one solution. In this case we have more than one stationary value of μ and consequently more than one stationary distribution for x given by Eq. (9). These different stationary distributions correspond to different macroscopic states —those given by the different values of μ— and they typically break some of the symmetries present in the microscopic force $f(x)$. We recover, in a perfect parallelism, all the features of the Weiss mean-field description of thermal ferromagnetic phase transitions.

At first sight one can mistake x for the macroscopic state of the system and interpret the fluctuations of the random process x as macroscopic ones. We stress that this is not correct. The stationary probability $P_{st}(x)$ is a *microscopic* property: it is the probability distribution for the value of the field *at a given site*. The macroscopic state is still given by expression (4) and therefore, in the thermodynamic limit, is a non-fluctuating quantity, precisely equal to $\langle x \rangle$. Therefore, here we do not have the interpretation problems that arise in zero-dimensional models, where the physical meaning of the moments is sometimes not completely clear, and much lesser their role in locating critical points. Here the macroscopic state is well defined, given by μ, and with this in mind it is possible to sharply define critical and transition points as those points where μ is not an analytical function of the parameters of the model.

2.2. THE LIMIT OF LARGE DIFFUSION

Before turning to the discussion of the self-consistent Eq. (10), we consider the limit of large diffusion, $D \to \infty$, which can be analyzed in a very simple way.

By averaging Eq. (8) (either using Novikov theorem or the corresponding Fokker-Planck equation) one obtains the following exact result:

$$\langle \dot{x} \rangle = \langle f(x) \rangle + \frac{\sigma_m^2}{2} \langle g(x)g'(x) \rangle. \tag{12}$$

In general, for nonlinear $f(x)$ and/or $g(x)$, this equation is not closed in the first moment. However, for D very large, every realization of the process $x(t)$ feels a strong force towards the mean value $\mu(t)$ of the process itself. As a consequence, the probability distribution of the process $x(t)$ will be sharply peaked around μ. If we neglect fluctuations, Eq. (12) reduces to

$$\langle \dot{x} \rangle = f(\langle x \rangle) + \frac{\sigma_m^2}{2} g(\langle x \rangle)g'(\langle x \rangle), \tag{13}$$

which is now a closed equation in $\langle x \rangle = \mu$. Note the familiar Stratonovich contribution coming from the multiplicative noise.

The stationary values μ_{st} are the solutions of

$$f(\mu_{st}) + \frac{\sigma_m^2}{2} g(\mu_{st})g'(\mu_{st}) = 0. \tag{14}$$

Consider, on the other hand, the zero-dimensional version of our model:

$$\dot{x} = f(x) + \xi_m g(x) + \xi_a. \tag{15}$$

It is well known that the equation for the extremes x_s of the stationary probability density reads

$$f(x_{st}) - \frac{\sigma_m^2}{2} g(x_{st})g'(x_{st}) = 0. \tag{16}$$

Note that the sign of the term coming from the noise is precisely the opposite! So will be the effects of external noise on zero-dimensional and spatially extended systems. Moreover, if one linearizes Eq. (5) around a steady point x_0 ($f(x_0) = g(x_0) = 0$), the stability of x_0 is given by that of the first moment $\langle x \rangle - x_0$. As it was advanced in the introduction, we have shown how in the mean-field models the linearization turns out to give a good estimation of the location of critical points (changes of stability) for D large enough, whereas for zero-dimensional systems, as the opposite sign evidences, gives even qualitatively wrong

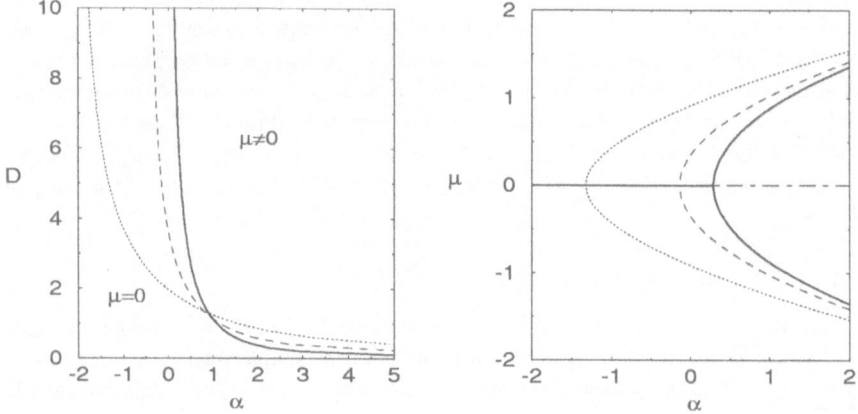

Figure 1: Phase diagram and order parameter (for $D = 5$) corresponding to the Ginzburg-Landau model defined by Eq. (18). Solid line: $\sigma_m^2 = 0$, dashed line: $\sigma_m^2 = 1$, dotted line: $\sigma_m^2 = 2$.

results. We expect the former assertion to be more general, valid for models in two and higher dimension with nearest-neighbor interactions, since there is strong evidence that the mean-field approximation becomes exact for these models in the limit $D \to \infty$.

3. The effects of external multiplicative noise

We present in this section three examples of concrete models where the external noise induces different effects. We have solved in the three cases the self-consistency equation (10) that gives the value of the macroscopic state or mean-field μ. We have also obtained the corresponding phase diagram in examples 3.1 and 3.2 where second order phase transition occurs, calculating the points where

$$\phi'(\mu = 0) = 1, \tag{17}$$

i.e., those points where the self-consistency equation (10) goes from having one solution to having two solutions.

3.1. NOISE CREATES ORDER

Consider the Ginzburg-Landau model with multiplicative noise [4]:

$$\dot{x} = -x^3 + \alpha x + \xi_m x + D(\mu - x) + \xi_a. \tag{18}$$

The potential model ($\sigma_m^2 = 0$) exhibits a second order phase transition at a value of α depending on the coupling D and the temperature σ_a^2 [5, 6, 7]. Analyzing the limit $D \to \infty$ one immediately finds that the critical point $\alpha_c = 0$ is shifted to $-\sigma_m^2/2$ by the external noise, increasing the zone of non-zero magnetization μ, or *ordered phase*.

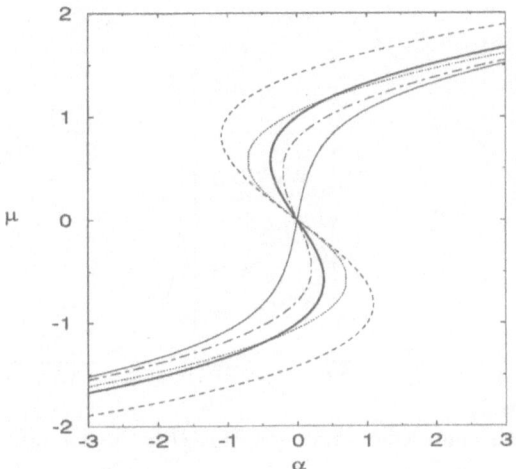

Figure 2: The effect of external noise on hysteresis cycles as described by Eq. (19), for $\sigma_m^2 = 0$ (deterministic system, solid line bold face), $\sigma_m^2 = 1$ and $D = \infty$ (dashed line), $D = 10$ (dotted line), $D = 3$ (dot-dashed line) and $D = 1$ (solid line).

In Fig. 1 we have depicted the phase diagram of the model for different values of σ_m^2 and for $\sigma_a^2 = 1$ in the plane D, α. We stress again the parallelism with the Weiss theory of paramagnetism: in the region where $\mu \neq 0$, the system is no longer ergodic. The local field x has two possible stationary distributions which are not connected in the thermodynamic limit. One can see immediately that these stationary distributions, as given by (9), are not symmetric around $x = 0$ or, in other words, the system has spontaneously broken the symmetry $x \rightarrow -x$ present in the dynamics.

Observe that, for D large enough but finite, the region corresponding to the ordered phase ($\mu \neq 0$) is enlarged. Thus we can rightfully assert that, for these values of D, the external noise *orders the system*. Note also that in the corresponding zero-dimensional model, if transitions are defined as a change in the shape of the stationary distribution, the shift of the critical point is the opposite or, in other words, the external noise destroys the ordered phase.

3.2. HYSTERESIS CYCLES

Consider now the equation

$$\dot{x} = (x^2 + 1)[-x(x^2 - 1) + \alpha + \xi_m] + D(\mu - x). \tag{19}$$

This is a simple model of a system exhibiting a hysteresis cycle when varying α (see Fig. 2). From the former example 3.1, on can note that, if the external noise term is linear on x, the noise only induces a shift on the control parameter. By including the non linear term $(1 + x^2)\xi_m$ we find in this example a more involved effect.

The solution of the corresponding self-consistency equation is shown in Fig. 2 for a fixed value of the external noise intensity $\sigma_m^2 = 1$. For strong enough coupling, the hysteresis cycle

164

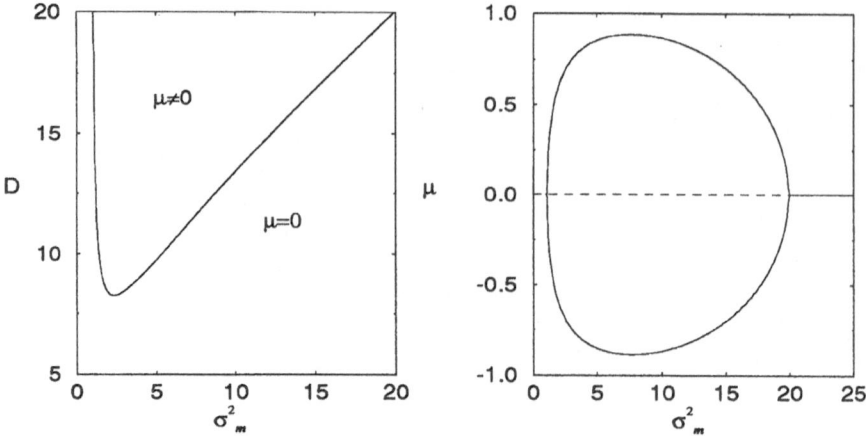

Figure 3: Phase diagram and oreder parameter (for $D = 20$) for the pure noise-induced model (20).

is enlarged. This is indeed the qualitative result found by De Kepper and Horsthemke [1, 2] in a stirred chemical reaction. The experiment probably corresponds to a strong diffusion of reactants, but its dynamics is much more complex than the one modeled by (19) (for instance, in the real experiment the upper branch of the cycle corresponds to an oscillatory stationary state). Nevertheless, an interesting and testable prediction that is suggested by model (19) is that the hysteresis cycle can be either enlarged or reduced depending on the values of D and σ_m^2. This effect is shown in Fig. 2, where μ is plotted against α in the deterministic case (bold face) and for $\sigma_m^2 = 1$ and different values of D. Observe that the cycle is reduced as D decreases and becomes smaller than the deterministic one for $D = 3$ and even disappears for $D = 1$. The same effect can be observed by varying σ_m^2: for small values of σ_m^2 the cycle is enlarged whereas it is reduced and disappears above a given value of σ_m^2 which depends on D.

3.3. PURE NOISE-INDUCED PHASE TRANSITIONS

The above example suggests the possibility of a *pure noise-induced transition*. If the external noise can enlarge a hysteresis cycle, could it produce such a cycle in a system not exhibiting any phase transition at all? A slight modification of the previous example gives us a positive answer to this question. Consider the equation [8]:

$$\dot{x} = (x^2 + 1)[-x(x^2 + 1) + \alpha + \xi_m] + D(\mu - x). \tag{20}$$

It is easy to check that, for $\sigma_m^2 = 0$, the local potential has only one minimum for any value of α. Adding a noise we are creating a hysteresis cycle exactly as in the former example we increased the existing cycles. If we fix α, then the appearance of the hysteresis cycle when varying σ_m^2 is a second-order phase transition with μ as the order parameter.

In Fig. 3 we show the phase diagram for $\alpha = 0$ in the plane D, σ_m^2. The ordered phase $\mu \neq 0$ is reached only for large enough values of D. Fixing D the system undergoes two second-order phase transitions when varying σ_m^2. Note also that, for $\alpha = 0$ the dynamics of the system is invariant under the transformation $x \to -x$ and that we again have the characteristic features of a genuine phase transition: the appearance of two not connected and non-symmetric stationary distributions.

4. Discussion

We have studied systems with a large number of degrees of freedom perturbed by a local or microscopic multiplicative noise, i.e., the noise affects independently each degree of freedom. These two aspects cause the vanishing of the fluctuations of the macroscopic state of the system in the thermodynamic limit, precisely as it happens in Equilibrium Statistical Mechanics.

So far, the effects of multiplicative noise have been studied only in systems with one or two degrees of freedom. These effects will apply to intrinsically zero-dimensional systems, such as the intensity of the resonant mode in a laser [9], and to systems with a large number of degrees of freedom but with the external noise acting in a global way. For instance, if one assumes for the systems presented in this paper that the noise $\xi_m^{(i)}$ is the same at any site i, in the limit $D \to \infty$ the system can be described by one variable μ_t which is an stochastic process obeying the zero-dimensional stochastic equation (15).

The mean field theory that we have presented is technically not much more difficult than the usual approach followed in the zero-dimensional theory of noise-induced transitions. Yet the results are very different both conceptually and qualitatively. In fact, we believe that the results of the zero-dimensional theory have little to do with the noise-induced phenomena in spatially distributed systems.

In the three examples presented in section 3 a cooperation between diffusion and external fluctuations is observed. For instance, let us assume, in the Ginzburg-Landau model (example 3.1), a negative value for α. Then the local deterministic potentials have only one minimum at every site. If the external noise is switched on, the actual value of α fluctuates and it can take a positive value at some sites i destabilizing $x_i = 0$. As a result of these fluctuations, small domains are created where the field is either negative or positive and the diffusion makes these domains to grow.

In the case of the pure noise-induced phase transition (example 3.3), there is a similar mechanism: depending on the sign of the noise, the local force points towards positive or negative values of the field. With a disordered initial condition, small domains are again created by the coincidence of the noise sign at some neighboring sites and, if the diffusion is strong enough, this domains grow and remain stable in spite of the action of further local fluctuations.

But observe the delicate compromise between the two mechanisms acting upon the system: diffusion and external fluctuations. This compromise is noticed in a more apparent way in example 3.3 where the ordered phase again disappears when increasing the external noise intensity σ_m^2 (see Fig. 3). It can also be observed in the other two examples: in the Ginzburg-Landau case 3.1., for weak diffusion the external noise destroys the order of the system (see Fig. 1). This occurs if D is below a given threshold which increases with the noise intensity σ_m^2. In example 3.2, we have seen that the hysteresis cycle enlarges with respect to the deterministic system only for strong enough diffusion (see Fig. 2) and that the effect is precisely the opposite for weak diffusion. If D is fixed and we continuously increase σ_m^2, we would observe first an enlargement and a subsequent shrinkage of the cycle until it finally disappears following a similar phase transition to that of example 3.3.

We have seen, using a simple mean-field theory, how the combination of diffusion and external fluctuations brings about a new and non trivial phenomenology that moreover fully connects the study of noise-induced transitions with phase transitions such as they are pictured in Equilibrium Statistical Mechanics. Some of the qualitative predictions obtained by means of this theory, such as order creation by noise and the existence of a phase transition purely induced by noise, have been confirmed numerically in two-dimensional systems [4, 8] as well as by a more refined analytical theory that appraises the dimensionality of the model [8]. We hope that these first results animate further research on spatially extended systems perturbed by external noise. We find of special interest the extension of the theory to other types of noise, more accurate analysis than the mean-field theory such as renormalization performed on the Langevin equation, and, from the experimental point of view, the practical implementation of this type of models using, for example, photosensitive reactions.

Acknowledgments

We thank the Program on Inter-University Attraction Poles, Prime Minister's Office, Belgian Government for financial support. J.M.R. Parrondo also acknowledges support from from Dirección General de Investigación Científica y Técnica (DGICYT) (Spain) Pro. No. PB91-0222 and Pro. No. PB91-0378 and C. Van den Broeck from the NFWO Belgium.

References

[1] W. Horsthemke and R. Lefever, *Noise-Induced Transitions* (Springer-Verlag, Berlin, 1984).

[2] P. de Kepper and W. Horsthemke, C. R. Acad. Sci. Ser. C **287**, 251 (1987).

[3] J. García-Ojalvo, A. Hernández-Machado, and J. M. Sancho, Phys. Rev. Lett. **71**, 1542 (1993).

[4] C. Van den Broeck, J.M.R. Parrondo, J. Armero, and A. Hernández-Machado, Phys. Rev. E **49**, 2639 (1994).

[5] Y. Onodera, Prog. Theor. Phys. **44**, 1477 (1970).

[6] M. Shiino, Phys. Rev. A **36**, 2393 (1987).

[7] R. Toral and A. Chakrabarti, Phys. Rev. B **42**, 2445 (1990).

[8] C. Van den Broeck, J.M.R. Parrondo, and R. Toral, *Pure noise-induced phase transitions*. In preparation.

[9] For a review, see M. San Miguel, in *Instabilities and Chaos in Quantum Optics II*, edited by N.B. Abraham, F.T. Arecchi, and L.A. Lugiato (Plenum Press, New York, 1988).

COMPETITIVE COEXISTENCE IN BIOLOGICAL SYSTEMS: EXACT ANALYTICAL RESULTS THROUGH A QUANTUM MECHANICAL ANALOGY

H.S. WIO,* M.N. KUPERMAN†
Centro Atómico Bariloche (CNEA) and Instituto Balseiro (UNC)
8400 Bariloche, Río Negro, Argentina
and
B. VON HAEFTEN, M. BELLINI and R.R. DEZA
Dpto. de Física, Fac. Cs. Exactas, U. Nac. de Mar del Plata
Funes 3350, 7600 Mar del Plata, Argentina
and
C. SCHAT‡
Dpto. de Física, TANDAR (CNEA) Buenos Aires, Argentina

ABSTRACT. We have studied an ecological system of two species (called *strong* and *weak*) competing for a single food resource, modelled as a reaction diffusion process. A whole family of exact analytical solutions has been found resorting to a quantum mechanical analogy. Such solutions indicate that in certain situations (and essentially as a consequence of the *weak* species mobility), the classical results on extinction and coexistence of Lotka-Volterra type equations are no longer valid. We have analyzed the stability of these solutions and discussed different possibilities for extending our results.

1. Introduction

One of the most important problems among those discussed by Mathematical Ecology is that related with the survival of species in coexistence. Of particular relevance are those cases where two or more species compete for the same food resource. The mathematical theory of competition was originated, among others in Volterra's 1927 work [1], where he showed that the coexistence of two or more species of predators limited to only one species of prey is impossible. Volterra's results have been discussed and extended, and today are included within the *Competitive Exclusion Principle* (or *ecological theorem*), that states: *N species that compete for n(< N) food resources, cannot coexist* [2]. This problem has been analyzed by several authors emphasizing, for instance, the conflict between the need to forage and the need to avoid competition; effects of diffusion-mediated persistence; global effects associated with the possibility of refuges; etc. [3,4,5].

*Member Carrera del Investigador CONICET, Argentina.
†Fellow CONICET, Argentina.
‡Fellow CNEA, Argentina.

E. Tirapegui and W. Zeller (eds.), Instabilities and Nonequilibrium Structures V, 167–181.
© 1996 Kluwer Academic Publishers.

In this work we study the possibility of coexistence of two species competing for a unique food resource. Even though this is a very simplified situation regarding the ecological reality, through its analysis we can gain in the knowledge of the role played by the different model parameters on the global behaviour before attempting a more complete analysis. The model to be discussed is similar to the one used by Eigen [6] in relation with problems of prebiological evolution and by Mikhailov [7] in order to describe a noise-induced transition in a biological system with diffusion. This model assumes that the *strong* species (the one that survives according to the indicated principle), as well as the food resource, are heavy and immobile; while the *weak* species (the one that becomes extinct) is light and mobile. In the case of spatially homogeneous distributions, the validity of the *Competitive Exclusion Principle* is verified, being impossible the coexistence of both species. Mikhailov's work [7] has shown that the existence of spatio- temporal fluctuations in the food resource, when larger than certain threshold value, makes possible the coexistence due exclusively to the mobility of the *weak* species. Here, we analize the alternative possibility of coexistence in the case where the food source is such that it acquires the characteristic of a *solitary wave* with constant velocity. Such a behaviour could have, for instance, an stational origin. This spatial inhomogeneity, that at variance with [7] is highly coherent, allows for the coexistence of both species, again as a consequence of the mobility of the *weak* species.

2. Mathematical Model

As indicated before, the model consist of two species competing for the same food resource. We indicate the population density of the *strong* species with the variable N, with n the population density of the *weak* species, and with M the food density. The set of differential equations we adopt in order to describe the behaviour of such a system for the spatially homogeneous case is :

$$
\begin{aligned}
\partial_t n(t) &= [b\,M(t) - a]\,n(t) \\
\partial_t N(t) &= [B\,M(t) - A]\,N(t) \\
\partial_t M(t) &= Q(t) - [G + c\,n(t) + C\,N(t)]\,M(t),
\end{aligned}
\tag{1}
$$

that are Malthusian-like *birth-death* equations [8] for each species : $b\,M(t)$ and $B\,M(t)$ indicate the growth rate while a and A indicate the death rate (assumed constant) of the species $n(t)$ and $N(t)$ respectively. For the food $M(t)$ the production is given by $Q(t)$ and is assumed to be independent of the population densities, while the decay is due not only to the natural degradation (rotting) with a rate G, but to the consumption by both predator species with rates $c\,n(t)$ and $C\,N(t)$. These equations have the following stationary solutions :

$$M_n = \frac{a}{b}, \qquad n_n = \frac{Qb - Ga}{ca}, \qquad N_n = 0$$

$$M_N = \frac{A}{B}, \qquad N_N = \frac{QB - GA}{CA}, \qquad n_N = 0$$

(2)

We assume the condition

$$\frac{A}{B} < \frac{a}{b}$$

(3)

which indicates that the threshold for the survival of the species N is lower than the threshold needed for the survival of the species n. A linear stability analysis in such a case shows that the first solution is unstable and the second one is stable. This corresponds to the initial assumption that N will have the characteristics of the *strong* species, the one that survives, while n will be the *weak* species, the one that becomes extinct. Hence, an arbitrary initial condition will move towards the attractor corresponding to the second of the indicated stationary solutions.

Now, we include the possibility of migration of the *weak* species modelled as a diffusive process [2,9]:

$$\begin{aligned}
\partial_t n(x,t) &= D_n \partial_x^2 n(x,t) + [bM(x,t) - a]\, n(x,t) \\
\partial_t N(x,t) &= [BM(x,t) - A]\, N(x,t) \\
\partial_t M(x,t) &= Q(x,t) - [G + cn(x,t) + CN(x,t)]\, M(x,t).
\end{aligned}$$

(4)

With the conditions indicated before, Eq.(3), and with a spatially homogeneous source $Q(t)$, the mobility of the *weak* species does not prevent its extinction.

3. Fluctuating Source

As it was indicated before, Mikhailov [7] considered the case where the food resource fluctuates in space and time. The model is again the same as in Eq.(4), but assuming that $Q(x,t) = Q(x,t) + f(x,t)$, with $f(x,t)$ a random function of space and time. Hence, $f(x,t)$ can be interpreted as an external random factor. Without loss of generality, it is possible to assume that the form of the fluctuation peaks are Gaussian, where the correlation function has the form

$$\langle f(x,t)\, f(x',t') \rangle = 2\,G\theta\, e^{-k_f |x - x'|} \delta(t - t')$$

(5)

The parameter k_f is related to the inverse of the characteristic radius of the peak fluctuation, while the intensity θ will be the parameter controlling the possibility of survival of the *weak* species. The study of the temporal evolution of such a system makes it necessary to resort to by now well known techniques for the treatment of stochastic systems.

The formalism used by Mikhailov resembles the one that is standard in the theory of second order phase transitions and consists in the following: First, to isolate the slow component of the field associated with the *weak* species and adopt it as the *order parameter*. Secondly, to adiabatically eliminate the rest of variables. Such a procedure leads us to a time dependent Ginzburg-Landau like equation for the indicated order parameter. In this equation a characteristic parameter arises: $\theta_c \propto D_n k_f^2$. Mikhailov has shown that if the value of θ (noise intensity) is smaller than the threshold value given by θ_c, the mean value of the population density of the *weak* species, $\langle n(x) \rangle$ goes asymptotically to zero at long times: the *weak* species becomes extinct. However, for $\theta > \theta_c$ the asymptotic value of $\langle n(x) \rangle$ is not zero, with the interpretation that the *weak* species survives (such a situation being a stationary solution of the problem). It is worth to remark that the fact that θ_c is proportional to D_n does not mean that $\theta_c \rightarrow 0$ when $D_n \rightarrow 0$, because in this limit several of the approximations are no longer valid. The meaning of this result is that, due to the mobility of the *weak* species, the increase of its population density in those regions where the fluctuations occurs, compensates for the decay outside these regions, rendering a *noise induced transition*. Mikhailov's corollary was: *in order to survive it is better to be clever, but sometimes it is enough to be mobile.*

4. Case of a Solitary Wave Like Source [10]

We start from the same original equations indicated in Eq.(4). In order to simplify the algebra we rewrite the equations making a scaling of variables and parameters. Also, and in order to simplify our analysis, we assume that $c = b$ and $C = B$, scale the variables n, N, M, Q, (multipliying each one by b/a), call: $G = g/a$, $\beta = B/b$, $\alpha = A/a$ and $q = bQ/a^2$, and also do the following change of spatial and temporal coordinates

$$t \rightarrow \tau = at, \quad x \rightarrow y = x\sqrt{\frac{a}{D_n}}$$

After these changes, our system adopts the simplified form

$$
\begin{aligned}
\partial_\tau n(y,\tau) &= \partial_y^2 n(y,\tau) + [M(y,\tau) - 1]\, n(y,\tau) \\
\partial_\tau N(y,\tau) &= [\beta M(y,\tau) - \alpha]\, N(y,\tau) \\
\partial_\tau M(y,\tau) &= q(y,\tau) - [g + n(y,\tau) + \beta N(y,\tau)]\, M(y,\tau)
\end{aligned} \tag{6}
$$

As our interest is to consider that the food source behaves like a *traveling wave* with a constant velocity c, it is more convenient to change to a reference frame that follows the wave, i.e. $y \rightarrow \xi = y - ct$, rendering

$$\partial_\tau n(\xi,\tau) = \partial_\xi^2 n(\xi,\tau) + c\partial_\xi n(\xi,\tau) + [M(\xi,\tau) - 1]\, n(\xi,\tau)$$

$$\partial_\tau N(\xi,\tau) = c\partial_\xi N(\xi,\tau) + [\beta M(\xi,\tau) - \alpha]\, N(\xi,\tau)$$
$$\partial_\tau M(\xi,\tau) = q(\xi,\tau) - [g + n(\xi,\tau) + \beta N(\xi,\tau)]\, M(\xi,\tau) + c\partial_\xi M(\xi,\tau). \qquad (7)$$

The possibility of solving analytically this problem hinges on the fact that, with an adequate transformation, the equation for $n(\xi,\tau)$ adopts the form of a Schrödinger like equation, where the potential is associated with the food wave. If we propose for $n(\xi,\tau)$

$$n(\xi,\tau) = e^{(-c\xi/2)}\, \phi(\xi,\tau), \qquad (8)$$

The resulting equation for $\phi(\xi,\tau)$ is

$$\frac{\partial^2 \phi}{\partial \xi^2} + \left(M(\xi) - 1 - \frac{c^2}{4}\right) \phi = \frac{\partial \phi}{\partial \tau}. \qquad (9)$$

We assume the stationary case: $\frac{\partial \phi}{\partial \tau} = 0$, that is to say that all the time dependence comes through the variable ξ. Comparing with the adimensional Schrödinger equation [11]

$$\frac{\partial^2 \psi}{\partial \xi^2} + (E - V(\xi))\, \psi = 0 \qquad (10)$$

we can identify the potential and the energy eigenvalue as

$$V(\xi) \to -M(\xi), \quad E \to -1 - \frac{c^2}{4}.$$

On the other hand, as we are considering population densities, n must be positive. Hence, the only relevant solution is the one corresponding to the ground state of the potential $-M(\xi)$, with a wave function $\psi_f(\xi)$ and energy E_f, determining a unique value for the velocity c, given by $c^2 = -4(1 + E_f)$. The form of the density $n(\xi)$ is fixed by Eq.(8), corresponding to the ground state wave function times an exponential decreasing factor that shifts the maximum of $\psi_f(\xi)$ backwards.

Hence, it is clear that the possibility of survival of the *weak* species as a wave like density is determined by an adequately chosen behaviour of the density $M(\xi,\tau)$. Here we have adopted for $M(\xi,\tau)$ a localized perturbation moving on top of the homogeneous distribution corresponding to the threshold $M_N = \alpha/\beta$ (i.e. the second value in Eq.(2) with the scaled parameters). Hence, we can expect that both $N(\xi)$ and $n(\xi)$ will grow inside the perturbated region, reaching some given values outside. The general result for N is

$$N_s(\xi) = N^s \exp\left[-\frac{1}{c} \int_\xi^\infty (\beta M(\xi') - \alpha) d\xi'\right] \qquad (11)$$

for $\xi \in [-\xi_0, \xi_0]$, $N_s(\xi) = N^s$ for $\xi > \xi_0$, and (on principle) a different constant value, to be determined, for $\xi < -\xi_0$.

Finally, the source q consistent with the stationary state $(M_s(\xi), N_s(\xi), n_s(\xi))$ we have just obtained, is given by

$$q_s(\xi) = (g + \beta N_s(\xi) + n_s(\xi)) M_s(\xi) + c \frac{\partial}{\partial \xi} M_s(\xi). \tag{12}$$

All the external influence will be included in the function $q_s(\xi)$ that now plays the role of an external parameter keeping the system far from equilibrium.

Here it is worth to make a comment concerning the form found for n: in the above indicated solution we expect the wave function $\phi(\xi, \tau)$ to decay exponentially outside the support of the *potential function* $\frac{\alpha}{\beta} - M(\xi)$. The rate of decay, as it is known from quantum mechanics, is $\gamma = \sqrt{-E_f}$, whereas from the indicated solution $\frac{c}{2} = \sqrt{-(1 + E_f)}$. Hence, the sign of the combined exponent $-\gamma | \xi | -\frac{c}{2}\xi$ is effectively negative as $\xi \to -\infty$, since $\gamma > \frac{c}{2}$. The fact that $n \neq 0$ in a localized region of the ξ-space is to be interpreted in the following way: before the food wave reaches the point x, there is a negligible number of individuals of the *weak* species (*extinction situation*); but because their ability to follow the wave (as opposite to the *strong* species) their population can grow to a non-negligible value, but only around the wavefront. Those individuals which are lagged behind, are condemned to extinction by competition with the *apter* species in the homogeneous situation.

For the form of the wave $M(\xi)$ mounted on top of the homogeneous value M_s we can choose several different possibilities. In [10] we have adopted a truncated parabola, however, there are several other possibilities corresponding to all known cases of exactly (or quasi-exactly) solvable potentials [14].

Here we have chosen to present the results for the following potential

$$M_s(\xi) = \begin{cases} \frac{\alpha}{\beta} & \xi \notin [-1, 1] \\ 3\sqrt{a_3}\xi^2 - a_3\xi^2(\xi^2 + \frac{a_2}{2a_3})^2 + \frac{\alpha}{\beta} & \xi \in [-1, 1] \end{cases} \tag{13}$$

i.e. a sextic potential which, although being symmetric like the one in [10], has more structure. The corresponding stationary Schrödinger like equation has the following groundstate wave function [14],

$$\phi(\xi) \propto e^{-\frac{a_2}{4\sqrt{a_3}}\xi^2 - \frac{\sqrt{a_3}}{4}\xi^4} \tag{14}$$

with the energy eigenvalue

$$E_f = -\frac{a_2}{4\sqrt{a_3}} + \frac{\alpha}{\beta} \tag{15}$$

This yields for the population density of the *weak* species n:

$$
n_s(\xi) \propto
\begin{cases}
R\,e^{\left(-\frac{c}{2} \pm \sqrt{-(E_f + \frac{\alpha}{\beta})}\right)\xi} & \xi \notin [-1,1] \\[2mm]
e^{-\left(\frac{c}{2}\xi + \frac{a_2}{4\sqrt{a_3}}\xi^2 + \frac{\sqrt{a_3}}{4}\xi^4\right)} & \xi \in [-1,1]
\end{cases}
\tag{16}
$$

The factor R in Eq.(16) assures the continuity of the function n at the points $\xi = \pm 1$ and is given by:

$$
R = e^{-\left(\frac{a_2}{4a_3} + \frac{\sqrt{a_3}}{4} + \sqrt{-(E_f + \frac{\alpha}{\beta})}\right)}
\tag{17}
$$

On the other hand, reeplacing the form of $M(\xi)$ into the expression for the *strong* species $N(\xi)$ we obtain:

$$
N_s(\xi) =
\begin{cases}
e^{\frac{\beta}{c}\left((\sqrt{a_3} + \frac{a_2}{6})(\xi^3 - 1) + \frac{a_3}{7}(\xi^7 - 1) + \frac{a_2}{5}(\xi^5 - 1)\right)} & \xi \in [-1,1] \\[3mm]
1 & \xi > 1 \\[3mm]
e^{\frac{\beta}{c}\left(2\sqrt{a_3} + \frac{a_3}{2} + \frac{2a_2}{5} + \frac{a_2}{3}\right)} & \xi < -1
\end{cases}
\tag{18}
$$

where $N_s = \text{cte.} = N_N = 1$. In Figure (1) we show the results for the indicated M_s, N_s and n_s for a particular set of parameters, while in Figure (2) we show the form of the associated source $q_s(\xi)$.

5. Conclusions

Even though the complexity of the ecological reality makes its complete modelization very hard, the study of simplified models could help in the understanding of the role played by some *enviromental* parameters in different situations. In particular, the model we have studied here could give some hints on the behaviour of systems of species in competence and the possibility of coexistence. Here we have analyzed an aspect that is usually neglected in models related with competence, that is the inclusion of spatial dependence and the dynamical aspects associated with it. Through a quantum analogy, we have shown that the existence of spatial variation of the shared food source (of seasonal or artificial origin) can be a possible origin of coexistence of competing species. However, the solution indicated in Eqs.(13), (16) and (18) would be relevant from an ecological point of view as far as it is stable against non-extremal variations of environmental parameters (for instance velocity or intensity of the food source). The full analytical stability study of the above indicated solutions is far from trivial and we have chosen to do a numerical analysis. In order to do such analysis, we have adapted an evolution scheme previously applied to another reaction-diffusion

174

systems [13]. We have considered as our initial condition the solution we have just found, and studied the evolution against variations in the propagation velocity c of the source q as well as in the potential parameter a_2. The results are indicated in Figures (3a,b,c). It is clear from these graphical results that the solution is robust against the indicated perturbations. This is due to the fact that the form of q, that plays the role of an external control parameter, is kept essentially fix.

So far we have only considered an immobile *strong* species and a mobile *weak* one. A natural question is: how could the inclusion of mobility for the *strong* species influence the above indicated results?. That means to consider again the scaled equations Eq.(7), that when we include diffusion of the *strong* species adopts the form

$$
\begin{aligned}
\partial_\tau n(y,\tau) &= \partial_y^2 n(y,\tau) + [M(y,\tau) - 1]\, n(y,\tau) \\
\partial_\tau N(y,\tau) &= d\partial_y^2 N(y,\tau) + [\beta M(y,\tau) - \alpha]\, N(y,\tau) \\
\partial_\tau M(y,\tau) &= q(y,\tau) - [g + n(y,\tau) + \beta N(y,\tau)]\, M(y,\tau)
\end{aligned}
\tag{19}
$$

with $d = D_N/D_n$.

The study of such a case, as well as other possible extensions of the present model (including contributions corresponding to the *struggle for life* for each species; inclusion of *cleaverness* of the *weak* species -i.e *chemotaxis*-; increasing the number of species; increasing the number of dimensions) will be discussed elsewhere [16]

Acknowledgments

The authors thanks to G. Abramson, D. Zanette and A. S. Mikhailov for fruitful discussions and to V. Grunfeld for a careful reading of the manuscript. Partial support from CONICET, Argentina is greatly acknowledged.

References

[1] Volterra, V., *Variazioni e fluttuazioni del numero d'individui in species animali conviventi*, R. Comitato Talassografico Italiano, Memoria **131**, pp 1-142 (1927).

[2] Murray, J. D., *Mathematical biology*, Springer–Verlag, 1989.

[3] Takeuchi, Yasuhiro, *Conflict between the need to forage and the need to avoid competition: persistence of two–species model* Math. Biosci. **99**, 181 (1990).

Takeuchi, Yasuhiro, *Diffusion mediated persistence in three species competition models with heteroclinic cycles*, Math. Biosci. **106**, 111 (1991).

Takeuchi, Yasuhiro, *Refuge mediated global coexistence of multiple competitors on a single resource* , WSSIA A **1**, 531, (1992).

[4] Muratori, S. and Rinaldi, S., *Remarks on competitive coexistence*, SIAM J. Appl. Math. **49**, 1462 (1989).

[5] Mimura, M. and Fife, P. C., *A 3 component system of competition and diffusion*, Hiroshima Math. J. **16**, 189 (1986).

[6] Eigen, M., Naturwissenschaften **58**, 465 (1971).

[7] Mikhailov, A. S., *Noise-induced phase transition in a biological system with diffusion*, Phys. Lett. **73A**, 143 (1979).

Mikhailov, A. S., *Effects of diffusion in fluctuating media: a noise induced phase transition*, Z. Physik B **41**, 277 (1981).

Mikhailov, A. S., *Selected topics in fluctuational kinetics of reactions*, Phys. Rep. **184**, 308 (1989).

[8] Malthus, T. R., *An essay on the principle of population, 1798*, Penguin Books, 1970.

[9] Fife, Paul C., *Mathematical aspects of reacting and diffusing systems*, Springer-Verlag, 1979.

[10] Schat, Carlos, *Nonequilibrium Dynamics in Reaction-Difussion Systems*, M.Sc.Thesis in Physics, Instituto Balseiro, 1991.

Schat, C. and Wio, H.S., *An Exact Analytic Solution of a Three Component Model for Competitive Coexistence*, Submitted to Math.Biosc. (1993).

[11] Landau, L. D. and Lifshitz, E. M., *Quantum Mechanics*, Pergamon Press, 1965.

[12] Simmons, G. F., *Differential Equations*, McGraw-Hill, 1972.

[13] Hassan, S. A., Kuperman, M. N., Wio, H. S., Zanette, D. H., *Evolution of reaction diffusion patterns in infinite and bounded domains*, accepted in Physica A (1993).

[14] Salem, L.D., *Solvable Models in Quantum Mechanics and their Relation with Algebraic Many Body Models*, Ph.D Thesis, Instituto Balseiro, 1992.

Salem, L.D. and Montemayor, R.; *A modified Ricatti approach to partially solvable quantum Hamiltonians*; Phys.Rev. A **43**, 1169 (1991).

[15] Tyson, J. J., Keener, J. P., *Singular perturbation theory of traveling waves in excitable media (a review)*, Physica D **32**,327 (1988)

[16] Kuperman, M., von Haeften, B. and Wio, H.S., *Competitive coexistence: Influence of the mobility of the strong species on the stability of exact solutions*, to be submitted (1994).

Wio, H.S., *Struggle for life: its effect on competitive coexistence*, in preparation.

Kuperman, M., Wio, H.S. and Rodriguez, M.A., *Influence of chemotactic like behaviour on competitive coexistence*, in preparation.

Strong Species N

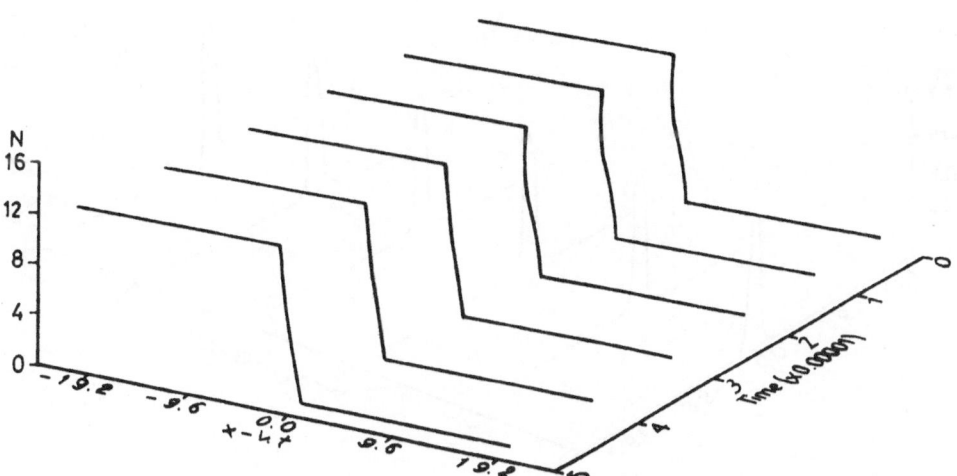

FIGURE 1 : Stationary solutions M_s, N_s and n_s for the set of parameters: $a_2 = 0.79$ and $a_3 = 1$, yielding $c = 1.366$ for the propagation velocity.

Weak Species

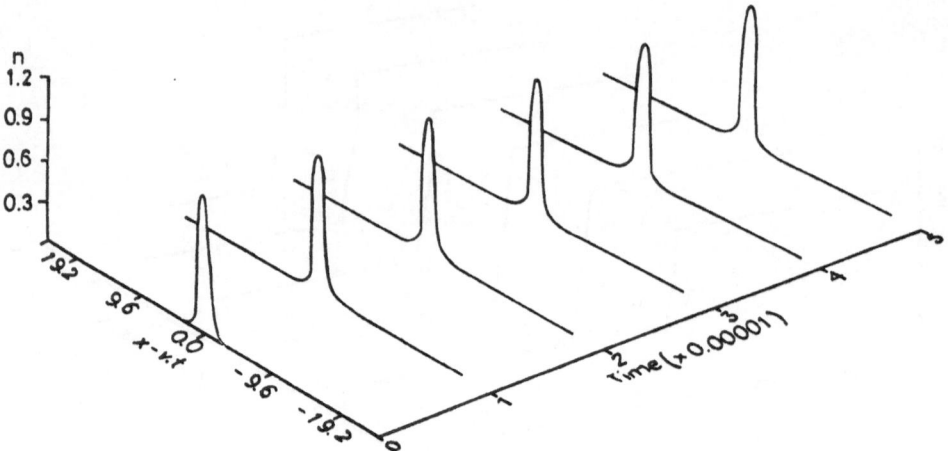

Food M

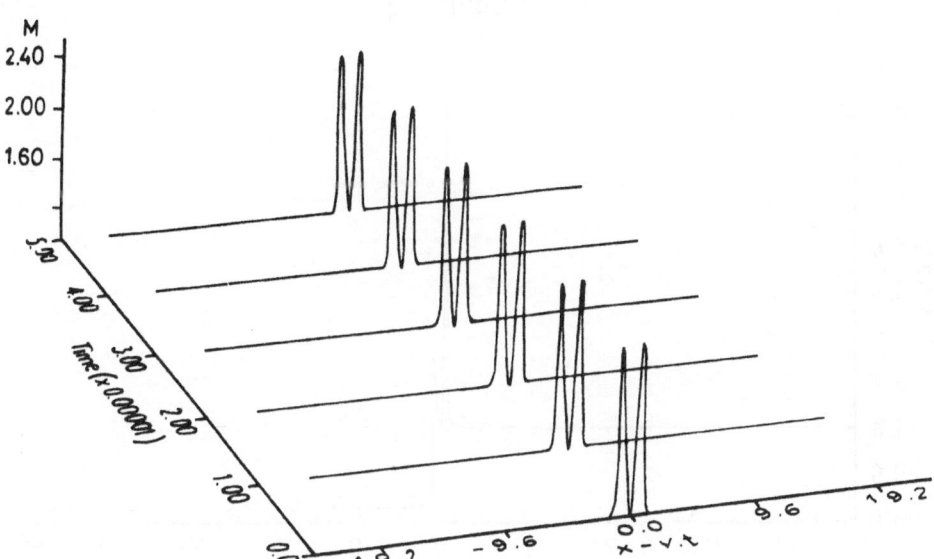

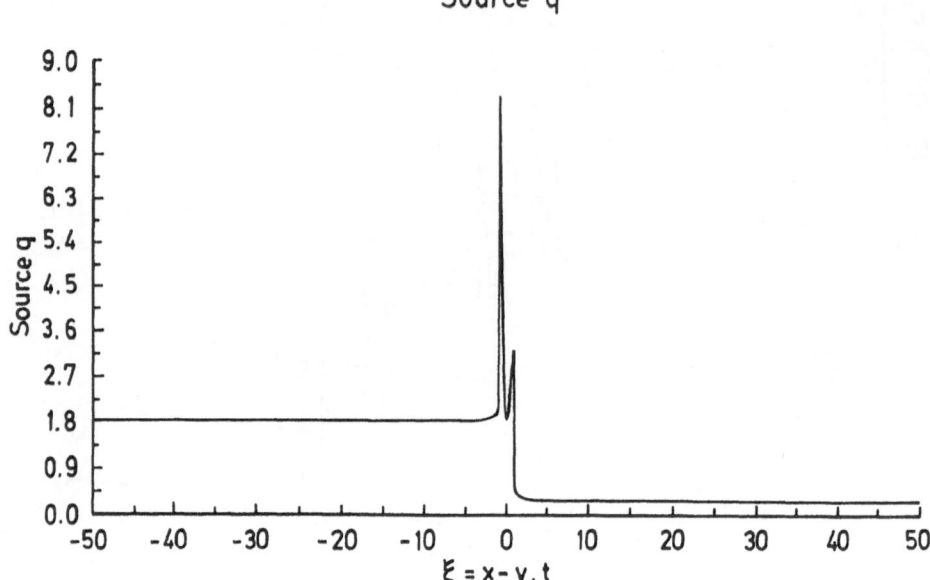

FIGURE 2 : Source $q_s(\xi)$ for the same set of parameters as in Fig.(1).

Initial Conditions

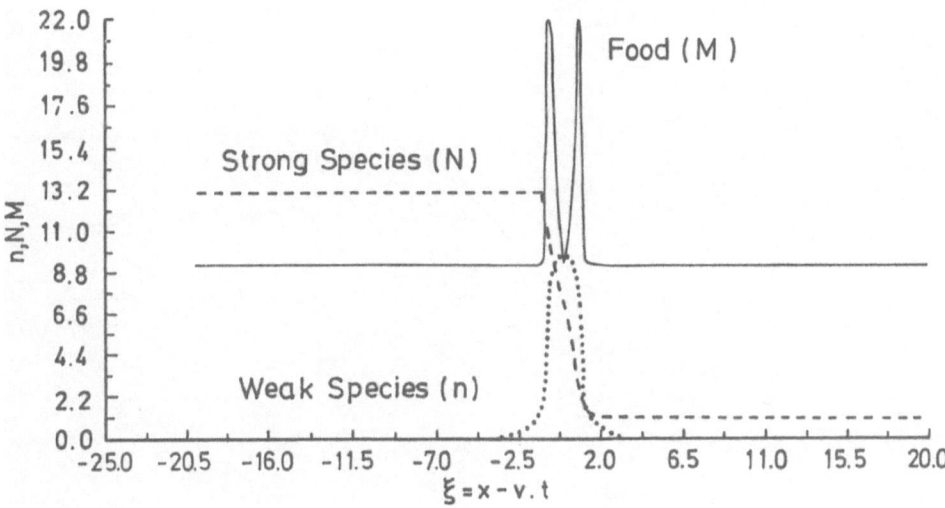

FIGURE 3: Evolution of the populations for M, N and n, for the initial conditions given by M_s, N_s and n_s as indicated in Fig.(1). We have considered the following variations of the parameters : $a_2 = 0.79 \pm 0.01$ and $c = 1.366 \pm 0.01$

Effects of disorder on reaction diffusion systems : An analytical aproach.

M.A. Rodriguez

Instituto Mixto de Astrofisica y estructura de la Materia,

C.S.I.C y Universidad de Cantabria,

Facultad de Ciencias, 39005 Santander, Spain

ABSTRACT . We show several effects produced by disorder using reaction diffusion models. When the system has not organized structures only effects due to the subdiffusion of the microscopic components appear. With organized structures the effects become more complex depending on the involved sizes and characteristic times. These facts are analytically illustrated with an annihilation reaction system and a Ballast resistor.

1. Introduction

Disorder is an essential component of many physical systems. Their effects are well known in linear systems [1,2] and are under investigation in non linear systems [3]. Examples of physical systems in which disorder is relevant are porous media, amorphous semiconductors, glasses, granular media [1,2] etc.. Transport phenomena associated to these systems are influenced by disorder. If disorder is strong enough the behavior ot such systems becomes anomalous. Tipical effects produced by disorder are subdiffusion of excitations and particles, localization of waves and particles, backscattering of light, weak electrical conduction, pinning of interfaces [1,2] etc..Some of these effects can be satisfactory explained by means of linear models whose evolution equations have inhomogeneous random coefficients (Random fields). Here it is worth to mention the models of random diffusion, usually random walks with random jump probabilities [4], wave propagation in random media [5] and Schrodinger equations with random potentials [6] . The main effect of disorder in such linear systems consists of a change in the diffusivity of the corresponding excitation. Matematically it can be made evident by averaging the evolution equations obtaining, in more or less rigorous way, equations with renormalized diffusion coefficients [4].

When the disorder is very strong the diffusion laws change dramatically giving rise to new phases in which the diffusion lenght varies in time with a characteristic exponent

183

E. Tirapegui and W. Zeller (eds.), Instabilities and Nonequilibrium Structures V, 183–192.
© 1996 *Kluwer Academic Publishers.*

determined by the intensity of disorder. In some cases it is possible to perform an overall analysis of these phases and to relate the exponent of the diffusion lenght with the degree of disorder [4].

In a nonlinear system the effect of disorder is not well known. On the one hand one expects to have a similar change in the diffusivity of each component that in a linear system. On the other hand the collective behavior, tipical of a nonlinear system, leads to the formation of structures that could interact with disorder in a more global manner .

The aim of this paper is to illustrate these different behaviors by means of reaction diffusion models with analytical solutions. In a first example we show disorder effects in the segregation proccess of an annihilation reaction. Since in this model there is only one relevant lenght, the diffusion lenght, the effect of disorder is, simply, to change this lenght. In a second example we analyse disorder effects in the propagation of a trigger wave in the ballast resistor. Here there are several lenghts and characteristic times. Under influence of disorder the trigger wave has different behaviors. It can be either like a particle in a positive velocity field or like a structure driven by a time depending noise.

2. Effects of disorder on segregation in the $A + B \to 0$ reaction in 1D

In this section we are going to consider a simple model of annihilation dominated by diffusion in a disordered medium. The analysis of such model has received a lot of attention in recent times[7]. This is due to the fact that this model is very sensitive to changes in the initial conditions, presence of sources and degree of correlations, among other changes. Here we restrict our attention to the study of disorder in a case with symmetrical and Poissonian initial conditions without sources. A more complete analysis of this model can be found in [8].It is well established that in these conditions segregation of like particles occurs for dimension less than four and the size of the segregated clusters growths in a diffusive manner [7].

We introduce disorder into the model by considering a random diffusion coefficient in the classical evolution equations:

$$\frac{\partial n_A(x,t)}{\partial t} = \partial_x \eta(x) \partial_x n_A(x,t) - kn_A n_B \tag{1}$$

$$\frac{\partial n_B(x,t)}{\partial t} = \partial_x \eta(x) \partial_x n_B(x,t) - kn_A n_B \tag{2}$$

where $n_{A,B}$ are the corresponding densities of reactants A and B, $\eta(x)$ is the random diffusion coefficient and k is the reactivity coefficient.

The important point in this model is that it is possible to analyze exactly the effect of disorder because the difference of densities, $\gamma(x,t) = n_A - n_B$, follows a linear equation:

$$\frac{\partial \gamma}{\partial t} = \partial_x \eta(x) \partial_x \gamma(x,t). \tag{3}$$

The size of segregation clusters can be obtained by calculating the correlation lenght associated to the correlation function $C_\gamma(x-y,t) = \langle \gamma(x,t)\gamma(y,t) \rangle$. Hence, the problem reduces to the calculation of this correlation function with different kinds of disorder. If disorder and initial conditions are uncorrelated the correlation function is given by :

$$C_\gamma(x-y,t) = \int dx_1 \int dy_1 \overline{G_2}(x,y,t/x_1,y_1)\{\gamma(x_1,0)\gamma(y_1,0)\} \tag{4}$$

where $\overline{G_2}$ is the averaged two point Green function:

$$\overline{G_2}(x,y,t/x_1,y_1,0) = \langle G(x,t/x_1,0)G(y,t,y_1,0) \rangle$$

and $\{\gamma(x_1,0)\gamma(x_2,0)\}$ is the average over initial conditions.From now on braquets and overlines will mean averages over disorder. The calculation of $\overline{G_2}$ in a model of random walk is a quite difficult task. It is related with the analysis of self averaging properties and it has been only done in more simple models of directed random walks [9]. At this moment we are developing a new method for the calculation of these quantities. Results obtained in the directed random walk and also unpublished results obtained with our method suggested the following ansatz:

$$\overline{G_2}(x,y,t/x_1,y_1,0) \sim \overline{G}(x,t/x_1,0)\overline{G}(y,t/y_1,0)$$

where $\overline{G}$, the averaged one point Green function, is governed by the following equation [10]:

$$\frac{\partial \overline{G}(x,t/x_1,0)}{\partial t} = \int_0^t D(t-t')\frac{\partial^2 \overline{G}(x,t'/x_1,0))}{\partial x^2}dt' + \int \int H(x,t/x',t')\overline{G}(x',t'/x_1,0) \tag{5}$$

$\overline{G}(x,0/x_1,0) = \delta(x-x_1)$. $D(t-t')$ is an effective diffusion coefficient that is determined by the following condition (in Laplace transform):

$$\langle \frac{\eta - D(s)}{1 - (\eta - D(s))G_{xx}(0,s)} \rangle = 0 \tag{6}$$

At long times the first term of (5) is dominant and as we mentioned in the introduction the averaged equation reduces to a diffusion equation with a renormalized coefficient. From the above conditions it is inmediate to calculate the equivalent diffusion coefficient $D(s)$ for the different types of disorder. Disorder can be clasified by its intensity as weak, marginal and strong depending of the behavior of the one point probability density, $P(\eta)$, when η goes to zero [10]. In the weak disordered case, $lim_{\eta \to 0} P(\eta) = 0$ and the equivalent diffusion coefficient is a constant. In this case the effect of disorder consists of a simple renormalization of the diffusion coefficient. In the marginal ($P(0) = constant$) and strong $(lim_{\eta \to 0} P(\eta) \sim \eta^{-\alpha})$ cases the diffusion coefficient is zero. Then it is neccesary to consider a frecuency dependent diffusion coefficient that presents logaritmics corrections in the marginal case and a potential power, $D(s) \sim s^{\frac{\alpha}{\alpha-2}}$, in the strong disordered case. This change in D implies a strong change in all physical properties of the system. Now the diffusion lenght growths in time as $t^{-\frac{1}{2+\alpha}}$. With uncorrelated initial conditions the lenght of segregated clusters also growths, from (4), as $t^{-\frac{1}{2+\alpha}}$. In conclusion, the effect of disorder in the annihilation reaction system is simple. Since each particle has a lower diffusivity due to disorder the corresponding segregation cluster growths slowly. No other effects happen since there are not organized structures.

3. Disordered Ballast resistor

In this section we study the propagation of a trigger wave in a disordered medium. The model is a Ballast resistor with random heat conductivity[11].The Ballast resistor consists of a superconducting wire inmersed in a bath with constant temperature T_B. If $T(x,t)$ is the temperature of the wire at site x and time t the evolution equation of this function will be given by:

$$c\partial_t T(x,t) = \partial_x \lambda(x)\partial_x T(x,t) - q(T(x,t) - T_B) + I^2 R(T) \tag{7}$$

where the first term of the right hand side corresponds to heat diffusion, $\lambda(x)$ is the heat conductivity, the second one represents the heat transfer to the bath and the last is the electrical power density. I is the electrical intensity, c is the specific heat and R(T) is the resistence as function of the temperature. In the case of a superconducting wire we take $R(T) = R_0\theta(T - T_c)$, being $\theta(x)$ the step function, T_c the critical temperature and R_0 a constant resistance. As initial condition we take $T(x,0) = T_0\theta(x)$, with $T_0 > T_c$. In these conditions, in an ordered case with $\lambda(x) = \lambda$ (constant), the further evolution of the system is well known. Moreover, the piecewise linear form of (7) allows an exact time assymptotic solution. The final result [12] is a trigger wave of width $W = (\frac{\lambda}{q})^{\frac{1}{2}}$ propagating with a velocity given by $v = \sqrt{\frac{q\lambda}{c^2}}$ [12]. We note that the trigger wave has the form of a kink. We take as widht of the wave the lenght in which the slope changes appreciably.

Let us focus on the analysis of the effects of disorder. Now $\lambda(x)$ is a random field with mean value λ. In order to avoid unnecessary parameters we write (7) in a reduced form doing $x \to \sqrt{\frac{\lambda}{q}}x, t \to \frac{c}{q}t$ and taking T_B as new origin of temperatures. Moreover we consider the wave in a moving frame with velocity v which will be adjusted in order to get fix the stationary wave. The resulting equation is:

$$\partial_t T_s(y,t) = \partial_{yy}T_s + \partial_y \xi_{y+vt}\partial_y T_s + v\partial_y T_s - T_s + T_h\theta(T_s - T_c) \tag{8}$$

where $T_s(y,t) = T(y+vt)$ is the function of the wave in the moving frame, $T_h = \frac{I^2 R_0}{q} - T_B$ and $\xi_x = \frac{\lambda(x)-\lambda}{\lambda}$ is the transformed random field with zero mean value. For characterizing ξ we take its correlation lenght l_c and intensity D in such a manner that the correlation function of the field can be writen as $\langle\xi_x\xi_{x'}\rangle = DH(x - x')$, being H a normalized function of width l_c. In the white noise limit, $l_c \to 0$, H is a delta function. Both parameters l_c and D as well as the one point probability density used in the above section are relevants in the further analysis.

Now we are in conditions of applying a method of analysis which gives exact results. It is based on the correct interpretation of the averaged nonlinear term $\langle\theta(T - T_c)\rangle$.

3.1 The averaged wave.

In general the effects of disorder in the propagation of a wave are twofold. One is the fluctuation of the position of the wave with respect to the origin of the moving frame. The other are fluctuations themselves in the shape of the wave. To separate both effects is a very difficult task since the average of the motion equations yields to solutions with mixed effects. Hence, by averaging the equation (8) one obtains an equation for $\overline{T(x,t)}$ which is difficult to interpret. Only in the case in which the approximation $\langle\theta(T - T_c)\rangle =$

$\theta(\overline{T} - T_c)$ holds it is possible to consider $\overline{T_s(x,t)}$ as a wave of the same type than the original wave. In principle this approximation is only valid when the effect of disorder is so weak that position fluctuations become negligible. However we can demostrate that with a correct interpretation the approximation is valid in all cases in which a stationary state is available. Indeed, instead of the original wave let us consider the normalized shape function $S(x,t) = \frac{1}{T_H} \partial_x T_s(x,t)$ which magnifies the sensitive zone of the wave. Using this function as a probability density it is possible to define a mean position and width of the original wave. From (8) we obtain the equation for $S(x,t)$:

$$\partial_t S(x,t) = \partial_{xx} S + v \partial_x S + \partial_{xx} \xi_{x+vt} S - S + \delta(x - x_c(t)) \tag{9}$$

where $x_c(t)$ is the point of the critical temperature $T(x_c,t) = T_c$. Taking now as reference of the wave the critical point and as origin of coordinates the mean value of this point, $\overline{x_c(t)} = 0$, and after averaging (9) we obtain:

$$\partial_t \overline{S} = L\overline{S} + P_c(x,t) \tag{10}$$

where L must be a linear operator with effective coefficients modified by the disorder and $P_c(x,t)$ is the probability of a shift of x in the position of the wave. In this equation we have in a separate form both effects of disorder. The averaged wave is the solution of (10) and appears as a convolution that combines both effects :

$$\overline{S(x,t)} = \int \int G(x,t/x',t') P_c(x',t') dx' dt' \tag{11}$$

Finally we can show how to separate these effects when the wave has a stationary evolution. The effect over the shape of the wave will be given by the stationary conditional probability of having a mean wave with a fixed position x_c, which can be defined as

$$S(x/x_c) = lim_{T \to \infty} \frac{1}{T P_c(x_c)} \int_0^T S(x,t) \delta(x_c - x_c(t)) dt. \tag{12}$$

From (9) and (12) it is easy to write the equation for this probability:

$$LS(x/x_c) = \delta(x - x_c). \tag{13}$$

This equation with $x_c = 0$ corresponds to do the approximation $\langle \theta(T - T_c) \rangle \sim \theta(\overline{T} - T_c)$ in the original equation (8). Hence, with this approximation we are exactly calculating the mean wave conditioned to have fixed the critical point at the origin.As a consecuence the conditions $\overline{S(0/0)} = \frac{T_c}{T_h}$ or, in the original equation, $\overline{T(0)} = T_c$ determine the velocity v.We remark that this is valid whenever a stationary evolution happens, independly of the intensity of the disorder.

3.2 Adiabatic propagation with long correlation lenght.

We define an adiabatic propagation as a case in which the wave formation is faster than the wave propagation. This condition can be analytically expressed by $\frac{L_F^2}{\lambda} \ll \frac{L_F}{v}$, where the first term is a estimation of the formation time of the wave as the time in which the diffusion is able to connect points over a lenght L_F. The term on the right is obviously the propagation time of the wave through the same distance. If moreover the correlation time is longer than L_F the wave propagates adiabatically following the variation of the field. $T_s(y,t)$ is then governed by (8) with $\partial_t T_s(y,t) = 0$, and the velocity and width can be exactly calculated, obtaining:

$$v(x) = (z - \frac{1}{z})\sqrt{(1+\xi_x)}$$

and

$$L_F(x) = z\sqrt{(1+\xi_x)}$$

with $z = (\frac{T_h}{T_c} - 1)^{\frac{1}{2}}$.The center of the wave moves like a directed random walk with velocity $v(x)$ and then the random phases produced by disorder are now governing by the behavior of the one point probability density of $\sqrt{1+\xi}$ when $\xi \to -1$ [9].Finally, the conditions for this case can now be more preciselly formulated as $\frac{T_h}{T_c} \ll \frac{3}{2}$ and $l_c \gg z\langle\sqrt{1+\xi}\rangle$.We note that this is an interesting case because it is possible to see analytically the effect of strong localization on the propagation of the wave. The wave behaves like a particle in a positive random velocity field.

3.3 Adiabatic propagation with short correlation lenght.

The difference with the previous case is that the wave does not follow adiabatically the fluctuations of the field but some average of them. This happens because the width of the wave is greater than the correlation lenght. An interesting point to be confronted with simulations is if whether the inertia of the wave is able to avoid the strong localization in some regions with strong disorder or, by the contrary , in these zones the shape of the wave becomes so wide that the adiabatic propagation, shown in the above subsection, happens.In a case of very short correlation lenght with weak disorder one expects to have an stationary evolution following the equation:

$$0 = \lambda_{eff}\partial_{yy}\overline{T_s} + v\partial_y\overline{T_s} - \overline{T_s} + T_h\theta(\overline{T_s} - T_c) \tag{14}$$

being λ_{eff} the same effective diffusion coefficient used in the above section and calculated from the disorder by the condition (6). The velocity and width of the wave are now $v = (z - \frac{1}{z})\sqrt{\lambda_{eff}}$ and $L_F = z\sqrt{\lambda_{eff}}$.

3.4 Fast propagation with weak disorder.

In a non adiabatic case the process of formation and propagation of the pulse are simultaneous. Then all terms in the equation (8) are important and the random field is now time depending. Here the correlation lenght also plays an important role because if it is larger than L_F the noise can be considered as global (only time depending). In this case it is also possible to get analytical predictions. Here we restrict ourselves to the case of weak intensity of the noise. By applying the method outlined in 3.1 we average (8) with $\langle\theta(T_s - T_c)\rangle \sim \theta(\overline{T_s} - Tc)$ in order to get a stationary equation for the conditioned wave, obtaining :

$$\partial_{yy}\overline{T_s} + \partial_y F(y) + v\partial_y\overline{T_s} - \overline{T_s} + T_h\theta(\overline{T_s} - T_c) = 0 \tag{15}$$

$F(y)$ can be calculated at the first order in the noise intensity D by means of the Novikov theorem [13] as:

$$F(y) = lim_{t\to\infty}\langle\xi_{y+vt}\partial_y T_s(y,t)\rangle \sim lim_{t\to\infty} D \int H(y + vt - y\prime)\langle\frac{\delta\partial_y T_s(y,t)}{\delta\xi_{y\prime}}\rangle dy\prime, \tag{16}$$

obtaining, after calculation of the stationary solution of the functional derivative, the equation:

$$D\frac{\partial^4}{\partial y^4}\int G_h(y - y')\overline{T_s(y')}dy\prime + \partial_{yy}\overline{Ts} + +v\partial_y\overline{T_s} - \overline{T_s} + T_h\theta(\overline{T_s} - T_c) = 0 \tag{17}$$

with $G_h(y) = \int G(y,t)H(y + vt)dt$, being $G(y,t)$ the Green function of the deterministic linear operator of (8). The first term of this equation represents the leading effect of the noise. The case $l_c \gg L_F$ corresponds to the global noise situation in which the noise acts as time depending. If moreover $v \gg l_c$ the noise is white and $G(y) \sim \frac{1}{v}\delta(y)$. In the oposite case, with $l_c \gg L_F$, $G(y) \sim \frac{1}{2v}\theta(-y)G(y,-y)$. In the white noise limit the first correction to the velocity can be easily calculated, obtaining:

$$v = z - \frac{1}{z} + D\frac{z^2([\frac{z(z+1)}{2}]^3 - 1)}{4(z^2 - 1)(z^5 + 1)}.$$

In other cases the velocity will depend of the explicit form of $H(x)$.

In conclusion, the effects of disorder in a case with a organized structure, like a trigger wave, are very dependent of the condition of propagation and of the relative size of the wave with respect to the correlation lenght of disorder. In this paper we have adressed such problem showing this dependence in an analytical form. In order to corroborate our predictions we are performing simulations with a generalized Poisson random field. Comparison of theory and simulations will be reported in a future paper.

Acknowledgments

Financial support from DGICYT Project NO. PS90-0098 (Spain) and Instituto de Cooperacion Iberoamericana is acknowledged.

References

[1] P.A. Lee and T.V. Ramakrishan, Rev. Mod. Phys. 57, 287 (1985). S. Alexander, J. Bernasconi, W.R. Schneider, R. Orbach, Rev. mod. Phys. 53,175 (1981).

[2] Chance and Matter, ed. Souletie, J. Vanimeus and R. Stora (Elsevier, Amsterdan,1987).

[3] W. Zimmermann, M. Seesselberg, F. Petruccione P.R.E. 48,2699 (1993).

[4] J.W. Haus and K.W. Kher, Phys. Rep. 150,263 (1987). J.P. Bouchaud, A. Georges, Phys. Reports 4-5,127,293 (1990)

[5] Wave Propagation and Scattering in Random Media, Vol I and II by A. Ishimaru, ed. Academic Press (1988).

[6] Introduction to the Theory of Disordered Systems. I.M. Lifshits, S.A. Gredeskul and L.A. Pastur, Ed. John Wiley (1988).

[7] K. Lindemberg, B.J. West and R. Kopelman in Noise and Chaos in Nonlinear Dynamical Systems, edited by F. Moss, L.A. Lugiato and W. Schleich. (Cambridge Univ. Press, Cambridge, england, 1990), p.142.

[8] H.S. Wio, M.A. Rodriguez, C.B. Briozzo and L. Pesquera Phys. Rev. A 44,R813 (1991).

[9] C. Aslangul, M. Barthelemy, N. Pottier, D. Saint James J. Stat. Phys. 61,403 (1990), J. Stat.Phys. 65,673 (1991).

[10] M.A. Rodriguez, E. Hernandez Garcia, L. Pesquera and M. San Miguel. Phys. Rev. B40,4212 (1989), Phys. Rev. B42, 10653 (1990).

[11] D. Bedeaux and P. Mazur. Physica A180, 295 (1992).

[12] C. Schat and H.S. Wio, Physica A180, 295 (1992).

[13] P. Hanggi in Stochastic Processes Applied to Physics, L. Pesquera and M.A. Rodriguez eds. (World Scientific, 1985, Singapore) p. 69.

CONCERNING THE NOISE STRENGTH IN PERIODICALLY DRIVEN PATTERN CONVECTION

M.O. CÁCERES

Centro Atómico Bariloche and Instituto Balseiro

Comisión Nacional de Energía Atómica, Universidad Nacional de Cuyo

8400 San Carlos de Bariloche, Río Negro, Argentina

ABSTRACT. In this paper we consider the influence of color noise on periodically driven pattern formation. The issue of the huge noise strength necessary to fit the experimental data, in periodically driven Rayleigh-Bénard convection, is studied by introducing a non-Markovian amplitude equation. Projector operator techniques are used to eliminate the fast variables, thus in first approximation an analytic expression for the order-disorder transition line is obtained. The fit with experiment is shown to be satisfactory (for a realistic thermal noise strength) provided the Langevin term has a correlation time τ_c of the order $1/\omega$.

1 Concept of a single-mode amplitude equation

Let us begin the discussion with the equations of fluid dynamics in the Oberbeck-Boussinesq approximation [1] , for a fluid bounded by infinite horizontal plates separated by a distance d and temperatures T_1 and $T_1 + \Delta T$ respectively. There will be three fields of interest: $V = V(x, y, z, t)$, the fluid velocity, $T = T_1 + \Delta T \, (z/d) + \Theta(x, y, z, t)$, the temperature, and $P = P(x, y, z, t)$, the pressure. The Oberbeck-Boussinesq equations are supplemented with Langevin noise terms to represent random molecular motion [2], and the noise terms are assumed to have a Gaussian White-Noise (W-N) distribution with strength satisfying the fluctuation-dissipation theorem[3] . By eliminating the pressure and introducing a "slow" order parameter

E. Tirapegui and W. Zeller (eds.), Instabilities and Nonequilibrium Structures V, 193–203.

$\psi(x, y, t)$, it is possible to write down a generalized free energy for ψ near the convective threshold $R \simeq R_c$, where the real field ψ is proportional to the vertical velocity V_z in such a way that the Nusselt number (to the lowest $\mathcal{O}(R - R_c)$) is given by:

$$(N - 1) = S^{-1} \int dr \ \psi(r, t)^2 \tag{1}$$

here S is the area of the cell and $dr = dx \ dy$. It is possible to show that ψ is a linear combination of V_z and Θ , thus corresponding to the slow eigenmode of the linear instability with normalization (1). Specifically, the field ψ satisfies the equation:

$$\tau_o \ \partial \psi(r, t)/\partial t = -\delta \mathcal{F}/\delta \psi + \zeta(r, t) \tag{2}$$

$$\mathcal{F} = -\int dr \ \left\{ \frac{1}{2} \tilde{a} \ \psi^2 - \frac{1}{4} \tilde{g} \ \psi^4 - \frac{1}{2} \xi_o^4 \left[(\nabla^2 + q_o^2) \ \psi \right]^2 \right\} \tag{3}$$

$$\langle\langle \zeta(r, t) \zeta(r', t') \rangle\rangle = 2 \tilde{F} \ \tau_o \ \delta(t - t') \ \delta(r - r') \tag{4}$$

which is the stochastic Swift-Hohenberg equation (S-H) [3], [4] in its functional form. This equation involves only the two-dimensional horizontal direction r, where t is the time, q_o is the critical wave-vector, and $\tilde{a}$ is the reduced Rayleigh number. The quantities τ_o, q_o , ξ_o , and $\tilde{g}$ can be evaluated for different boundary conditions[4] . The noise strength coefficient $\tilde{F}$ (see (4)) can be shown to scale with the ratio of the thermal fluctuation energy $(k_B T)$ to the characteristic dissipative energy of convection $(\rho d^3)(\nu/d)^2$, for free or rigid boundary conditions[4], [5], [6]. Therefore the Bénard system near R_c is reduced to the study of a relaxation model characterized by the free energy (3). An infinite system of rolls perpendicular to the x direction will be characterized by the order parameter:

$$\psi(r, t) = \sqrt{2} Re \left[A(r, t) \ \exp(i \ q_o \ x) \right] \tag{5}$$

where the complex amplitude $A(r, t)$ introduces small deviations from a pattern of parallel rolls. For the case of concentric rolls in a cylinder, the order parameter is written in the form:

$$\psi(r,t) = \sqrt{2} Re \left[\sqrt{L/r} \ A(r,\phi,t) \ \exp(i \ q_o \ r) \right] \tag{6}$$

In general, the evolution equation for the amplitude $A(r,t)$ is obtained by introducing the order parameter ψ into the Swift-Hohenberg equation. This amplitude equation will be in partial derivatives. An approximate reduction to an effective single-mode description of the problem has been discussed by various authors[6], [7]; this approximation is equivalent to considering an effective mode $\mathcal{A}(t)$ (space-independent). The single-mode model has the advantage that a direct solution of the stochastic equation is feasible. Therefore rather than solving the full S-H model, integrating over space to get the Nusselt number, and fitting to the experiment, it is much simpler to solve an ordinary stochastic differential equation for some effective mode $\mathcal{A}(t)$.

In several experiments, Meyer et al.[8], [6] gave convincing evidence that stochastic effects influence the initial pattern formation in their Rayleigh-Benard convection cell. When only one mode fits into the container the effective mode $\mathcal{A}(t)$ satisfies the Stochastic Amplitude Equation (SAE):

$$\tau_o \ d\mathcal{A}/dt = \left[a - g \ \mathcal{A}^2 \right] \ \mathcal{A} + \xi(t) \tag{7}$$

$$\langle \xi(t)\xi(t') \rangle = 2\tau_o \ \mathbf{F} \ \delta(t - t')$$

with renormalized coefficients [6] . The dimensionless strength of the thermal noise at convective threshold $\mathbf{F}$ turns to be roughly four orders of magnitude smaller than required to fit the experiments . This huge difference has not been explained up to now . Here we report that the factor of 10^4 can be accounted for if we relax the hypothesis of the W-N Langevin term. Such a new model describes much better the adiabatically eliminated modes which are, in some respects, represented in eq.(2) .

2 Order-Disorder transition in Rayleigh-Benard convection

When the Rayleigh number $a = (R - R_c)/R_c$ is periodically modulated in time

$$a = a(t) = \epsilon_o + \delta \, \sin(\omega t) \tag{8}$$

stochastic effects turn out to be of crucial importance in describing what has been called the Order-Disorder Transition Line (O-DTL). The roll patterns appear and disappear periodically, the change from "ordered" (consecutive patterns strongly correlated) to "disordered" (weakly correlated) behavior occurs rather suddenly with some critical value of ϵ_o (for fixed δ) and can be described by a line $\epsilon_o = \epsilon_o(\delta)$ in the phase space of the parameters. But the noise intensity, in periodically driven systems still remains an open question . Recently [9] the O-DTL has been obtained and a good fit to the experiments was achieved with a noise strength $\mathbf{F} = 10^4$ x $\mathbf{F}_{th}$.

2.1 The non-Markovian amplitude equation model

Let the non-Markovian SAE be:

$$\tau_o \, d\mathcal{A}/dt = \left[a(t) - g \, \mathcal{A}^2\right] \, \mathcal{A} + f(t) \tag{9}$$

the constants are the same as in Ref.[8], but here $f(t)$ is a Gaussian noise with $\langle f(t_1)f(t_2)\rangle = 2\mathbf{F} \, \tau_o/\tau_c \exp(-2 \mid t_1 - t_2 \mid /\tau_c)$ and $< f(t) >= 0$ (τ_c is the correlation time of the Langevin force). By scaling time in units ω and $\mathcal{A}$ in units $\sqrt{\omega/g}$ one is left with four dimensionless parameters $\alpha = \epsilon_o/\omega\tau_o$, $\beta = \delta/\omega\tau_o$, $\tilde{\epsilon} = 2g\mathbf{F}/(\omega\tau_o)^2$, $\tau = \omega\tau_c$. Note that the process (9) can be cast in the 2-dimensional Markovian form (in dimensionless units):

$$dA/dt = \left[a - g \, A^2\right] \, A + f(t) \equiv \mathcal{F}(A,t) + f(t) \tag{10}$$

$$df/dt = -1/\tau \, f + \sqrt{\tilde{\epsilon}}/\tau \, \xi(t) \tag{11}$$

here $\xi(t)$ is a zero mean value Gaussian W-N. These equations include all the dynamics, necessary to get the O-DTL. Instead of solving (10), we can go one step further and get some approximation for the leading order in the parameter τ. This is done by working out some effective SAE from (10) and (11) . The approximate equation obtained in this way may be regarded as defining a Markov process which approximates the actual non-Markovian (9).

2.2 The effective stochastic amplitude equation

From (10) and (11), the evolution equation for the 2-dimensional Markovian propagator $\mathcal{P} \equiv \mathcal{P}(A, f, t \mid A_o, f_o, t)$, is the Fokker-Planck Equation:

$$\partial_t \mathcal{P} = -\partial_A \left[\mathcal{F}(A, t) + f \right] \mathcal{P} + 1/\tau \left[\partial_f f + \tilde{\epsilon}/\tau \, \partial_f^2 \right] \mathcal{P} \tag{12}$$

If τ is small we are concerned with the limit $< f^2 >_{st} = \tilde{\epsilon}/\tau \gg 1$. Hence in order to introduce a perturbation in τ we have to rescale f by setting $f = \eta \sqrt{\tilde{\epsilon}/\tau}$ to exhibit the powers of τ. Therefore we rewrite (12) (defining the operators $\mathcal{L}_o$, $\mathcal{L}_1$, and $\mathcal{L}_2$) in the form:

$$\partial_t \mathcal{P}(A, \eta, t) = \left[\tau^{-1} \mathcal{L}_o + \tau^{-1/2} \mathcal{L}_1 + \mathcal{L}_2 \right] \mathcal{P}(A, \eta, t) \tag{13}$$

Even when $\mathcal{L}_2$ is explicitly time-dependent, projector operator techniques[10] can be used without any difficulty. From (13) we see that the largest term is the one with $\mathcal{L}_o$, therefore we select the projection operator $\mathbf{P} = \mathbf{1} - \mathbf{Q}$ defined by

$$\mathbf{P} \, \mathcal{P}(A, \eta, t) = G(\eta) \int \mathcal{P}(A, \eta', t) \, d\eta' \,, \qquad G(\eta) = (2\pi)^{-1/2} \, \exp(-\eta^2/2)$$

Then Eq. (13) can be worked out to get an effective Fokker-Planck equation for the marginal probability $P(A, t) = \mathbf{P} \, \mathcal{P}(A, \eta, t)$. The calculation is straightforward and the leading order in τ gives:

$$\partial_t P(A, t) = \left[-\partial_A \, \mathcal{F}(A, t) + \tilde{\epsilon} \, \partial_A^2 \left(1 + \tau \, \mathcal{F}'(A, t) \right) \right] P(A, t) \tag{14}$$

With these assumptions, up to the $\mathcal{O}(\tau)$, the effective SAE is:

$$dA/dt = \mathcal{F}(A, t) + \sqrt{\tilde{\epsilon}} \, \left(1 + \tau \, \mathcal{F}'(A, t) \right)^{1/2} \xi(t) \tag{15}$$

here $F'(A) \equiv (a(t) - 3A^2)$ and $a(t) \equiv \alpha + \beta \sin(t)$. Since small τ means $\tau = \omega \tau_c$ $\ll 2\pi$, the correlation time τ_c must be shorter than the period $T = 2\pi/\omega$ (the deterministic time scale). In order to solve (15) we will use the quasi-deterministic method presented in Ref. [9].

2.3   The quasi-deterministic approximation

We want to calculate the random switching times between the deterministic attractors of Eq.(15) . To this end we split the dynamics into a non-linear deterministic $A(Ao, t', to)_{\text{det}}$ and a linear stochastic one. In the linear regime the problem corresponds to a time-dependent generalized Ornstein-Uhlenbeck process [11].

$$dA/dt = a(t) \ A + \sqrt{\tilde{\epsilon}} \ (1 + \tau \ a(t))^{1/2} \ \xi(t) \tag{16}$$

This process can be solved rigorously by a Gaussian probability distribution with variance

$$\sigma(t) = e^{2I(t_o,t)} \ \tilde{\epsilon} \int_{t_o}^{t} dt' \ D(t') \ e^{-2I(t_o,t')} \tag{17}$$

where $D(t) \equiv (1 + \tau \ a(t))$ is proportional to the time-dependent diffusion coefficient and

$$I(t_o,t) = \int_{t_o}^{t} dt' \ a(t')$$

By introducing the new variable

$$Z = A/\sigma(t) \qquad for \qquad t > t_o \tag{18}$$

the probability of finding Z at time t if $A(to) = Ao$ is given by

$$P(Z, t \mid Z_o, t_o) = (2\pi)^{-1/2} \ \exp\left(-(Z - Z_o)^2/2\right) \tag{19}$$

where

$$Z_o^2 = \left[\tilde{\epsilon} \int_{t_o}^{t} dt' \ A_o^{-2} \ D(t') \ e^{-2I(t_o,t')}\right]^{-1} \tag{20}$$

The quasi-deterministic approximation consists in regarding the trajectories which have already come sufficiently close to the origin, because those trajectories are the important ones in the calculation of the transition probability [12]. Therefore we approximate the mean value in the linear-regime by the non-linear deterministic solution, which implies: $A_o^{-2} \exp[-2I(t_o, t')] \simeq A^{-2}(Ao, t', to)_{\det}$. Then it is possible to obtain a simple expression for the transition probability between the attractors:

$$P^{tr} = Prob[that \ A(t) < 0 \quad if \quad A(0) > 0] \tag{21}$$

where the Z_o, given by (20), is approximated as follows. The integral in Eq.(20) can be done analytically by the steepest-descent method if we introduce the approximation: $I(t_o, t) - 1/2 \ ln[D(t)] \simeq I(t_o, t) - \frac{\tau}{2} a(t)$. Therefore the matching time between the linear and non-linear dynamics is characterized by the dominant contribution of the integrand in Eq.(20). This characteristic time $\hat{t}$ is given by the solution of

$$a(\hat{t}) = \frac{\tau}{2} \dot{a}(\hat{t}) \tag{22}$$

with $\dot{a}(\hat{t}) - \tau/2 \ \ddot{a}(\hat{t}) > 0$ (we also ensure that $D(t) > 0$). The next approximation in order to get an analytic expression for Z_o is to consider strong convergence of the trajectories within one period towards their attractors. Then we can substitute the deterministic solution by its asymptotic attractor:

$$A^2(A_o, t, t_o)_{\det} \simeq (e^{4\pi\alpha} - 1)/h(t - T, t) \tag{23}$$

where

$$h(t_o, t) = 2g \int_{t_o}^{t} dt' \ e^{2I(t_o, t')} \tag{24}$$

Approximating $h(t-T, t)$ by another application of the method of steepest descent we finally arrive at the formulae

$$Z_o^2 = (1 - e^{-4\pi\alpha}) \exp \left(4 \left[\alpha \ ar \cos(\alpha/\beta) - [1 + \cos(\Delta)]/2 \ (\beta^2 - \alpha^2)^{1/2} \right] \right) \times$$

$$\times \exp \left(2\alpha [\Delta - \sin(\Delta)] - \tau \left\{ \alpha [1 - \cos(\Delta)] + (\beta^2 - \alpha^2)^{1/2} \ \sin(\Delta) \right\} \right) \times \tag{25}$$

$$\times \left[(\beta^2 - \alpha^2)/\cos(\Delta) \right]^{1/2} \ /(2\pi \ \hat{t})$$

where $tg(\Delta) = \tau/2$. If $\tau=0$ we recover the result of Ref.[9] . The small τ approximation is bounded by $\tau\alpha/2 \leq (\beta^2 - \alpha^2)^{1/2}$ so that transition processes are completely dominated by trajectories in the neighborhood of $\tilde{t}$. Physically speaking the short correlation τ $(= \omega\tau_c)$ will introduce a shift in the matching time $\tilde{t}$ which ultimately is the one responsible of the switching time between attractors (from (22) we see that: $\alpha + \beta \sin(\tilde{t} - \Delta) = 0$, with $\cos(\tilde{t} - \Delta) > 0$).

2.4   The Order-Disorder transition line

A remarkable result about Eqs.(21) and (25) is the fact that if we choose the thermal noise strength[8]$_{a,b}$, [13] $\mathbf{F}_{th}$ (i.e.: $\tilde{\epsilon} = 2g\mathbf{F}_{th} /(\omega\tau_o)^2 = 10^{-4}$ x $\epsilon_{exp.} = 3.39$ x10^{-8}) it is possible to fit the experimental data by using a noise correlation $\tau \simeq 2.5$. The O-DTL is obtained by imposing the condition [13], [9], [14] :

$$P^{tr} = 1/4 \tag{26}$$

Therefore by using (21) and (25) in (26) (at the leading $\mathcal{O}(\tau)$) we show in Fig.1 the O-DTL for several τ (=0, 2.5, 4). We see that the agreement is quite good despite all the approximations made (which is reflected in a not so good fitting for $\alpha < 1$) . The analysis for larger τ and for every α, and β can also be made from (10) and (11) but other techniques have to be used, for example the Kolmogorov Eigenvalue Theory , work along this line is in progress . Even when τ is smaller than 2π (the dimensionless deterministic time scale) the amazing result is the fact that τ_c is $\mathcal{O}(1/\omega)$, this means that the time scale of the adiabatically eliminated modes, represented in the Langevin term in (9), cannot be treated as very fast variables, this is why our non-Markovian SAE is important. This result is not entirely unexpected because the elimination of variables (modes) in a non-autonomous system (driven with a frequency ω) will introduce a stroboscopic correlation time of $\mathcal{O}(1/\omega)$. It is clear from the agreement between experiment and theory that the O-DTL could be accounted for by a SAE model with thermal noise strength, provided the noise has a nonzero time correlation

τ_c. We are convinced that the introduction of a non-Markovian SAE illustrated here by a simple example will also be very important for the treatment of complicated periodically-driven systems with more excitable modes.

The author wishes to thank Prof. L. Kramer and Dr. A. Becker for useful discussions, and to Dr. V. Grunfeld for the English critical reading. of the manuscript. CONICET grant PIA Nro: 0314/92 is acknowledged.

References

[1] Chandrasekhar S. (1961) 'Hydrodynamic and Hydromagnetic Stability', Clarendon, Oxford.

[2] Landau L.D. and Lifshitz E.M. (1959) 'Fluid Mechanics' Addison-Wesley, Reading, MA.

[3] Swift J.B. and Hohenberg P.C. (1977) 'Hydrodynamic fluctuations at the convective instability', Phys. Rev. A, 15, 319.

[4] Hohenberg P.C. and Swift J.B. (1992) 'Effects of external noise at the onset of Raleigh-Bwnard convection', Phys. Rev. A, 46, 4773.

[5] Graham R.(1974) 'Hydrodynamic fluctuation near the convective instability', Phys. Rev. A; 10,1762.

[6] Ahlers G., Cross M.C., Hohenberg P.C., and Safran S.(1981) 'The amplitude equation near the convective threshold: application to time-dependent experiments', J. Fluid Mech. 110,297.

[7] van Beijeren H. and Cohen E.G.D. (1988) 'The effects of thermal noise in a Rayleigh-Bwnard cell near its first convective instability' , J. Stat. Phys. 53, 77.

[8] (a) Meyer C.W., Cannell D.S. and Ahlers G. (1992) 'Hexagonal and Roll flow patterns in temporally modulated Rayleigh-Bwnard convection', Phys. Rev. A, 45,

8583; (b) Meyer C.W., Ahlers G.and Cannell D.S. (1991), 'Stochastic influences on pattern formation in Rayleigh-Bwnard convection: Ramping experiments', Phys. Rev. A 44, 2514; (c) Meyer C.W., Ahlers G.and Cannell D.S. (1987) 'Initial stages of pattern formation in Rayleigh-Bwnard convection', Phys. Rev. Lett. 59, 1577.

[9] Stiller O., A. Becker, and L. Kramer (1992) 'Noise-induced transition between attractors in time-periodically driven systems' Phys. Rev. Lett. 68, 3670.

[10] (a) van Kampen N.G. (1985) 'Elimination of fast variables', Phys. Rep. 124, 69-160, (1985); (b) idem (1989) 'Langevin-Like equation with colored noise', J. Stat. Phys. 54, 1289.

[11] Gardiner C.W. (1983) 'Handbook of stochastic methods', Springer-Verlag, Berlin.

[12] de Pasquale F., Rack Z., San Miguel M., and Tartaglia P. (1984) 'Fluctuations and limit of metastability in a periodically driven unstable system', Phys. Rev. B, 30, 5228.

[13] Swift J.B. and Hohenberg P.C. (1988) 'Comment on Initial stages of pattern formation in Rayleigh-Bwnard convection', Phys. Rev. Lett. 60,75.

[14] Becker A., Caceres M.O. and Kramer L. (1992) 'Correlation function in stochastic periodically driven instabilities', Phys. Rev. A, 46, R4463.

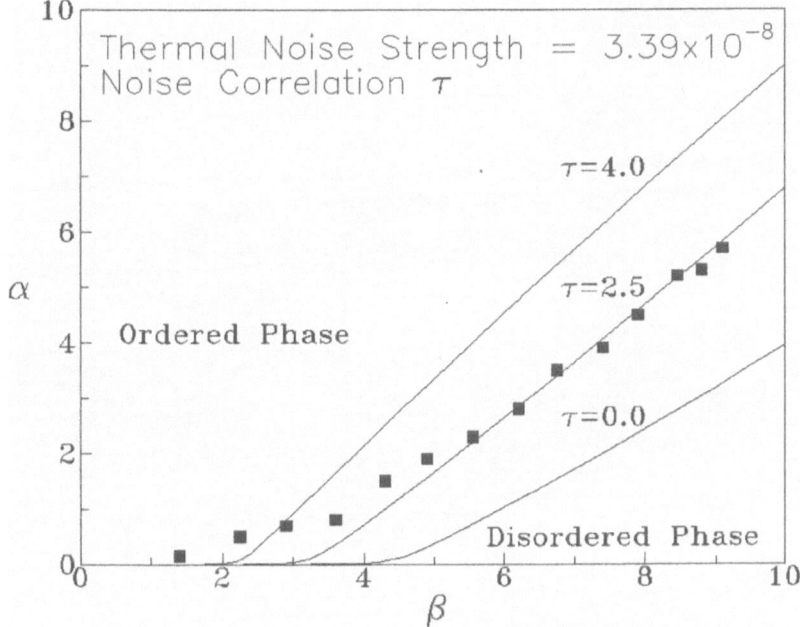

Figure 1. The O-DTL in the α, β plane calculated from Eq.(25) and for several values of τ. The thermal noise strength is taken from Refs. [8]. [13]. $2g\,\mathbf{F}_{th}/(\omega\,\tau_c) = 3.39 \times 10^{-8}$. The squares are data of Ref. [8c]

QUASI-STATIONARY DISTRIBUTION AND
GIBBS MEASURE OF EXPANDING SYSTEMS

Pierre Collet
C.N.R.S., Physique Théorique
Ecole Polytechnique
91128 Palaiseau Cedex, France

Servet Martínez
Departamento de Ingeniería Matemática
Facultad de Ciencias Físicas y Matemáticas
Universidad de Chile
Casilla 170-3 Correo 3, Santiago, Chile

Bernard Schmitt
Département de Mathématiques
Faculté de Sciences Mirande
Université de Bourgogne
BP-138, 21004 Dijon Cedex, France

ABSTRACT. Let T be an expanding $C^{1+\gamma}$ transformation defined on $A = \bigcup_{i=1}^{p} A_i$, $(A_i)_{i=1}^{p}$ being a finite collection of connected open bounded subsets of $\mathbb{R}^n$, such that TA contains strictly A and T is Markovian. We prove the existence of a quasi-stationary distrition for T. We show that the T-invariant probability on the limit Cantor set is Gibbsian with potential $\mathrm{Log}|DT|$. Using the Hilbert projective metric we prove that both distributions are weak limits of conditional laws of probabilities, the speed of convergence being exponential. These results develop a previous work by G. Pianigiani and J.A. Yorke.

0. Introduction

In a pionering work, G. Panigiani and J.A. Yorke [P-Y] investigated the statistical behavior of the orbits of some dynamical systems having a property of metastability. They gave as examples of motivation the well known example of the Lorenz differential equations [L] near the transition between chaotic and non chaotic behavior [K-Y], [R], [Y-Y], and the example of an energy conservating billiard table with smooth obstacles giving instable trajectories, in which a small hole was cut out. In both cases one observes at first a chaotic behavior of the trajectories but after a random time, they are either attracted by a stable fixed point (first example) or fall in the hole (second example).

E. Tirapegui and W. Zeller (eds.), Instabilities and Nonequilibrium Structures V, 205–219.

Therefore it is of interest to study the statistics of the "survivor": let $P_t(E)$ be the conditional probability of the trajectories to be in a measurable set E at time t knowing that they avoided a neighbourhood of a stable fixed point (first example) or the hole (second example) for at least time t. Does there exist a limit for $P_t(E)$ when t goes to infinity ?

G. Panigiani and J.A. Yorke gave a positive answer to this question in the following case: T is a C^2-expansive transformation defined on a bounded open set $A \subset \mathbf{R}^n$ with $TA \supset \neq A$ and T is topologically transitive. More precisely they proved the existence of a "conditionally invariant" probability μ on A, i.e. a measure satisfying $\mu(T^{-1}E/T^{-1}A) = \mu(E)$ for every Borel set $E \subset A$; this measure has moreover regular density relatively to the Lebesgue measure λ on A and is the weak limit of the sequence of probabilities $\lambda(T^{-n} \bullet /T^{-n}A)$ when n goes to infinity.

This notion of conditionally invariant measure was introduced at first for countable state Markov chains $(X_n)_{n \in \mathbf{N}}$ with an absorbing state S in Vere-Jones [V-J] and recently developed by P. Ferrari, H. Kesten, S. Martinez and P. Picco [F-K-M-P] in relation with a geometric absorption property. For finite Markov chains Darroch and Seneta [D-S] have shown that there exists a unique conditional invariant measure which also turns out to be the weak limit of the sequence of the probability distributions $\mathbf{P}[X_n \in \bullet /X_n \notin S]$. When it exists, this last limit is called "limiting conditional distribution" and was first studied by Yaglom [Y] for the branching processes.

As usual, the statistical results obtained for Markov chains is a strong motivation for discovering similar properties for some hyperbolic dynamical systems. In this paper our aim consists in revisiting the Panigiani-Yorke results, which is the so called Markov case, that is when the transformation T has a finite number of inverse branches which are well defined on the range of T. We prove the existence of a conditionally invariant absolutely continuous probability measure; we show that it is the limiting conditional distribution and we estimate the speed of convergence. Moreover we study the statistics of the orbits staying in A, prove the existence of a singular invariant probability measure concentrated on $\bigcap_{n \geq 0} T^{-n}A$ and establish its Gibbsian property.

The main technical ingredient is the property of contraction of the transfer operator relatively to a Hilbert projective metric.

Similar results are proved in [C-M-S], but the mathematical approach is different; moreover we prove in the present paper that the speed of convergence toward the conditionally invariant measure is exponential.

1. Notations and Main Result

Let $A = \bigcup_{i=1}^{\ell} A_i$ be an open subset of $\mathbf{R}^n$, where the subsets A_i are open, pairwise disjoints and connected.

We consider a transformation $T : \overline{A} \to \mathbf{R}^n$ which is assumed to verify the following conditions:

(C_1) $A \subset TA$ and T can be extended to $\overline{A}$, being $C^{1+\gamma}$ on $A \cap T^{-1}A$ $(0 < \gamma \leq 1)$
(C_2) $A \cap T(\partial A) = \emptyset$
(C_3) There exists $\beta > 1$ such that $DT(x)$ is β-expansive for all $x \in A \cap T^{-1}A$,   i.e.:

$$\inf \left\{ \|DT_x(v)\| \; ; \; \|v\| = 1 \right\} \geq \beta, \quad \forall x \in A \cap T^{-1}A.$$

Let us denote by λ the Lebesgue measure on $\mathbf{R}^n$; by $C_1)$ the transformation T is non singular relatively to λ, hence for each $h \in L^1_\lambda(A)$ we can define the Radon-Nikodym derivative $\dfrac{d(h\lambda \circ T^{-1})}{d\lambda} \in L^1_\lambda(TA)$. ($d(h\lambda \circ T^{-1})$ is the measure defined on TA by:

$$\int_B d(h\lambda \circ T^{-1}) = \int_{T^{-1}B} h \, d\lambda$$

for all borelian sets B in TA). We define a positive operator P $P : L^1_\lambda(A) \longrightarrow L^1_\lambda(TA)$ by

$$Ph = \frac{d\,(h\lambda \circ T^{-1})}{d\lambda},$$

which is called the transfer operator. It is clear that for any $h_1 \in L^1_\lambda(A)$ and $h_2 \in L^1_\lambda(TA)$ we have:

$$\int_A h_1 \, h_2 \circ T \, d\lambda = \int_{TA} Ph_1 \, h_2 \, d\lambda.$$

Identifying the functions of $L^1_\lambda(A)$ with the functions of $L^1_\lambda(TA)$ vanishing outside A obtain:

$$\int_A h_1 \, h_2 \circ T \, d\lambda = \int_A Ph_1 \, h_2 \, d\lambda, \quad \forall h_1, \, h_2 \in L^1_\lambda(A).$$

If we restrict Ph_1 to A we can then iterate P and obtain:

$$\int_A h_1 \, h_2 \circ T^n \, d\lambda = \int_1 P^n h_1 . h_2 \, d\lambda, \quad \forall h_1, \, h_2 \in L^1_\lambda(A), \quad \forall n \in \mathbf{N}.$$

By $C_3)$ the transformation T admits local inverse branches on A. More precisely let us consider the restriction of T to $A \cap T^{-1}A$. Then this restriction maps $A \cap T^{-1}A$ onto A; if x belongs to A, it has at least one preimage y in $A \cap T^{-1}A$. Since $DT_y \neq 0$, there exists a neighbourhood U of x in A and a neighbourhood V of y in $A \cap T^{-1}A$ such that $T : V \longrightarrow U$ is a diffeomorphism; a local inverse branch of T on U is the diffeomorphism $\varphi : U \to V$ such that $T \circ \varphi = Id_U$ and $\varphi \circ T = Id_V$.

We would like to show that under the condition $C_2)$, the number of inverse branches is constant on A_i.

We notice that $C_2)$ implies that: $TA_j \cap A_i \neq \emptyset \Longrightarrow TA_j \supset A_i$ (Markov condition). Therefore T maps $\left(\bigcup_{j \; ; \; TA_j \supset A_i} A_j \right) \cap T^{-1}A_i$ onto A_i, so that for analysing the number of inverse branches of T in any $x \in A_i$, it is sufficient to consider the restriction of T to $A_j \cap T^{-1}A_i$ (for $A_j \cap T^{-1}A_i \neq \emptyset$) which maps $A_j \cap T^{-1}A_i$ onto A_i.

Let i, j be such that $A_j \cap T^{-1}A_i \neq \emptyset$ and $x \in A_i$; let us suppose that the number of inverse branches of T at x in A_j is infinite; A being bounded (condition $C_1)$) is relatively compact so that we can build a sequence $(y_n)_{n \in \mathbb{N}}$, $y_n \in T^{-1}A_i \cap A_j$, $y_n \longrightarrow y$ and $Ty_n = x$; by continuity of T, $Ty = x$ and by $C_2)$ $y \in A_j \cap T^{-1}A_i$. This is impossible because of the existence of a local inverse branch defined on a neighbourhood of x onto a neighbourhood of y.

Let us define: $T^{-1}\{x\} \cap A_j = \{y_1, y_2, \cdots, y_k\}$, $y_t \in A_j \cap T^{-1}A_i$, $t = 1, 2, \cdots, k$. There exists neighbourhoods $U \subset A_i$ of x, $V_t \subset A_j \cap T^{-1}A_i$ of y_t and inverse branches φ_t of T, $\varphi_t : U \longrightarrow V_t$, and assume that the number of inverse branches is not constant on U. Then for any neighbourhood W of x, $W \subset U$, there exists $x_W \in W$, $\mathbf{Z}_W \in T^{-1}\{x_W\}$ such that $\mathbf{Z}_W \neq \varphi_t(x_W)$ $(t = 1, 2, \cdots, k)$ so that x_W and $\mathbf{Z}_W$ create a new inverse branch. By recurrence we can build a sequence $(x_n)_{n \in \mathbb{N}}$, $x_n \in A$, $x_n \to x$ such that $T^{-1}\{x_n\} \cap (A_j \cap T^{-1}A_i)$ contains z_n and the inverse branch between x_n and z_n is a new branch; as above $(z_n)_{n \in \mathbb{N}}$ has an accumulation point z in $A_j \cap T^{-1}A_i$ by $C_2)$ such that $Tz = x$; we conclude as before that this is a contradiction.

The number of inverse branches is then locally constant on A_i, and then constant on A_i by connectedness.

Let x be in A_i; we denote by $\varphi_{j_i} (j_i = 1, 2, \cdots, p_i)$ the inverse branches of T defined on A_i; an easy computation gives:

$$\forall x \in A_i, \quad \forall h \in L^1_\lambda(A_i) \; ; \; Ph(x) = \sum_{j_i=1}^{P_i} | \det D\varphi_{j_i}(x) | \cdot h \circ \varphi_{j_i}(x) \qquad (1)$$

which implies

$$\forall x \in A, \quad \forall h \in L^1_\lambda(A), \quad Ph(x) = \sum_{1}^{\ell} \left(\sum_{j_i=1}^{P_i} | \det D\varphi_{j_i}(x) | h \circ \varphi_{j_i}(x) \right) 1_{A_i}(x).$$

We remark that P acts on $C(A)$, the space of the continuous functions defined of A.

Definition 1.1. A probability measure μ on A is conditionally invariant if $\mu \circ T^{-1} = \alpha\mu$ on A for some $\alpha > 0$.

Remarks.
1. If μ is conditionally invariant necessarily $\alpha = \mu[T^{-1}(A)]$. Therefore μ is conditionally invariant if $\mu(T^{-1}E/T^{-1}A) = \mu(E)$ for any Borel set E in A (where $\mu(\cdot/T^{-1}A)$ denotes the probability induced by $\mu \circ T^{-1}$ on $T^{-1}A$), obviously $0 < \alpha \leq 1$.

2. If μ is conditionally invariant and absolutely continuous with a density $h \in L^1_\lambda(A)$, then for every Borel set E contained in A we have:

$$\mu(T^{-1}E) = \int_A 1_E \circ T \, h d\lambda = \int_A 1_E \, Ph \, d\lambda = \alpha \int_A 1_E \, h d\lambda.$$

It follows that $Ph = \alpha h$; the converse is trivial. Therefore in order to prove the existence of a conditionally invariant absolutely continuous probability for T, it is necessary and sufficient to prove the existence of an eigenfunction in $(L^1_\lambda(A))^+$ associated to an eigenvalue $0 < \alpha \le 1$.

We will prove more; we will find an eigenfunction in the space of Hölder-continuous functions defined on $\overline{A}$. On this space the iterates of P will converge exponentially fast to a rank one operator; actually this rank one operator will be the projection π into an eigenspace of dimension one.

In order to prove this result we will assume that T is transitive on components, that is:

$$\forall (i,j) \in \{1, 2, \cdots, \ell\}^2 , \ \exists \, n(i,j) : \ T^{n(i,j)} A_j \supset A_i.$$

It is a classical result from the theory of Markov chains that if T is transitive on components, it can be decomposed in irreducible components on which an iterate of T is irreducible. We say that T is irreducible on components if:

$$\forall (i,j) \in \{1, 2, \cdots, \ell\}^2 , \ \exists n : \ T^n A_j \supset A_i.$$

Then the existence of a conditionally invariant measure for the irreducible components of T implies the existence of a conditionally invariant measure for T. [cf [B] for example for more details].

Moreover if T is irreducible on components and expanding, then an iterate of T is onto, that is:

$$\forall i \in \{1, 2, \cdots, \ell\} , \ \exists N : \ T^N A_i \supset \bigcup_{j=1}^{\ell} A_j.$$

Following a classical method used in the area of the study of the thermodynamical properties of differentiable dynamical systems, we can deduce the convergence of the iterates of P from the convergence of the iterates of P^N ([B]). Therefore it is not a restriction to assume that $N = 1$ and we will make this hypothesis for the simplification of the proofs:

C_4) $\forall i \in \{1, 2, \cdots, \ell\} : \ TA_i \supset \bigcup_{j=1}^{\ell} A_j.$

We will consider the following cones of positive γ-Hölder functions defined on $\overline{A}$ and depending on a real number $\eta > 0$ to be fixed later on

$$\Lambda_L = \left\{ f \in C^+(\overline{A}) ; \ \forall i \in \{1,2,\cdots,\ell\} ; \ \forall (x,y) \in \overline{A}_i^2, \ \|x-y\| \le \eta : f(x) \le e^{L\|x-y\|^\gamma} f(y) \right\}$$

(where $\| \ \|$ is the usual norm on $\mathbf{R}^n$).

We will show in the next paragraph that $P\Lambda_L \subset \Lambda_L$ for an adequate L and that P contracts strictly the cone Λ_L for an adapted metric.

Finally we denote by P^* the adjoint operator of P as acting on $C^+(\overline{A})$; the operator P^* acts on the set $\mathcal{M}(\overline{A})$ of the Borel measures on $\overline{A}$ as follows:

$$\forall f \in C(\overline{A}) , \ \int_{\overline{A}} f \cdot P^* dm = \int_{\overline{A}} Pf \cdot dm.$$

We have the following theorems.

Theorem 1. Assume that $T : A \longrightarrow \mathbf{R}^n$ verifies the conditions $C_1)$, $C_2)$, C_3, $C_4)$, then

i) $\exists h_0 \in C^+(\overline{A})$, $\exists m \in \mathcal{M}^+(\overline{A})$, $\exists \ 1 \ge \alpha > 0$ such that:

$$Ph_0 = \alpha h_0 ; \ P^* m = \alpha m ; \ \int_{\overline{A}} h_0 dm = 1.$$

ii) h_0 is γ-Hölder continuous on $\overline{A}$ and:

$$\forall h \in C(\overline{A}) ; \ \left\| \frac{P^n h}{\alpha^n} - h_0 \cdot \int h dm \right\|_u \xrightarrow[n \to \infty]{} 0$$

$\left(\| \ \|_u \text{ is the uniform norm on } C^+(\overline{A}) \right)$.

The convergence is exponential on $\displaystyle\bigcup_{L>L_0} \Lambda_L$ for some $L_0 > 0$.

iii) The pair (α, m) is uniquely determined by the condtions: $\alpha > 0$, $m \in \mathcal{M}^+(\overline{A})$, $P^* m = \alpha m$.

iv) h_0 is the unique γ-Hölder function defined on $\overline{A}$ verifying: $h_0 > 0$; $m(h_0) = 1$ and $Ph_0 = \alpha h_0$.

Theorem 2. Under the conditions of the theorem 1

i) $\mu = \dfrac{h_0 d\lambda}{\int_{\overline{A}} h_0 d\lambda}$ is the unique conditionally invariant probability with a γ-Hölder density function.

ii) μ is the weak-limit of the probability measures $\lambda(T^{-n} \bullet /T^{-n}A)$.

iii) m is the weak limit of the probability measures $\mu(\bullet /T^{-n}A)$.

iv) The measure $\nu = h_0 m$ is T-invariant and ergodic with support $\bigcap\limits_{n \geq 0} T^{-n}\overline{A}$. Furthermore, it is a Gibbs measure associated to the potential $\mathrm{Log}\,|\det DT\,|$.

The definition of a Gibbs measure associated to a given potential will be given at the end of section 2.

The idea of the proof of theorem 1 follows a general formulation given by P. Walters [W]; in our case (due to the expansive property) we obtain the exponential speed of convergence ii) for the γ-Hölder functions.

2. Proofs of the Theorems

a) An estimation of Ph, h ∈ Λ_L.

We will make first the following elementary remarks.

1) Let h be a function in Λ_L which is not identically zero and such that there exists $i \in \{1, 2, \cdots, p\}$ and $x \in A_i$ verifying $h(x) > 0$. Then h is strictly positive on A_i.

2) By (1) and mixing on components we deduce that if $h \in \Lambda_L$ and $h \not\equiv 0$, then $Ph > 0$ on A.

3) By the hypothesis C_1) and C_3), there exists $\varepsilon > 0$ such that:

$$\forall i, j \in \{1, 2, \cdots, p\}\ \forall x, y \in A_j \cap T^{-1}A_i,\ \|x - y\| < \varepsilon :\ \|Tx - Ty\| \geq \beta \|x - y\|.$$

In the definition of Λ_L we will fix $\eta = \varepsilon$.
Let h be in Λ_L, $h \not\equiv 0$, and $x, y \in A_i$, $\|x - y\| < \eta$ for some i, then

$$Ph(x) = \sum_{j_i=1}^{pi} |\det D\varphi_{j_i}(x)|\cdot h \circ \varphi_{j_i}(x).$$

By the remark 3, $\|\varphi_{j_i}(x) - \varphi_{j_i}(y)\| \leq \frac{1}{\beta} \cdot \eta < \eta$ and $\varphi_{j_i}(x), \varphi_{j_i}(y) \in A_j$. Since $h \in \Lambda_L$ we have:

$$h(\varphi_{j_i}(x)) \leq \exp\left(L\|\varphi_{j_i}(x) - \varphi_{j_i}(y)\|^\gamma\right)\cdot h(\varphi_{j_i}(y)). \tag{2}$$

On the other hand since T is $C^{1+\gamma}$ and DT is bounded on $\overline{A}$, we get:

$$\exists\, a > 0,\ \inf_{x \in \overline{A}} |\det D\,\varphi_{j_i}(x)| \geq a\ ,\quad \forall i = 1, 2, \cdots, p,\ \forall j_i = 1, 2, \cdots, p_i. \tag{3}$$

$\exists \, 0 < b < \infty \; ; \; \forall x, y \in \overline{A} \; | \; | \det D \, \varphi_{j_i}(x) \, | - | \; \det D \, \varphi_{j_i}(y) \, | \, | \leq b \|\varphi_{j_i}(x) - \varphi_{j_i}(y)\|^{\gamma}.$

$$(4)$$

From (3) and (4) we obtain that for all x and y in some A_i with $\|x - y\| < \eta$

$$\frac{| \det D \, \varphi_{j_i}(x) |}{| \det D \, \varphi_{j_i}(y) |} \leq \exp \left[\frac{b}{a} \, \|\varphi_{j_i}(x) - \varphi_{j_i}(y)\|^{\gamma} \right]. \qquad (5)$$

Then (2) and (5) imply:

$$\forall x, y \in A_i : \;\; \|x - y\| < \eta \, , \;\; \frac{Ph(x)}{Ph(y)} \leq \exp \left[(\frac{a}{b} + L) \cdot \sup_{j_i} \|\varphi_{j_i}(x) - \varphi_{j_i}(y)\|^{\gamma} \right]. \quad (6)$$

The inequality (6) makes sense because of the remark 2). Moreover by the choice of η, we have:

$$\|\varphi_{j_i}(x) - \varphi_{j_i}(y)\| \leq \frac{1}{\beta} \, \|x - y\|.$$

Therefore we obtain the following property:

$$\forall i = 1, 2, \cdots, p, \;\; \forall x, y \in A_i, \;\; \|x - y\| < \eta : \qquad (7)$$

$$\frac{Ph(x)}{Ph(y)} \leq \exp \left[(\frac{1}{\beta})^{\gamma} (\frac{b}{a} + L) \, \|x - y\|^{\gamma} \right].$$

Thus we have proved:

$$P\Lambda_L \subset \Lambda_{(\frac{1}{\beta})^{\gamma}(\frac{b}{a}+L)}.$$

From this relation we get

Lemma 2.1 *For any* $L > \dfrac{b}{a(\beta^{\gamma} - 1)} = L_0, \; P\Lambda_L \subset \Lambda_L.$

b) Hilbert's projective metric; the property of contraction.

For any $L > 0$ let us consider the projective space $\dot{\Lambda}_L$ of Λ_L. The space $\dot{\Lambda}_L$ can be represented as follows:

$$\dot{\Lambda}_L = \left\{ h \in \Lambda_L \; ; \; \int_{\underline{A}} h(x) \, d\lambda(x) = 1 \right\}.$$

The action of P on $\dot{\Lambda}_L$ will be also denoted by P and it is $P\dot{h} = \dfrac{Ph}{\int Ph d\lambda}$.

If $L > L_0$ lemma 2.1 asserts that $P\dot{\Lambda}_L \subset \dot{\Lambda}_L$. In a pioneering paper [Bi], G. Birkhoff studied the action of a positive operator on an invariant cone in a Banach lattice and introduced Hilbert's projective metric. We give its definition in the context of our problem.

Let h_1 and h_2 be two non zero elements of the closed, convex, bounded cone Λ_L. Then there exists a largest real $u_1 > 0$ and a smallest one $v_1 > 0$ such that:

$$\begin{cases} u_1\, h_1 \le h_2 \le v_1\, h_1 \\ h_2 - u_1\, h_1 \in \Lambda_L \\ v_1\, h_1 - h_2 \in \Lambda_L \end{cases} \tag{8}$$

We define: $\theta_{\Lambda_L}(h_1,h_2) = \operatorname{Log} \frac{v_1}{u_1}$; it is clear that $\theta_{\Lambda_L}(h_1,h_2)$ depends only on the projective classes of h_1 and h_2; moreover it belongs to $\overline{\mathbf{R}}^+$.

The reader will find the proof of the following proposition in [Bi], [Fe,Sc] or [Ko].

Proposition 2.1.

i) θ_{Λ_L} is a pseudo metric on $\dot{\Lambda}_L$.

ii) The set $\Pi_{\Lambda_L}(1) = \left\{ h \in \Lambda_L;\ \theta_{\Lambda_L}(1,h) < +\infty \right\}$ is a complete metric space for the metric θ_{Λ_L} (1 is the function identically equal to 1).

iii) If $\|\ \|_u$ denotes the uniform norm on $\overline{A}$, then for all pair $h_1,\ h_2$ in $\dot{\Lambda}_L$ we have:

$$\|h_1 - h_2\|_u \le \left(e^{\theta_{\Lambda_L}(h_1,h_2)} - 1 \right)\|h_i\|,\ i = 1,2.$$

For $L > L_0$, it is possible to define:

$$\operatorname{diam}_{\Lambda_L}(P\Lambda_L) = \operatorname*{Sup}_{h_1,h_2 \in \Lambda_L} \theta_{\Lambda_L}(Ph_1, Ph_2).$$

Remark. If $\operatorname{diam}_{\Lambda_L}(P\Lambda_L) < \infty$, then by using (8) we obtain the existence of a positive real u_1' associated to h_1, h_2 such that:

$$u_1' \cdot Ph_1 \le Ph_2 \le u_1' \cdot \exp\left[\operatorname{diam}_{\Lambda_L}(P\Lambda_L)\right] \cdot Ph_1.$$

In his paper, G. Birkhoff proved the crucial inequality of contraction:

$$\forall h_1, h_2 \in \Lambda_L,\quad \theta_{\Lambda_L}(Ph_1, Ph_2) \le th\left[\frac{1}{4}(\operatorname{diam}_{\Lambda_L}(P\Lambda_L))\right] \cdot \theta_{\Lambda_L}(h_1,h_2). \tag{9}$$

We are going to prove that for $L > L_0$, $\operatorname{diam}_{\Lambda_L}(P\Lambda_L) < +\infty$. This will imply, by using (9), that P is a strict contraction on Λ_L relatively to the metric θ_{Λ_L}. To prove the finiteness of $\operatorname{diam}_{\Lambda_L}(P\Lambda_L)$ we have to give the explicit expression of the projective metric; an easy computation using (8) gives

$$\begin{cases} \forall h_1, h_2 \in \Lambda_L,\ \theta_{\Lambda_L}(h_1,h_2) = \operatorname{Log}\left(M(h_1,h_2) \cdot M(h_2,h_1) \right),\ \text{where:} \\ M(h_1,h_2) = \operatorname*{Sup}_{i=1,2,\cdots,\ell}\ \operatorname*{Sup}_{\substack{x,y \in A_i \\ \|x-y\|<\eta}} \dfrac{e^{L\|x-y\|^\gamma}\, h_1(x) - h_2(y)}{e^{L\|x-y\|^\gamma}\, h_2(x) - h_2(y)}. \end{cases} \tag{10}$$

Let h_1, h_2 be in Λ_L and let us estimate $\frac{e^{L\|x-y\|^\gamma}}{e^{L\|x-y\|^\gamma}} \frac{Ph_1(x)-Ph_1(y)}{Ph_2(x)-Ph_2(y)}$, for $x, y \in A_i$, $\|x-y\| < \eta$. Since $L > L_0$, $P\Lambda_L \subset \Lambda_L$ and:

$$\frac{e^{L\|x-y\|^\gamma}}{e^{L\|x-y\|^\gamma}} \frac{Ph_1(x) - Ph_1(y)}{Ph_2(x) - Ph_2(y)} \le \frac{\left(e^{(L+L')\|x-y\|^\gamma} - 1\right) Ph_1(y)}{\left(e^{L\|x-y\|^\gamma} - e^{L'\|x-y\|^\gamma}\right) Ph_2(x)},$$

where $L' = (\frac{1}{\beta})^\gamma (\frac{b}{a} + L) < L$.

The quantity $\left(\frac{e^{(L+L')\|x-y\|^\gamma} - 1}{e^{L\|x-y\|^\gamma} - e^{L'\|x-y\|^\gamma}}\right)$ is clearly bounded on each A_i, $i = 1, 2, \cdots, p$, by a constant which we will denote by K_1.

On the other hand the local bound $\frac{Ph_1(x)}{Ph_1(y)} \le e^{L\|x-y\|^\gamma}$ for $x, y \in A_i$, $\|x-y\| < \eta$, gives us (by precompactness of each A_i and continuity of Ph_1) that there exists a constant

K_2 such that: $\dfrac{\underset{\overline{A_i}}{\text{Sup }} Ph_1}{\underset{\overline{A_i}}{\text{Inf }} Ph_1} \le K_2$, the constant K_2 depending only on L and of the minimal

cover of $\overline{A_i}$ by balls of radius $\eta \cdot (i = 1, 2, \cdots, p)$.

Therefore $M(h_1, h_2) \cdot M(h_2, h_1) \le K_1^2 K_2^2$ and $\text{diam}_{\theta_{\Lambda_L}} (P\Lambda_L) < 2\text{Log} (K_1, K_2)$. Actually we have proved more:

Lemma 2.2 *If $L > L_0$ then for all $k \ge 1$, $\ell \ge 1$, $\forall h \in \Lambda_L$, then $\theta_{\Lambda_L} (P^k h, P^k \ell)$ is uniformly bounded by $2 \text{Log } K_1 K_2$.*

Let us denote by: $\gamma_L = th (\frac{1}{4} 2\text{Log } K_1 K_2) < 1$. Then we have from Birkhoff's estimate

(9)

$$\forall L > L_0, \quad \forall h_1, h_2 \in \Lambda_L,$$

(11)

$$\theta_{\Lambda_L} (Ph_1, Ph_2) < \gamma_L \, \theta_{\Lambda_L}(h_1, h_2).$$

c) Proof of the theorems.

Proposition 2.2. For $L > L_0$ there exists a unique fixed point $\dot{h}_0$ in $\dot{\Lambda}_L$ such that for all $\dot{h} \in \dot{\Lambda}_L$: $\theta_{\Lambda_L} (P^n \dot{h}, \dot{h}_0) \le c^{te} \cdot \gamma_L^n$.

Since $L > L_0$, then for any $\dot{h} \in \dot{\Lambda}_L$ we deduce from (11) and the lemma 2.2 that the sequence $(P^n h)_{n \ge 1}$ is a Cauchy sequence in the complete metric space $\left(\Pi_{\Lambda_L}(1), \theta_{\Lambda_L}\right)$ (proposition 2.1. ii)). This sequence converges to a fixed point $\dot{h}_0$ for P in $\dot{\Lambda}_L$ and:

$$\theta_{\Lambda_L} (P^n \dot{h}, \dot{h}_0) \le c^{te} \gamma_L^n, \tag{12}$$

the constant depending on $\dot{h}$ and L.

If $\dot{h}_1$ is another function in $\dot{\Lambda}_L$, then by iterating (11) we obtain:

$$\theta_{\Lambda_L} \left(P^n \, \dot{h}, \, P^n \, \dot{h}_1 \right) \leq \gamma_L^{n-1} \, \theta_{\Lambda_L} \left(P\dot{h}, \, P\dot{h}_1 \right).$$

Then, inequality (12) implies proposition 2.2.

Let us consider the cone $\Lambda = \bigcup_{L > L_0} \Lambda_L$. For any $\dot{h} \in \Lambda$ there exists $L_1 > L_0$ such that $\dot{h} \in \Lambda_{L_1}$. The estimation (7) gives $P\Lambda_{L_1} \subset \Lambda_{\frac{1}{\beta}(\frac{b}{a}+L_1)}$ with $\beta > 1$; therefore there exists an integer $m \geq 0$ for which $P^m \, \Lambda_{L_1} \subset \Lambda_L$ for any fixed $L > L_0$, the integer m depending on L_1 and L. We have:

Corollary 2.1. For $L > L_0$ and for any $\dot{h} \in \Lambda$ we have:

$$\theta_{\Lambda_L} \left(P^n \dot{h}, \, \dot{h}_0 \right) \leq c^{te} \, \gamma_L^m,$$

the constant depending on $\dot{h}$ and L.

We conclude that P admits an unique fixed point in Λ and that the iterates $\dot{P}^n h$ converge exponentially fast in the θ_{Λ_L} metric to $\dot{h}_0$, for any $\dot{h} \in \Lambda$.

In order to prove theorem 1 it remains to relate the projective metric and the uniform one. The method is classical and we will summarize it below.

At first, the projective equality $P\dot{h}_0 = \dot{h}_0$ implies the existence of a real $\alpha > 0$ and a Hölder- function $h_0 > 0$ in Λ_L such that:

$$Ph_0 = \alpha h_0.$$

Second we shall use property (8). We fix some $L > L_0$ and $h \in \Lambda_L$. For any integer n, there exists a largest real u_n and a smallest v_n such that:

$$\begin{cases} u_n \, h_0 \leq \dfrac{P^n h}{\alpha^n} \leq v_n \, h_0 \\ r_n = \dfrac{P^n \, h}{\alpha^n} - u_n \, h_0 \in \Lambda_L \\ s_n = v_n \, h_0 - \dfrac{P^n \, h}{\alpha^n} \in \Lambda_L \end{cases} \tag{13}$$

We have $u_n \leq v_n$. Since P is a positive operator on Λ_L, the sequence u_n is non increasing while the sequence v_n is non decreasing. The functions Pr_n, Ps_n and h_0 belong to Λ_L (since $L > L_0$); then by using again (8) we get the existence of ξ_n and β_n such that:

$$\begin{aligned} \xi_n h_0 \leq Pr_n \leq \xi_n \exp\left(\operatorname{diam}_{\Lambda_L} P\Lambda_L \right) h_0 \\ \beta_n \, h_0 \leq Ps_n \leq \beta_n \exp\left(\operatorname{diam}_{\Lambda_L} P\Lambda_L \right) h_0 \end{aligned} \tag{14}$$

From lemma 2.2. $\operatorname{diam}_{\Lambda_L} (P\Lambda_L)$ is bounded. Combining inequalities (13) and (14) it is easy to see that:

$$0 \leq \left(v_{n+1} - u_{n+1} \right) \leq \left(v_n - u_n \right) \cdot \left(1 - \exp\left(-\operatorname{diam}_{\Lambda_L} (P\Lambda_L) \right) \right). \tag{15}$$

The sequences $(u_n)_{n \in \mathbf{N}}$ and $(v_n)_{n \in \mathbf{N}}$ are adjacent and define a common limit $m(h)$. Thus, we deduce from (13) and (15) that:

$$\left\|\frac{P^n h}{\alpha^n} - h_0 \, m(h)\right\|_u \leq (v_1 - u_1)\left(1 - \exp\left(-\operatorname{diam}_{\Lambda_L}(P\Lambda_L)\right)\right)^{n-1} \|h_0\|_u \quad , \qquad (16)$$

where $\| \ \|_u$ denotes the uniform norm on $\overline{A}$.

Corollary 2.1. allows to get the same estimation (16) for $h \in \Lambda = \bigcup_{L > L_0} \Lambda_L$, and then for any function $h \in \Lambda - \Lambda$.

The mapping $h \to m(h)$ is well defined and it is linear on $\Lambda - \Lambda$; the density of $\Lambda - \Lambda$ in $C(\overline{A})$ for the uniform norm and the uniform boundness of the family of operators $\left(\frac{P^n}{\alpha^n}\right)_{n \in \mathbb{N}}$ imply that we can extend the linear functional m to a Borel measure on $\overline{A}$; inequality (16) remains valid on $C(\overline{A})$ but we loose the exponential speed of convergence. This proves the part ii) of the theorem 1.

To obtain i) it is enough to replace m by $\dfrac{m}{m(h_0)}$ and h_0 by $m(h_0).h_0$

The other statements of theorem 1 can be easily proved. This can also be done using the results of P. Walters [W].

The proof of theorem 2 is now easy. Part (i) follows immediatly from theorem 1. Moreover for any continuous function g on $\overline{A}$ we have:

$$\int_{\overline{A}} g \circ T^n \cdot 1_{\overline{A}} \circ T^n \, d\lambda = \int_{\overline{A}} g \cdot P^n \, 1 \, d\lambda.$$

Then:

$$\frac{\int_{\overline{A}} g \circ T^n \cdot 1_{\overline{A}} \circ T^n \, d\lambda}{\lambda(T^{-n}\overline{A})} = \frac{\frac{1}{\alpha^n} \int_{\overline{A}} g \circ T^n \, 1_{\overline{A}} \circ T^n \, d\lambda}{\frac{1}{\alpha^n} \int_{\overline{A}} 1_{\overline{A}} \circ T^n \, d\lambda} = \frac{\frac{1}{\alpha^n} \int_{\overline{A}} g \cdot P^n \, 1 d\lambda}{\frac{1}{\alpha^n} \int_{\overline{A}} P^n \, 1 \, d\lambda}.$$

By using theorem 1 ii) we see that the limit of the above quantity is:

$$\frac{\int_{\overline{A}} g \cdot h_0 d\lambda}{\int_{\overline{A}} h_0 d\lambda} = \int_{\overline{A}} g \, d\mu$$

which proves part ii) of theorem 2.

With the same notation:

$$\frac{\int_{\overline{A}} g \cdot 1_{\overline{A}} \circ T^n \, d\lambda}{\int_{\overline{A}} 1_{\overline{A}} \circ T^n \, d\lambda} = \frac{\int_{\overline{A}} P^n g d\lambda}{\int_{\overline{A}} P^n 1 \, d\lambda} \xrightarrow{n \to \infty} \frac{\int_{\overline{A}} h_0 \, m(g) \, d\lambda}{\int_{\overline{A}} h_0 \, d\lambda} = m(g)$$

which proves part iii) of the theorem 2.

It is clear that the measure $\nu = h_0 m$ has its support in $\bigcap_{n \geq 0} T^{-n} \overline{A}$. We leave to the reader the proof of the invariance of ν by T, and of the ergodicity of ν. Moreover it can be proved that the ν-measure of any open set in A is strictly positive. [See for instance [B]; [W]].

The Gibbs Property

We summarize the construction of a Markov partition for T. The method is the same as Bowen's construction [B] of Markov partitions for Axiom A diffeomorphisms.

We have seen that there exists $\varepsilon > 0$ and $\beta > 1$ such that $\|Tx - Tx'\| \geq \beta\|x - x'\|$ when $\|x - x'\| < \varepsilon$. Then T has the pseudo orbit shadowing property, that is: $\forall \xi > 0$, $\exists \zeta > 0$ such that if $(x_i)_{i \in \mathbf{N}}$ is a sequence of points in $\overline{A}$ with $\|Tx_i - x_{i+1}\| < \zeta$ $\forall i \in \mathbf{N}$, then there exists $x \in \overline{A}$ verifying: $\|x_i - T^i x\| < \xi$, $\forall i \in \mathbf{N}$.

The point x is unique if 2ξ is less than the expansive constant β of T.

We choose 4ξ smaller than the expansive constant β of T and we take ξ as in the shadowing property. Note that $K < \xi$, $K < \dfrac{\zeta}{2}$ and $\|x - x'\| < K$ implies $\|Tx - Tx'\| < \zeta/2$. Let $C = \{x_1, x_2, \cdots, x_r\}$ be a K-dense subset of $\overline{A}$ and put: $\Sigma(C) = \left\{ q \in \displaystyle\prod_0^\infty C \; ; \right.$ $\left. \|Tq_i - q_{i+1}\| < K, \forall i \in \mathbf{N} \right\}$. We define $\theta : \Sigma(C) \to \overline{A}$ by taking $\theta(q)$ the unique point given by the shadowing property; θ is continuous and onto.

Let $P_i = \{\theta(q) \; ; \; q_0 = x_i\}$; $\{P_1, P_2, \cdots, P_r\}$ is a closed cover of $\overline{A}$ and $\overline{A} \backslash \displaystyle\bigcup_{i=1}^r \partial P_i$ is open and dense. We consider the collection of open sets $\{D_1, D_2, \cdots, D_m\}$ of the form int $(P_i \cap P_j)$ and int $(P_k \backslash P_\ell)$. One can show that $\{R_1, R_2, \cdots, R_m\}$ where $R_i = \overline{D}_i$ verifyes the following conditions:

- $\overline{A} \backslash \displaystyle\bigcup_{i=1}^m \partial R_i$ is open and dense

- diam $R_i < \xi$, $i = 1, 2, \cdots, n$ and $T\left(\displaystyle\bigcup_i \partial R_i \right) \subset \displaystyle\bigcup_i \partial R_i$

- int $R_i \cap T(\text{int } R_j) \neq \emptyset \implies R_i \subset T(R_j)$

- $T|_{R_i}$ is one to one and $\displaystyle\bigcap_{n=0}^\infty T^{-n} R_{i_n}$ contains at most one point for any sequence $(i_0, i_1, \cdots, i_n, \cdots)$.

Moreover if $\partial R = \displaystyle\bigcup_{i=1}^m \partial R_i$, then $\partial R \subset T^{-1}(\partial R)$ so that, by ergodicity of ν, $\nu(\partial \xi) = 0$ or 1. But $\overline{A}/\partial R$ is open and dense so that $\nu(\partial \xi) = 0$.

The collection $\mathcal{R} = \{R_1, R_2, \cdots, R_m\}$ is called a Markov partition for T. Let $\mathcal{R}_0^{n-1} = \displaystyle\bigcap_{i=0}^{n-1} T^{-i} \mathcal{R}$ be the partition associated to T^n. Of course T^n is invertible on each atom I_n of $\mathcal{R}_0^{n-1}$. We will denote by f_i the inverse branch of $T|_{R_i}$ $(i = 1, 2, \cdots, m)$. If

$$I_n = \bigcap_{i=0}^{n-1} T^{-i} R_{j_i}, \text{ then the inverse branch of } T^n|_{I_n} \text{ is of the form}$$

$$f_{j_1} \circ f_{j_2} \circ \cdots \circ f_{j_n} \left((j_1, j_2, \cdots, j_n) \in \{1, 2, \cdots, m\}^n \right).$$

The nth iterate of P satisfies:

$$\forall x \in \overline{A}, \ P^n h(x) = \sum_{(j_1, \cdots, j_n)} \frac{h(f_{j_1} \circ f_{j_2} \circ \cdots \circ f_{j_n}(x))}{|\ \text{Det } DT^n(f_{j_1} \circ \cdots \circ f_{j_n}(x)\ |} 1_{T^n I_n}.$$

Take $h = 1_{I_n}$ where $I_n = \bigcap_{i=0}^{n-1} T^{-i} R_{j_i}$ is some atom. Observe that $T^{n-1} I_n = R_{j_0}$. On the other hand, from the mixing and Markov properties there exists $\ell > 0$ such that $T^\ell(R_j) \supset \overline{A}$ for every j. We have:

$$P^{n-1} 1_{I_n}(x) = \frac{1_{I_n} (f_{j_1} \circ f_{j_2} \circ \cdots \circ f_{j_{n-1}}(x))}{|\ \text{Det } DT^{n-1}(f_{j_1} \circ \cdots \circ f_{j_{n-1}}(x))\ |} 1_{R_{j_0}}.$$

Hence:

$$\begin{aligned}
\nu(I_n) &= \int_A 1_{I_n} h_0 \, dm = \frac{1}{\alpha^{\ell+n-1}} \int_A P^{\ell+n-1}(h_0 \cdot 1_{I_n}) dm \\
&= \frac{1}{\alpha^{\ell+n-1}} \int_A P^\ell \left(\frac{h_0(f_{j_1} \circ \cdots \circ f_{j_{n-1}})}{|\ \text{Det } T^{n-1}(f_{j_1} \circ \cdots \circ f_{j_{n-1}}\ |)} 1_{R_{j_0}} \right) dm.
\end{aligned} \tag{17}$$

But h_0 and $|\ \text{Det } DT^n(f_{j_1} \circ \cdots \circ f_{j_n})\ |$ are Hölder functions and bounded away from zero and infinity. Therefore we have the distorsion property (cf. (5)):

$\exists M_1, M_2, N_1, N_2$ positive constants such that

$$M_1 \leq \frac{h_0(f_{j_1} \circ \cdots \circ f_{j_n}(x))}{h_0(f_{j_1} \circ \cdots \circ f_{j_n}(x_0))} \leq M_2$$

$$N_1 \leq \frac{|\ \text{Det } T^n(f_{j_1} \circ \cdots \circ f_{j_n}(x))\ |}{|\ \text{Det } T^n(f_{j_1} \circ \cdots \circ f_{j_n}(x_0))\ |} \leq N_2 \ ,$$

for any $x, x_0 \in \overline{A}$.

Therefore:

$$\frac{1}{\alpha^{\ell+n-1}} \frac{M_1}{N_2} \frac{h_0(f_{j_1} \circ \cdots \circ f_{j_{n-1}}(x_0))}{|\ \text{Det } DT^{n-1}(f_{j_1} \circ \cdots \circ f_{j_{n-1}}(x_0))\ |} \leq \nu(I_n)$$

$$\leq \frac{1}{\alpha^{\ell+n-1}} \frac{M_2}{N_1} \frac{h_0(f_{j_1} \circ \cdots \circ f_{j_{n-1}}(x_0))}{|\ \text{Det } DT^{n-1}(f_{j_1} \circ \cdots \circ f_{j_{n-1}}(x_0))\ |},$$

$\forall x_0 \in \overline{A}, \quad \forall n \geq n_0$. Using (17) and since the numbers $\int_A P^\ell(1_{R_j}) dm$ are bounded above and below by strictly positive quantities κ and κ^{-1} we get:

$$\alpha^{-m}\kappa^{-1} \underset{\overline{A}}{\text{Inf}} \; h_0 \cdot \frac{M_1}{N_2}$$

$$\leq \frac{\nu(I_n)}{\exp\left(-n\log\alpha\right)\mid \text{Det}\left(DT^{n-1}(f_{j_1}\circ\cdots\circ f_{j_{n-1}}(x_0))\right)\mid^{-1}}$$

$$\leq \alpha^{-m}\kappa\text{Sup}\; h_0 \cdot \frac{M_2}{N_1}.$$

This property is known as the Gibbs property; it means that the measure ν is strongly related to the volume measure. (We refer to R. Bowen [B] and D. Ruelle [R] for definitions and properties of Gibbs measures).

Acknowledgements.

The authors thank partial support from program ECOS-CONICYT, grant CEE # CI1*-CT920046. S.M. was partially financed by grant FONDECYT 1940405.

References

[P-Y] G. Panigiani, J.A. Yorke (1979), Expanding Maps on Sets which are almost Invariant: Decay and Chaos, Trans. Amer. Math. Soc., 252, 351-366.

[L] E. Lorenz (1963), Deterministic Nonperiodic Flow, J. Atmospheric Sci., 20, 130-141.

[K-Y] J. Kaplan, J. Yorke (1978), Preturbulence: A Regime observed in a Fluid Flow Model of Lorentz, Preprint.

[R] K. Robbins (1977), A New Approach to Subcritical Instability and Turbulent Transitions in a Simple Dynamo, Math. Proc. Cambridge Philos. Soc., 82, 309-325.

[Y-Y] J. Yorke, E. Yorke (1978), Metastable Chaos: the Transition to Sustamed Chaotic Behavior in the Lorentz Model, Preprint.

[V-J] D. Vere-Jones (1962), Geometric Ergodicity in Denumerable Markov Chains, Quart. J. Math. Oxford, 13, ser. 2, 7-28.

[D-S] J.N. Darroch, E. Seneta (1965), On Quasi-Stationnary Distributions in Absorbing Discrete time Finite Markov Chains, J. Appl. Prob., 2, 88-100.

[F-K-M-P] P.A. Ferrari, H. Kesten, S. Martínez and P. Picco (1993), Existence of Quasi Stationary Distributions. A Renewal dynamical Approach, Annals of Probability, to appear.

[R] D. Ruelle (1978), Thermodynamic Formalism, Addison Wesley.

[B] R. Bowen (1975), Equilibrium States and the Ergodic Theory of Anosov Diffeomorphisms, Lect. Notes in Math., 470, Springer-Verlag.

[W] P. Walters (1978), Invariant Measures and Equilibrium States for Mappings which Expand Distances, T.A.M.S., 236, 121-153.

[Y] A.M. Yaglom (1947), Certain Limit Theorems of the Theory of Branching Stochastic Processes, (in Russian) Dokl. Akad. Nank. SSSR (n.s), 56, 795-798.

DETAILED BALANCE AND REVERSED
PROCESS FOR MACROSCOPIC SYSTEM

by

Felipe Barra[1,2], Marcel Clerc[1], Cristián Huepe[1,2], Enrique Tirapegui[1]

[1]:Facultad de Ciencias Fisicas y Matematicas, Universidad de Chile, Casilla487-3, Santiago.
[2]:Centro de Física No Lineal y Sistemas Complejos de Santiago.

Abstract

We discuss the detailed balance symmetry for Markov processes with stationary probability describing macroscopic systems far from equilibrium and we show that one can always put these processes in generalized detailed balance adding a suitable set of parameters. Our analysis and construction uses as an essential tool the notion of reversed process of a Markov process which gives a new insight of the problem.

1   Introduction: Review of detailed balance

We shall present here a discussion of the detailed balance symmetry for Markovian processes. In most situations of interest in macroscopic physics a modelization with a Markov process is possible[1]. To cover a general case let $x = (x_1, \cdots, x_l)$ be a set of macroscopic variables which take continuous real values. The Markov process in the stationary state is defined by the conditional probability $P_+(x, t|x', t'; \sigma)$, $t > t'$, and the stationary probability $\rho(x, \sigma)$, where σ stands for a set of parameters $\sigma = (\sigma_1, \cdots, \sigma_m)$. The joint probability densities $W^{(n)}$ are then given by $(t_1 > t_2 > \cdots > t_n)$[1]

$$W^{(n)}(x_1, t_1; x_2, t_2; \cdots; x_n, t_n; \sigma) =$$
$$P_+(x_1, t_1|x_2, t_2; \sigma) P_+(x_2, t_2|x_3, t_3; \sigma) \cdots P_+(x_{n-1}, t_{n-1}|x_n, t_n; \sigma) \cdot \rho(x_n, \sigma) \quad (1)$$

We remark that $W^{(n)}$ depends only on the time differences $(t_i - t_j)$ since $P_+(x, t|x', t'; \sigma) = P_+(x|x', t - t'; \sigma)$.

The generalized detailed balance (GDB) symmetry is defined in terms of the two points joint probability density $W^{(2)}$ and is related to a time reversal transformation $x \to \tilde{x}$, $\sigma \to \tilde{\sigma}$, such that applied twice gives back the original values of the macrovariables x and parameters σ, i.e. $\tilde{\tilde{x}} = x$, $\tilde{\tilde{\sigma}} = \sigma$. We shall consider macrovariables and parameters which are even or odd under the time reversal transformation. This is always possible[2] and we have

$$
\begin{aligned}
x_\mu \to \tilde{x}_\mu &= \epsilon^\mu x_\mu, & \epsilon^\mu = \pm 1, & \quad \mu = 1, \ldots, l \\
\sigma_j \to \tilde{\sigma}_j &= \epsilon'_j \sigma_j, & \epsilon'_j = \pm 1, & \quad j = 1, \ldots, m
\end{aligned}
\quad (2)
$$

221

E. Tirapegui and W. Zeller (eds.), Instabilities and Nonequilibrium Structures V, 221–230.
© 1996 Kluwer Academic Publishers.

If there is a microscopic theory from which our macrovariables and parameters are constructed through coarse-graining procedures[1,3] then the physical transformation $(x \rightarrow \tilde{x}, \sigma \rightarrow \tilde{\sigma})$ is known since it can be derived from the time reversal invariance of the microscopic system which is described either by a Schrödinger equation if quantum effects are important or by Hamilton's equations if it is a classical system. We insist that GDB is defined with respect to a transformation $(x \rightarrow \tilde{x}, \sigma \rightarrow \tilde{\sigma})$, which needs not be the physical one. The definition is [1,2,4]: the Markov process in the stationary state with conditional probability $P_+(x,t|x',t';\sigma)$ and stationary probability $\rho(x,\sigma)$ is in GDB with respect to the given transformation $(x \rightarrow \tilde{x}, \sigma \rightarrow \tilde{\sigma})$ of the form (2) if

$$W^{(2)}(x_1,t_1;x_2,t_2;\sigma) = W^{(2)}(\tilde{x}_2,t_1;\tilde{x}_1,t_2;\tilde{\sigma}) \tag{3}$$

Integrating (3) over x_2 we find

$$\rho(x_1,\sigma) = \rho(\tilde{x}_1,\tilde{\sigma}) \tag{4}$$

which is a necessary condition on the stationary probability to have GDB. We can always consider a Markov process in the stationary state as defined for times t between $(-\infty)$ and $(+\infty)$ and we shall use this fact often. Relation (3) is the obvious thing one could expect to have in the microscopic system. For a Markov process the conditional probability density $P_+(x,t|x',t';\sigma)$ obeys a differential equation [4] which we take in the form

$$\partial_t P_+(x,t|x',t';\sigma) = \sum_{n \geq 1} \frac{(-1)^n}{n!} \partial_{\mu_1} \dots \partial_{\mu_n} A_{\mu_1 \dots \mu_n}(x,\sigma) P_+(x,t|x',t';\sigma) \tag{5}$$

where $\partial_\mu = \frac{\partial}{\partial x_\mu}$ (here repeated indices imply summation, a notation we use in what follows) and initial condition $P_+(x,t|x',t';\sigma) = \delta(x - x')$ for $t \rightarrow t'_+$. This general form in which the right hand is a formal infinite series covers the Fokker-Planck case of diffusion processes (the sum stops for $n = 2$) and the general case for which we have the whole series. In the case of jump processes with variables $(X_1, X_2, \dots, X_n)$ taking discrete values one usually has a small parameter $\eta = \Omega^{-1}$, where Ω is the volume of the system, and then $x_j = \frac{X_j}{\Omega}$ can be considered as a continuous variable for big Ω and the conditional probability obeys equation (5) changing there $\partial_t \rightarrow \eta \partial_t$, $\partial_\mu \rightarrow \eta \partial_\mu$ (see[1,5]).

2 Reversed process and detailed balance

We shall use for our discussion of GDB the notion of reversed process in the stationary state. As we shall see the reversed process correspond to look the realization of the process changing $t \rightarrow -t$. This notion is defined with an advanced conditional probability $P_-(x,t|x',t';\sigma)$ where now $t < t'$ (this is the reason for using the notations P_+ and P_-). The conditional probability $P_-(x,t|x',t';\sigma)$ is the probability that the stochastic variable $X(t)$ takes the value x knowing that in a future time $t' > t$ it will take the value $X(t') = x'$. The reversed process has a retarded conditional probability $\hat{P}_+(x,t|x',t';\sigma)$ defined by $(t > t')$

$$\hat{P}_+(x,t|x',t';\sigma) \equiv P_-(x,-t|x',-t';\sigma) \tag{6}$$

Since we are reversing time in the stationary state we have

$$
\begin{aligned}
W^{(2)}(x, -t : x', -t'; \sigma) &= P_-(x, -t|x', -t'; \sigma)\rho(x', \sigma) \\
&= P_+(x', -t'|x, -t; \sigma)\rho(x, \sigma)
\end{aligned}
\tag{7}
$$

From (6),(7) and recalling the invariance by time translation (all joint probability functions and conditional probabilities depend only on time differences as we have stated after eq.(1)) we obtain for $\hat{P}$ the equivalent expression

$$
\hat{P}_+(x, t|x', t'; \sigma) = P_+(x', t|x, t'; \sigma)\frac{\rho(x, \sigma)}{\rho(x', \sigma)}
\tag{8}
$$

From (3) and (8) it is immediate to see that the original process (1) is in GDB with respect to the transformation $(x \to \tilde{x}, \sigma \to \tilde{\sigma})$ if and only if

$$
\hat{P}_+(x, t|x', t'; \sigma) = P_+(\tilde{x}, t|\tilde{x}', t'; \tilde{\sigma})
\tag{9}
$$

The stationary probability $\hat{\rho}(x, \sigma)$ of the reversed process is easily seen from (8) to coincide with that of the original process

$$
\hat{\rho}(x, \sigma) = \rho(x, \sigma)
\tag{10}
$$

and one should remark that reversing twice takes one back to the initial process

$$
\hat{\hat{P}}_+(x, t|x', t'; \sigma) = P_+(x, t|x', t'; \sigma)
\tag{11}
$$

The differential equation for $\hat{P}_+$ can be calculated and is of the form (5)

$$
\partial_t \hat{P}_+(x, t|x', t'; \sigma) = \sum_{n \geq 1} \frac{(-1)^n}{n!} \partial_{\mu_1} \cdots \partial_{\mu_n} \hat{A}_{\mu_1 \ldots \mu_n}(x, \sigma)\hat{P}_+(x, t|x', t'; \sigma)
\tag{12}
$$

In order to calculate the functions $\hat{A}_{\mu_1 \ldots \mu_n}(x, \sigma)$ we recall that in general one has for any process satisfying (5) that [6] $(d(r) = dr_1 \cdots dr_n)$

$$
A_{\mu_1 \ldots \mu_n}(x, \sigma) = \lim_{\tau \to +0} \frac{1}{\tau} \int d(r) r_{\mu_1} \cdots r_{\mu_n} P_+(x + r, t + \tau|x, t; \sigma)
\tag{13}
$$

We calculate $\hat{A}_{\mu_1 \ldots \mu_n}(x, \sigma)$ using (13) with P_+ replaced by $\hat{P}_+$ given by (8). One has (we omit the σ here)

$$
\hat{A}_{\mu_1 \ldots \mu_n}(x) = \lim_{\tau \to +0} \frac{1}{\tau} \int d(r) r_{\mu_1} \cdots r_{\mu_n} P_+(x|x + r, \tau)\frac{\rho(x + r)}{\rho(x)}
\tag{14}
$$

We replace $P_+(x|x + r, \tau)$ up to $O(\tau)$ using the backward equation [4] $(\partial'_\mu = \frac{\partial}{\partial x'_\mu})$

$$
\partial_\tau P_+(x|x', \tau) = \sum_{n \geq 1} \frac{1}{n!} A_{\mu_1 \ldots \mu_n}(x') \partial'_{\mu_1} \cdots \partial'_{\mu_n} P_+(x|x', \tau)
\tag{15}
$$

This gives $(m \geq 1)$

$$\hat{A}_{\nu_1 \ldots \nu_m}(x) = \frac{1}{\rho(x)} \sum_{n \geq 1} \frac{1}{n!} \int d(r) r_{\nu_1} \cdots r_{\nu_m} A_{\mu_1 \ldots \mu_n}(x+r) \rho(x+r) \frac{\partial}{\partial r_{\mu_1}} \cdots \frac{\partial}{\partial r_{\mu_n}} \delta(r) \quad (16)$$

We develop now $A_{\mu_1 \ldots \mu_n}(x+r)\rho(x+r)$ in powers of r in a Taylor series and after partial integration and using that $A_{\mu_1 \ldots \mu_n}(x)$ is symmetric in all its indices we obtain the final formula

$$\hat{A}_{\nu_1 \ldots \nu_m}(x, \sigma) = \frac{(-1)^m}{\rho(x, \sigma)} \sum_{s \geq 0} \frac{(-1)^s}{s!} \partial_{\alpha_1} \cdots \partial_{\alpha_s} A_{\nu_1 \ldots \nu_m \alpha_1 \ldots \alpha_s}(x, \sigma) \rho(x, \sigma) \quad (17)$$

One can see directly using (17) that reversing twice gives $\hat{\hat{A}}_{\nu_1 \ldots \nu_m}(x, \sigma) = A_{\nu_1 \ldots \nu_m}(x, \sigma)$ We define now the quantities

$$D_{\mu_1 \ldots \mu_n}(x, \sigma) \equiv \frac{1}{2}(A_{\mu_1 \ldots \mu_n}(x, \sigma) + \hat{A}_{\mu_1 \ldots \mu_n}(x, \sigma)) \quad (18a)$$

$$R_{\mu_1 \ldots \mu_n}(x, \sigma) \equiv \frac{1}{2}(A_{\mu_1 \ldots \mu_n}(x, \sigma) - \hat{A}_{\mu_1 \ldots \mu_n}(x, \sigma)) \quad (18b)$$

in terms of which the differential equations for the conditional probabilities of the original (direct) process and the reversed process take the form

$$\partial_t P_+(x, t | x', t'; \sigma) = \sum_{n \geq 1} \frac{(-1)^n}{n!} \partial_{\mu_1} \cdots \partial_{\mu_n} (D_{\mu_1 \ldots \mu_n}(x, \sigma) + R_{\mu_1 \ldots \mu_n}(x, \sigma)) P_+(x, t | x', t'; \sigma) \quad (19a)$$

$$\partial_t \hat{P}_+(x, t | x', t'; \sigma) = \sum_{n \geq 1} \frac{(-1)^n}{n!} \partial_{\mu_1} \cdots \partial_{\mu_n} (D_{\mu_1 \ldots \mu_n}(x, \sigma) - R_{\mu_1 \ldots \mu_n}(x, \sigma)) \hat{P}_+(x, t | x', t'; \sigma) \quad (19b)$$

which motivates the definition (18) since we see that passing from the direct to the reversed process amounts to change the sign of $R_{\mu_1 \ldots \mu_n}(x, \sigma)$. In order to have a clear view of what represents the reversed process in the stationary states we consider the time interval $[-T, T]$ and the curves

$$\gamma : x = f(t), f(-T) = y, f(T) = z \qquad \hat{\gamma} : x = \tilde{f}(t) \equiv f(-t), \tilde{f}(-T) = z, \tilde{f}(T) = y \quad (20)$$

The curve $\hat{\gamma}$ is obtained from γ by time reversal $t \to -t$. If the initial conditions both for the original process and for the reversed one are taken at a time $t_0 << -T$ (say $t_0 \to -\infty$) then we can consider that in the interval $[-T, T]$ both processes are in the stationary states and the definition of the reversed process (6) is such that the probability that the curve γ is a realization of the initial process is equal to the probability that $\hat{\gamma}$ is a realization of the reversed process. We remark also that if the original process is in GDB with respect to $x \to \tilde{x}$, $\sigma \to \tilde{\sigma}$, i.e. (9) is true, then the usual irreversible drift $d_\mu(x, \sigma)$ and the reversible drift $r_\mu(x, \sigma)$ defined by[7]

$$d_\mu(x, \sigma) = \frac{1}{2}(A_\mu(x, \sigma) + \varepsilon^\mu A_\mu(\tilde{x}, \tilde{\sigma})) \quad (21a)$$

$$r_\mu(x,\sigma) = \frac{1}{2}(A_\mu(x,\sigma) - \varepsilon^\mu A_\mu(\tilde{x},\tilde{\sigma})) \tag{21b}$$

turn out to coincide with $D_\mu(x,\sigma)$ and $R_\mu(x,\sigma)$ defined by (18) for $n = 1$. The proof is simple since (9) implies that

$$\hat{A}_{\mu_1\ldots\mu_n}(x,\sigma) = \varepsilon^{\mu_1}\cdots\varepsilon^{\mu_n} A_{\mu_1\ldots\mu_n}(\tilde{x},\tilde{\sigma})) \tag{22}$$

and using (22) we see that

$$d_\mu(x,\sigma) = D_\mu(x,\sigma) \qquad r_\mu(x,\sigma) = R_\mu(x,\sigma) \tag{23}$$

Using the explicit expression (17) for $\hat{A}_{\mu_1\ldots\mu_n}(x,\sigma)$ we can see that relation (22) is the condition for GDB given in[8]. On the other hand the decomposition of $A_\mu(x,\sigma)$ as

$$A_\mu(x,\sigma) = D_\mu(x,\sigma) + R_\mu(x,\sigma) \tag{24}$$

which results from (18) coincides in the Fokker-Planck cased with the one done by Jauslin in[2] (in this last paper a classification of diffusion processes is proposed based on the previous separation of the drift $A_\mu(x,\sigma)$). Since the direct and reversed processes have the same stationary probability $\rho(x,\sigma)$ this function is a solution of both (19a) and (19b) and then one also has

$$\sum_{n\geq 1} \frac{(-1)^n}{n!}\partial_{\mu_1}\cdots\partial_{\mu_n} D_{\mu_1\ldots\mu_n}(x,\sigma)\rho(x,\sigma) = 0 \tag{25a}$$

$$\sum_{n\geq 1} \frac{(-1)^n}{n!}\partial_{\mu_1}\cdots\partial_{\mu_n} R_{\mu_1\ldots\mu_n}(x,\sigma)\rho(x,\sigma) = 0 \tag{25b}$$

3 Construction of processes in generalized detailed balance

We proceed now to construct from the original process (1) a new process depending on a new odd parameter λ which will be in GDB with respect to the transformation $(x \to \tilde{x}, \sigma \to \tilde{\sigma}, \lambda \to \tilde{\lambda})$. The new process will be defined by a conditional probability density $P_+(x,t|x',t' : \sigma,\lambda)$ which satisfies the equation

$$\partial_t P_+(x,t|x',t';\sigma,\lambda) = \sum_{n\geq 1} \frac{(-1)^n}{n!}\partial_{\mu_1}\cdots\partial_{\mu_n} B_{\mu_1\ldots\mu_n}(x,\sigma,\lambda)P_+(x,t|x',t';\sigma,\lambda) \tag{26}$$

where:

$$B_{\mu_1\ldots\mu_n}(x,\sigma,\lambda) = D_{\mu_1\ldots\mu_n}(x,\sigma) + \lambda R_{\mu_1\ldots\mu_n}(x,\sigma) \tag{27}$$

It is easy to see that for $|\lambda| \leq 1$ equation (26) gives a positive probability density (in the Fokker-Planck case there is no restriction on λ). Notice that for $\lambda = 1$ the process defined by

(26) coincide with the original process while for $\lambda = -1$ it coincide with the reversed process $\hat{P}_+(x,t|x',t';\sigma)$. Due to the form (27) of $B_{\mu_1\dots\mu_n}(x,\sigma,\lambda)$ and to equations (25) one has that the stationary probability of the process defined by (26) is $\rho(x,\sigma)$ and is then independent of λ. In order to prove that the new process is in GDB with respect to $(x \to \tilde{x} = x,\ \sigma \to \tilde{\sigma} = \sigma,\ \lambda \to \tilde{\lambda} = -\lambda)$ as announced we construct the reversed process $\hat{P}_+(x,t|x',t';\sigma,\lambda)$ defined by (see (8))

$$\hat{P}_+(x,t|x',t';\sigma,\lambda) = P_+(x',t|x,t';\sigma,\lambda)\frac{\rho(x,\sigma)}{\rho(x',\sigma)} \tag{28}$$

This conditional probability will satisfy a differential equation of the form (26) with $B_{\mu_1\dots\mu_n}(x,\sigma,\lambda)$ replaced by $\hat{B}_{\mu_1\dots\mu_n}(x,\sigma,\lambda)$ given by

$$\hat{B}_{\nu_1\dots\nu_m}(x,\sigma,\lambda) = \frac{(-1)^m}{\rho(x,\sigma)}\sum_{s\geq0}\frac{(-1)^s}{s!}\partial_{\alpha_1}\dots\partial_{\alpha_s}B_{\nu_1\dots\nu_m\alpha_1\dots\alpha_s}(x,\sigma,\lambda)\rho(x,\sigma) \tag{29}$$

Using the definition (27) of $B_{\nu_1\dots\nu_m}(x,\sigma,\lambda)$ and also (18) we see that (29) is the sum of four terms, since (we omit the indices)

$$B(x,\sigma,\lambda) = \frac{1}{2}(A(x,\sigma) + \hat{A}(x,\sigma)) + \frac{\lambda}{2}(A(x,\sigma) - \hat{A}(x,\sigma)) \tag{30}$$

We use now the property that reversing twice gives back the initial process. This implies from (30) and using (17) that

$$\hat{B}(x,\sigma,\lambda) = \frac{1}{2}(A(x,\sigma) + \hat{A}(x,\sigma)) - \frac{\lambda}{2}(A(x,\sigma) - \hat{A}(x,\sigma)) \tag{31a}$$

$$\hat{B}_{\nu_1\dots\nu_m}(x,\sigma,\lambda) = D_{\nu_1\dots\nu_m}(x,\sigma) - \lambda R_{\nu_1\dots\nu_m}(x,\sigma) \tag{31b}$$

This last property shows then that $\hat{P}_+(x,t|x',t';\sigma,\lambda)$ satisfies the same differential equation as $P_+(x,t|x',t';\sigma,\lambda)$ except for the change $\lambda \to -\lambda$. Since at $t \to t'_+$ both conditional probabilities have the same initial condition $\delta(x - x')$ they are identical. We have then

$$\hat{P}_+(x,t|x',t';\sigma,\lambda) = P_+(x,t|x',t';\sigma,-\lambda) \tag{32}$$

and comparing this equality with (9) we see that the process $P_+(x,t|x',t';\sigma,\lambda)$ is in GDB with respect to the transformation $(x \to \tilde{x} = x,\ \sigma \to \tilde{\sigma} = \sigma,\ \lambda \to \tilde{\lambda} = -\lambda)$. We remark that the construction of the process $P_+(x,t|x',t';\sigma,\lambda)$ we have done here is the generalization to a general Markov process of the construction done by R. Graham in[7]. As we have already mentioned for a macroscopic system obtained by coarse-graining from a microscopic one the time reversal transformation of the macroscopic variables and parameters, $x \to \tilde{x}$, $\sigma \to \tilde{\sigma}$, are known. One can ask then the question if it is possible to put this system in GDB adding parameters as we have done above but keeping the original transformations of x and σ. The answer is yes and the construction is based on a new reversed process $\tilde{P}_+(x,t|x',t';\sigma)$. In a first step we suppose that the original process (1) has the property that its stationary probability satisfies

$$\rho(x,\sigma) = \rho(\tilde{x},\tilde{\sigma}) \tag{33}$$

with respect to the physical transformations $x \to \tilde{x}$, $\sigma \to \tilde{\sigma}$ derived from the coarse-graining. If (33) is valid we can define $\tilde{P}_+$ by

$$\tilde{P}_+(x,t|x',t';\sigma) = P_-(\tilde{x},-t|\tilde{x}',-t';\tilde{\sigma}) \tag{34}$$

As in the preceding case one obtains

$$\tilde{P}_+(x,t|x',t';\sigma) = P_+(\tilde{x}',t|\tilde{x},t';\tilde{\sigma})\frac{\rho(x,\sigma)}{\rho(x',\sigma)} \tag{35}$$

and from (33) we have again that the stationary probability of (34) is

$$\tilde{\rho}(x,\sigma) = \rho(x,\sigma) \tag{36}$$

The differential equation satisfied by $\tilde{P}_+$ is of the form (5) with $A_{\mu_1\ldots\mu_n}(x,\sigma)$ replaced by $\tilde{A}_{\mu_1\ldots\mu_n}(x,\sigma)$ and one easily finds that

$$\tilde{A}_{\mu_1\ldots\mu_n}(x,\sigma) = \varepsilon^{\mu_1}\cdots\varepsilon^{\mu_n}\hat{A}_{\mu_1\ldots\mu_n}(\tilde{x},\tilde{\sigma}) \tag{37}$$

The relation between $\tilde{P}_+$ and $\hat{P}_+$ can be obtained from (35)

$$\tilde{P}_+(x,t|x',t';\sigma) = \hat{P}_+(\tilde{x},t|\tilde{x}',t';\tilde{\sigma}) \tag{38}$$

Once again if we reverse twice we go back to the original process. This initial process (1) is in GDB with respect to $(x \to \tilde{x}, \sigma \to \tilde{\sigma})$ if and only if (from (9) and (38)) $\tilde{P}_+$ coincides with P_+

$$\tilde{P}_+(x,t|x',t';\sigma) = P_+(x,t|x',t';\sigma) \tag{39}$$

We define again

$$\tilde{D}_{\mu_1\ldots\mu_n}(x,\sigma) \equiv \frac{1}{2}(A_{\mu_1\ldots\mu_n}(x,\sigma) + \tilde{A}_{\mu_1\ldots\mu_n}(x,\sigma)) \tag{40a}$$

$$\tilde{R}_{\mu_1\ldots\mu_n}(x,\sigma) \equiv \frac{1}{2}(A_{\mu_1\ldots\mu_n}(x,\sigma) - \tilde{A}_{\mu_1\ldots\mu_n}(x,\sigma)) \tag{40b}$$

If P_+ is in GDB with respect to $(x \to \tilde{x}, \sigma \to \tilde{\sigma})$ one has from (39,40) that

$$\tilde{D}_{\mu_1\ldots\mu_n}(x,\sigma) = A_{\mu_1\ldots\mu_n}(x,\sigma) \tag{41a}$$

$$\tilde{R}_{\mu_1\ldots\mu_n}(x,\sigma) = 0 \tag{41b}$$

We define a new process which depends on a new odd parameter $\lambda \to \tilde{\lambda} = -\lambda$ by

$$\partial_t P_+^{(2)}(x,t|x',t';\sigma,\lambda) = \sum_{n\geq 1}\frac{(-1)^n}{n!}\partial_{\mu_1}\cdots\partial_{\mu_n}B_{\mu_1\ldots\mu_n}^{(2)}(x,\sigma,\lambda)P_+^{(2)}(x,t|x',t';\sigma,\lambda) \tag{42}$$

where:

$$B^{(2)}_{\mu_1\ldots\mu_n}(x,\sigma,\lambda) = \tilde{D}_{\mu_1\ldots\mu_n}(x,\sigma) + \lambda\tilde{R}_{\mu_1\ldots\mu_n}(x,\sigma) \tag{43}$$

As before the conditional probability defined by (42) is positive for $|\lambda| < 1$ in the general case and for any λ in the Fokker-Planck case. For $\lambda = 1$ one obtains the original process and for $\lambda = -1$ the reversed process $\tilde{P}_+$. The stationary probability of (42) does not depends on λ and is again $\rho(x,\sigma)$. If we calculate the reversed process $\tilde{P}^{(2)}_+$ of $P^{(2)}_+$ using (35) one finds

$$\tilde{P}^{(2)}_+(x,t|x',t';\sigma,\lambda) = P^{(2)}_+(x,t|x',t';\sigma,\lambda) \tag{44}$$

which means by (39) that $P^{(2)}_+$ is in GDB with respect to the transformation $(x \to \tilde{x}$, $\sigma \to \tilde{\sigma}$, $\lambda \to \tilde{\lambda} = -\lambda)$. This construction answers our question in the case when (33) is verified. If this condition is not true the first thing we have to do is to introduce a minimal set of odd parameters $(\lambda_1,\ldots,\lambda_s;\mu_1,\ldots,\mu_k)$ in the following way. Let $x = (x_1,\ldots,x_p,x_{p+1},\ldots,x_l)$, $\sigma = (\sigma_1,\ldots,\sigma_q,\sigma_{q+1},\ldots,\sigma_m)$ where the first p variables are odd $(\varepsilon^\mu = -1, \mu = 1,\ldots,p)$ and the first q parameters are odd $(\varepsilon'_j = -1, j = 1,\ldots,q)$. Then a maximal choice of new odd parameters (λ_μ,μ_j) will be to define a new stationary probability

$$\rho'(x_1\ldots x_l;\sigma_1\ldots\sigma_m;\lambda_1\ldots\lambda_p;\mu_1\ldots\mu_q) \equiv \prod_{\mu=1}^{p}|\lambda_\mu|\rho(\lambda_1 x_1,\ldots,\lambda_p x_p,x_{p+1},\ldots,x_l;$$
$$\mu_1\sigma_1,\ldots,\mu_q\sigma_q,\sigma_{q+1},\ldots,\sigma_m) \tag{45}$$

and this will correspond to s=p, k=q . One obviously has now

$$\rho'(x;\sigma;\lambda_\mu;\mu_j) = \rho'(\tilde{x};\tilde{\sigma};\{\tilde{\lambda}_\mu\};\{\tilde{\mu}_j\}) \tag{46}$$

but in many cases one can obtain (46) with $s < p$, $k < q$ (this is what we mean by minimal choice of new parameters). We can now define a new conditional probability density P'_+ by

$$P'_+(x,t|x',t';\sigma;\{\lambda_\mu\};\{\mu_j\}) \equiv \prod_{\mu=1}^{s}|\lambda_\mu|P_+(\lambda_1 x_1,\ldots,\lambda_s x_s,x_{s+1},\ldots,x_p;x_{p+1},\ldots,x_l|$$
$$\lambda_1 x'_1,\ldots,\lambda_s x'_s,x'_{s+1},\ldots,x'_l;$$
$$\mu_1\sigma_1,\ldots,\mu_k\sigma_k,\sigma_{k+1},\ldots,\sigma_q,\sigma_{q+1},\ldots,\sigma_m) \tag{47}$$

Taking in (47) the limit $(t - t') \to \infty$ we see that we have (46) and we can apply the previous method and define the reserved process $\tilde{P}'_+$ by (34) since now (33) is verified with the introduction of new parameters $(\lambda_1,\ldots,\lambda_s;\mu_1,\ldots,\mu_k)$, $s \le p$, $k \le q$ .The final result is that adding one more odd parameter λ we shall be able to construct a process in GDB with respect to the transformation $(x \to \tilde{x}$, $\sigma \to \tilde{\sigma}$, $\lambda_\mu \to \tilde{\lambda}_\mu = -\lambda_\mu$, $\mu_j \to \tilde{\mu}_j = -\mu_j$, $\lambda \to \tilde{\lambda} = -\lambda)$. This construction answers completely our question and one should notice that again when all the new parameters take the value one the new process coincides with the original one . We shall now take a simple illustrative example to show a possible interpretation of our constructions. Consider the diffusion process describe by the Fokker-Planck equation

$$\partial_t P_+(x,t|x',t') = \partial_\mu(-A_\mu(x) + \frac{1}{2}g^{\mu\nu}\partial_\nu)P_+(x,t|x',t') \tag{48}$$

where $g^{\mu\nu}$ is a positive definite constant matrix. Suppose that the physical transformation of all the variables $x = (x_1, \ldots, x_l)$ is $x_\mu \to \tilde{x}_\mu = x_\mu$ and let us apply the first construction based on the reversed process $\hat{P}_+(x,t|x',t')$. It is simple to see that in this case one has from (17) that

$$\hat{A}_\mu = -A_\mu(x) + \frac{1}{\rho(x)}g^{\mu\nu}\partial_\nu\rho(x) \tag{49a}$$

$$\hat{g}^{\mu\nu} = g^{\mu\nu} \tag{49b}$$

and the reversed process will obey the Fokker-Planck equation

$$\partial_t \hat{P}_+(x,t|x',t') = \partial_\mu(-\hat{A}_\mu(x) + \frac{1}{2}g^{\mu\nu}\partial_\nu)\hat{P}_+(x,t|x',t') \tag{50}$$

The new process $P_+(x,t|x',t';\lambda)$ in GDB with λ an odd parameter will satisfy the equation

$$\begin{aligned}
\partial_t P_+(x,t|x',t';\lambda) &= \partial_\mu(-(D_\mu(x) + \lambda R_\mu(x)) + \frac{1}{2}g^{\mu\nu}\partial_\nu) \\
&\quad P_+(x,t|x',t';\lambda)
\end{aligned} \tag{51}$$

with D_μ and $R_\mu(x)$ defined by (18) and will be in GDB with respect to the transformation $(x \to \tilde{x} = x, \ \lambda \to \tilde{\lambda} = -\lambda)$. In general a parameter like λ is a variable representing the state of the environment of the system described by $x = (x_1, \ldots, x_l)$ which varies in a much slower time scale than the variables x. More precisely let τ_0 be the greatest of the characteristic times of the initial system $P_+(x,t|x',t')$. By this we mean that if we take initial conditions at $t = 0$ for times $t >> \tau_0$ the system will be in the stationary state. A possible way of representing this situation is to represent the system (24) in terms of the Langevin equations

$$\dot{x}_\mu = D_\mu(x) + \lambda R_\mu(x) + \xi_\mu(t) \tag{52}$$

where $\xi_\mu(t)$ are gaussian white noises with zero mean and correlation $< \xi_\mu(t)\xi_\nu(t') > = g^{\mu\nu}\delta(t-t')$ supplemented by an equation for $\lambda(t)$ of the form

$$\dot{\lambda} = -\frac{1}{\tau}\lambda(t) + \sqrt{\eta}\xi(t) \tag{53}$$

where $\xi(t)$ is a gaussian white noise with zero mean ,δ correlated and with vanishing correlations with the $\xi_\mu(t)$, $\tau \gg \tau_0$ and $\eta \ll 1$ is a small parameter measuring the intensity of the noise on $\lambda(t)$. The conditional probability $P_+(x,\lambda,t|x',\lambda',t')$ of the diffusion process defined by (52) and (53) satisfies

$$\partial_t P_+(x,\lambda,t|x',\lambda',t') = [\partial_\mu(-(D_\mu + \lambda R_\mu) + \frac{1}{2}g^{\mu\nu}\partial_\nu)$$

$$+\partial_\lambda(\frac{1}{\tau}\lambda + \frac{\eta}{2}\partial_\lambda)]P_+(x,\lambda,t|x',\lambda',t') \tag{54}$$

and the stationary probability of (49) will be

$$\rho(x,\lambda) = \rho_1(x)\rho_2(\lambda) \tag{55}$$

where $\rho_1(x)$ is the stationary probability of the system defined by (52) which is independent of λ and $\rho_2(\lambda)$ the stationary probability of (53) given by

$$\rho_2(\lambda) = \frac{1}{\sqrt{2\pi\eta\tau}}e^{-\frac{\lambda^2}{\eta\tau}} \tag{56}$$

For times $\tau_o \approx t \ll \tau$ the probability $P_+(x,\lambda,t|x',\lambda',t')$ will be approximately given by

$$P_+(x,\lambda,t|x',\lambda',t') \approx P_+(x,t|x',t';\lambda')\delta(\lambda - \lambda') \tag{57}$$

since λ will have had no time evolve from its initial condition λ'. For times $\tau_o \ll t \ll \tau$ the system (52) will be in its stationary state $\rho_1(x)$, but the stationary probability will still be of the form $\rho_1(x)\delta(\lambda - \lambda')$, i. e. very different from the real stationary probability given by (55) which corresponds to times $t \gg \tau$ .It is simple to verify that the system defined by (54) is in detailed balance without changing any parameter but this refers to the stationary state defined by (55) for which the mean value of λ is zero and this means that R_μ will play no role in (54). If we observe the system (54) for times $t \ll \tau$ then λ will be constant and the system will not be in detailed balance. We see then that in this interpretation the absence of detailed balance can be understood as the fact that the far from equilibrium original system has a stationary state that is in fact a slow varying state evolving in a time scale much shorter than the characteristic time of evolution of the parameters which represent the state of the environment which are taken as constants in the time scale in which the nonequilibrium system evolves.

References

[1]:N.G.Van Kampen, Stochastic Processes in Physics and Chemestry (North Holland,1981)

[2]:H.R.Jauslin, J.Stat.Phys.42,573 (1986).

[3]:G.Nicolis, S.Martinez, E.Tirapegui, Chaos, Solitons and Fractals 1,25 (1991)

[4]:C.W.Gardiner, Handbook of Stochastic Methods (Springer-Verlag,1989)

[5]:E.Tirapegui in Proceedings of the Latin American School Of Physics 1987. ed by J.J.Giambiagi et al. (World Scientific,1988)

[6]:P.Hanggi, H.Thomas Phys.Reports 88 N4, 207 (1982)

[7]:R.Graham, Z.physik B40, 149 (1980)

[8]:P.Hanggi, Phys.Rev A25 1130 (1982)

PART IV
STATISTICAL MECHANICS
AND APPLICATIONS

SUBSTANTIAL IRREVERSIBILITY IN QUANTUM COSMOLOGY

Mario Castagnino[a,b], Fabián Gaioli[a], and Daniel Sforza[b]

[a] *Departamento de Física, Facultad de Ciencias Exactas y Naturales,*
Universidad de Buenos Aires, Ciudad Universitaria, 1428 Buenos Aires — Argentina
[b] *Instituto de Astronomía y Física del Espacio,*
Casilla de Correo 67, Sucursal 28, 1428 Buenos Aires — Argentina

ABSTRACT: In this work we study the relationship between the thermodynamics and the cosmological arrows of time in the framework of Quantum Cosmology. We apply the formalism developed by the Austin (Sudarshan and co-workers) and the Brussels (Prigogine and co-workers) groups, concerning the decaying law of unstable quantum systems using rigged Hilbert spaces, to a cosmological unstable model. We obtain an intrinsic irreversible evolution of the very early Universe by means of a diagonalization of the Hamiltonian of the system whose eigenstates evolve in a "probabilistic time."

1. Introduction

We are going to show that an irreversible behavior appears in the quantum stage of the Universe. (As Cosmology is not the core of this meeting, we will provide a brief summary in the next section.) The motivation of this work is twofold:

On the one hand, we know that there exists a clear distinction between past and future in nature, generally associated with the thermodynamics arrow of time as expressed by the Second Principle of Thermodynamics. At the cosmological level we also know that the Universe is in expansion and the direction of this expansion is called the cosmological arrow of time. The connection between both arrows is one of the most interesting problems of Physics.

On the other hand, it was suggested [1] that a resonant behaviour between the scale factor and the scalar field is responsible for an instability in the system [a Friedmann-Robertson-Walker (FRW) Universe]. This is why chaos appears in classical gravity [1, 2] where the route to chaos is reached by the breaking of the resonant tori, and a dissipative process in semiclassical gravity [3] related to entropy generation by the particle creation mechanism. We propose a very simplified model in order to understand the origin of such an instability at the quantum gravity level.

The very early Universe is the unstable system par excellence, then it is tempting to apply the well known machinery of unstable quantum systems [4]. In this line of research we use the reformulation of the problem carried out by the Austin [5] and the Brussels [6] groups who have introduced the rigged Hilbert space formulation of Quantum Mechanics [7, 8] in order to take into consideration generalized eigenvectors of the Hamiltonian corresponding to complex eigenvalues [9]. In particular, these groups have studied this formalism in the Friedrichs model [10]. Our model will be reduced to the latter in the short time approximation.

233

E. Tirapegui and W. Zeller (eds.), Instabilities and Nonequilibrium Structures V, 233–245.
© *1996 Kluwer Academic Publishers.*

In this work we study the evolution of a closed FRW Universe, conformally coupled to a real massive scalar field Φ, self-interacting with a potential $V(\Phi)$. We organize the paper as follows: in section II we briefly explain the main ideas behind Quantum Cosmology. In section III we analyze the cosmological model, and section IV is devoted to solve it. Finally, in section V we draw our conclusions.

2. What is Quantum Cosmology about?

Quantum Cosmology (QC) involves the quantization of gravity coupled to matter using the as-yet incomplete formalism of Quantum Gravity [11]. In QC the system under consideration is the whole Universe. One represents the quantum state of the Universe by a wave function $\psi[h_{ij}, \Phi]$, a functional on superspace, the space of all three-metrics h_{ij} and matter field configurations Φ on a three-surface. Since it is a single system the concept of probability has no place in standard QC. Many different interpretations were suggested but the problem is still open.

The full superspace formalism of QC is very difficult to deal with in practice, because it involves differential equations in an infinite number of variables. Then, the attention has been focused on simplified models whose configuration space is a finite-dimensional approximation to superspace, called minisuperspace model, and only some modes of the matter field are taken into account while the rest are frozen. For example, one has the state characterized by a (scale factor of a FRW metric) and Φ (the homogeneous mode of the matter field).

The Hamiltonian version of General Relativity (previous step to the canonical quantization of the theory) requires to separate the space-time in space and time. One starts cutting the space-time with a spatial hipersurface and reducing the four-metric to an induced three-metric on the three-surface. The standard $3 + 1$ [12] form of the metric is

$$ds^2 = -(N^2 - N_i N^i)dt^2 + 2N_i dx^i dt + h_{ij} dx^i dx^j. \tag{2.1}$$

where N and N_i are known as lapse function and displacement vector, respectively. We assume $N = N(t)$ and $N_i = 0$. The next step in the derivation of the field equations is to write the superspace Einstein-Hilbert action of the gravitational field in terms of h_{ij}, N, and N_i. This reads

$$S = \int \mathcal{L}dt = \int d^4 x \sqrt{-g} N \{ \frac{m_P^2}{12} R - \frac{1}{2} [g^{\mu\nu} \partial_\mu \Phi \partial_\nu \Phi + \xi R \Phi^2 + V(\Phi)] \}, \tag{2.2}$$

where $g^{\mu\nu}$ is the 4-metric tensor, $g = det(g^{\mu\nu})$, R is the curvature scalar, ξ is the coupling factor between gravity and matter (in our case $\xi = \frac{1}{6}$, conformal coupling), and m_P is the Planck mass. If we now restrict the metric and only retain the nonfrozen modes of the matter field the action (2.2) (in this minisuperspace model) is reduced to

$$S = \int L dt = m_P^2 \int dt N \left[\frac{1}{2N^2} f_{ab}(q^c) \dot{q}^a \dot{q}^b - V(q) \right]. \tag{2.3}$$

where f_{ab} is a metric on minisuperspace and $q^a(t)$ represents certain components of the three-metric h_{ij} (e.g., in the case of a FRW Universe $h_{ij} = a^2 \delta_{ij}$, where a is the scale factor) and certain modes of the matter field (e.g., the homogeneous one, Φ). We easily see that the action (2.3) has the form of the usual action of a relativistic spinless particle in a curved background, with a kinetic and a potential term. The lapse function $N(t)$ plays the role of a Lagrange multiplier expressing the reparametrization invariance of S (that guarantees the general covariance).

Variation with respect to q^a yields the Einstein field equations. Variation with respect to N yields the constraint,

$$\frac{1}{2N^2} f_{ab} \dot{q}^a \dot{q}^b + V(q) = 0,$$

which is the 0-0 Einstein equation.

Canonical momenta are defined as usual,

$$p_a = \frac{\partial L}{\partial \dot{q}^a} = m_P^2 f_{ab} \frac{1}{N} \dot{q}^b.$$

The Hamiltonian is given by

$$H = p_a \dot{q}^a - L = N \left[\frac{1}{2m_P^2} f^{ab} p_a p_b + m_P^2 V(q) \right]. \tag{2.4}$$

Since N is arbitrary the Hamiltonian vanishes,

$$H = 0. \tag{2.5}$$

Using the substitution $p_a \longrightarrow -i \nabla_{q^a}$ (∇ stands for covariant derivative), the Hamiltonian constraint is canonically quantized yielding the Wheeler-DeWitt equation [13],

$$\left[-\frac{1}{2m_P^2} \Box + m_P^2 V \right] \psi(q) = 0. \tag{2.6}$$

The solution of this equation is called the wave function of the Universe. Equation (2.6) has not the form of the Schrödinger equation, the time dependence of ψ has disappeared. Thus, there also exists the difficulty to find an adequate Hilbert space for ψ. These facts are the origin of the problem of time in Quantum Gravity [14]. To recover this concept in this minisuperspace model we can build up a "cosmological time" breaking the invariance under temporal reparametrizations, fixing the lapse function N. In this way a Schrödinger equation naturally arises instead of the Wheeler-DeWitt one. This can be achieved using the Unruh-Wald [15] proposal, where N is taken to be a^{-3} in a FRW Universe [16], or a more general one [17], where N does not appear as a Lagrange multiplier since the constraint (2.5) is removed. In the first case we are dealing with a unimodular Einstein's theory in the classical limit, and in the second case we have general relativity on-shell and general relativity with an arbitrary classical potential in the off-shell classical theory. In all cases the wave function of the Universe depends on a cosmological time τ, $\psi(\tau, q^a)$.

For the sake of briefness we leave the Schrödinger equation derived by using this concept of time for the next section, where we shall study our concrete model.

3. The model

Let us consider a closed FRW Universe (isotropic and homogeneous) with a real scalar matter field. In such a case, the Einstein equations can only be satisfied if the quantum matter field is also homogeneous (as in minisuperspace models). To obtain a more realistic and consistent model, with inhomogeneous matter fields, it would be necessary to introduce inhomogeneous metric degrees of freedom (which, in a first approximation, can be described by gravitons). Therefore, we would have only two degrees of freedom (a and Φ), but we must go beyond this minisuperspace in order to have a statistical system. What we can do is to fix the value of $N_i = 0$ to retain all the inhomogeneous modes of the scalar field [3] and to consider the inhomogeneous graviton modes as having the same qualitative effects as any other matter field. In this way we reobtain an infinite-dimensional space, called midisuperspace [18].

Decomposing the matter field Φ in terms of the eigenfunctions of the spatial Laplacian as

$$\Phi(t, x) = \frac{1}{a} \sum_n \phi_n(t) Q_n(x),$$

we obtain the action for our model,

$$S = \frac{1}{2} \int dt \{ -m_P^2 \frac{a\dot{a}^2}{N} + m_P^2 N a + \sum_n [\frac{a}{N}\dot{\phi}_n^2 - \frac{N}{a}(n^2 + m^2 a^2)\phi_n^2] \}, \qquad (3.1)$$

from which we derive the Hamiltonian

$$H = \frac{N}{2a} \{ -\frac{1}{m_P^2}\pi^2 - m_P^2 a^2 + \sum_n [p_n^2 + (n^2 + m^2 a^2)\phi_n^2] \}, \qquad (3.2)$$

where m is the mass of the matter field and the canonical momenta associated with a and ϕ_n are $\pi = (-m_P^2/N)\dot{a}$ y $p_n = (a/N)\dot{\phi}_n$, respectively. The mass acts as a coupling constant whose intensity is a measure of the interaction between the subsystem given by the scalar factor part and the bath given by the matter field part. (Note the minus sign in the first oscillator term. This is a common feature of all cosmological models and it shows that the energy obtained in particle creation processes by the matter field is provided by the gravitational term, as it happens, for example, in semiclassical gravity.)

After quantization and using the "cosmological time" mentioned above we obtain a Schrödinger equation (in "coordinate" representation) which reads

$$i\frac{\partial \psi}{\partial \tau} = \{\frac{1}{2m_P^2}\partial_a^2 - \frac{m_P^2}{2}a^2 + \frac{1}{2}\sum_n[-\partial_{\phi_n}^2 + (n^2 + m^2a^2)\phi_n^2]\}\psi. \tag{3.3}$$

In the case of paper [17] all that matter is the "on-shell" theory where the factor N/a of Eq. (3.2) is irrelevant. For convenience we will take a Hamiltonian equivalent to the one derived from the choice $N = a$ in the "off-shell" theory. The Hamiltonian can be rewritten in terms of the usual creation and annihilation operators as

$$H = -\omega_0 a^\dagger a + \int_0^\infty d\omega \omega b_\omega^\dagger b_\omega + \lambda \int_0^\infty d\omega g(\omega)(a + a^\dagger)^2(b_\omega + b_\omega^\dagger)^2 + cte \tag{3.4}$$

where $\lambda g(\omega) = \frac{m^2}{8\omega}$, $\omega_0 = 1$, and we have regularized the expression of H and we have slightly modified it without substantially affecting the qualitative behavior of the system [19].

It is very hard to remove the interaction term. However, we can obtain a diagonalization of the Hamiltonian by means of a qualitative analysis.

It is easy to check that the interaction term does not conserve the number of particles. From a physical point of view, cosmological particle production can be schematically split into three stages [20]. First, the quantum creation regime, where particles are produced from vacuum fluctuations and from n-particle states, both spontaneously and by stimulation. Second, the particle interaction regime, where particles decay or scatter with each other and possibly settle into thermal equilibrium. Third, the classical regime, where matter becomes collision-dominated while undergoing expansion like a relativistic fluid. The division of these regimes is not precise but depends on the natural frequency of the normal modes of the system, the interaction rate of particles, and the dynamics of the Universe. These processes can also overlap with each other, e.g., particles are produced not only from the vacuum but also from particle interactions, and they are subjected to redshifting all the time during expansion. Entropy generation from the classical regime is well understood and has been calculated using finite temperature field theory. It is also easy to understand the cause of entropy generation in the interaction regime, as the scattering or decaying of particles change the correlation of the initial states. However, at the opening stage of quantum vacuum creation it was not at all clear which mechanism, if any, is directly responsible for entropy generation. Since the evolution of the system strictly follows quantum mechanical laws which were considered to be time-reversal invariant, without the introduction of interactions or of any measure to alter the correlation, no entropy change in the system was expected.

We propose a simple model where particle creation appears as a consequence of a resonance between the modes a and b_ω's. Taking into account the previous discussion, for very small times (namely for the quantum creation regime), we can consider small occupation numbers only because we can consider that the spontaneous creation of particles is dominant with respect to the stimulated one. Then, for the very early Universe we can consider only pair creation in each mode and for the lowest occupation numbers [21]. Under these conditions the Hamiltonian of the model is reduced to

$$
\begin{pmatrix}
-\omega_0 + 3\lambda \int g(\omega)d\omega & \cdots & 3\sqrt{2}\lambda g(\omega) & \cdots \\
\vdots & \ddots & & \\
3\sqrt{2}\lambda g(\omega) & & -\omega_0 + 2\omega + 3\lambda g(\omega) + 3\lambda \int g(\omega)d\omega & \\
\vdots & & & \ddots
\end{pmatrix}, \qquad (3.5)
$$

where this matrix representation stands for the states
$$|1,0,...,0\rangle \equiv |1\rangle \quad \text{and} \quad |1,...,2_\omega,...\rangle \equiv |\omega\rangle,$$

where in the first position we have put the oscillator and in the remaining ones the bath modes.

Redefining

$$
\bar{\omega}_0 = -\omega_0 + 3\lambda \int g(\omega)d\omega \tag{3.6}
$$

$$
\bar{\omega} = \bar{\omega}_0 + 2\omega + 3\lambda g(\omega) \tag{3.7}
$$

we obtain a Friedrichs-like model [10] for this subspace, in the form

$$
\begin{pmatrix}
\bar{\omega}_0 & \cdots & 3\sqrt{2}\lambda g(\omega) & \cdots \\
\vdots & \ddots & & \\
3\sqrt{2}\lambda g(\omega) & & \bar{\omega} & \\
\vdots & & & \ddots
\end{pmatrix}. \tag{3.8}
$$

The Hamiltonian (3.8) corresponds to the one-particle sector of an oscillator in a bosonic heat bath [22], which can be rewritten as

$$
H = \omega_0|1\rangle\langle 1| + \int_0^\infty d\omega\, \omega|\omega\rangle\langle\omega| + 3\sqrt{2}\lambda \int_0^\infty d\omega\, g(\omega)(|1\rangle\langle\omega| + |\omega\rangle\langle 1|), \tag{3.9}
$$

where $\omega_0 \in \mathcal{R}_{\geq 0}$ and its Hilbert space is $\mathcal{H} = \mathcal{C} \oplus \mathcal{L}^2(0, \infty)$. For the sake of simplicity we have eliminated all bars (we must also consider, $g(\omega) \sim g(\bar{\omega})$, because we neglect the terms of superior order in λ). We will do that from now on.

The orthonormality relations are

$$\langle 1|1 \rangle = 1, \quad \langle \omega|\omega' \rangle = \delta(\omega - \omega'), \quad \langle 1|\omega \rangle = \langle \omega|1 \rangle = 0. \tag{3.10}$$

and the closure relation is

$$|1\rangle\langle 1| + \int_0^\infty d\omega |\omega\rangle\langle \omega| = 1. \tag{3.11}$$

The eigenvalue problem can be solved in an exact manner [23]. This well known solution includes an unstable state ($|1\rangle$ in our case) which decays towards the future in the time τ, promoting transitions to states with increasing number of particles (pair creation).

4. A brief sketch of the solution

Solving the eigenvalue problem of (3.9) we obtain eigendistributions that must be interpreted in an adequate extension of the Hilbert space. The eigenvectors read

$$|\tilde{\omega}\rangle = |\omega\rangle + \frac{\lambda g(\omega)}{\alpha_+(\omega)}(|1\rangle + \int_0^\infty d\omega' \frac{\lambda g(\omega')}{\omega - \omega' + i\epsilon}|\omega'\rangle).$$

$$\tag{4.1}$$

$$\langle \tilde{\omega}| = \langle \omega| + \frac{\lambda g(\omega)}{\alpha_-(\omega)}(\langle 1| + \int_0^\infty d\omega' \frac{\lambda g(\omega')}{\omega - \omega' - i\epsilon}\langle \omega'|).$$

where $\alpha_\pm(\omega) \equiv \alpha(\omega \pm i\epsilon)$ and $\alpha(z) = z - \omega_0 - \lambda^2 \int_0^\infty d\omega \frac{g^2(\omega)}{z - \omega}$ is the reduced resolvent of H, and we have absorbed in λ all numerical factors.

The spectral decomposition of H becomes

$$H = \int_0^\infty d\tilde{\omega} \ \tilde{\omega}|\tilde{\omega}\rangle\langle \tilde{\omega}|. \tag{4.2}$$

To make an exponential decaying law explicit we can extend the problem analytically to the complex plane. The function $\alpha(z)$ is not analytic in all complex plane since it has a positive real cut. By deforming the contour of integration to perform the calculation, considering $g(\omega)$ has an analytical extension to the lower half-plane, we can extend $\alpha(z)$ to the half-plane corresponding to the new curve Γ (this curve is in the lower half-plane below the poles of the integrand of Eq. (4.2) and it has the same end points as the positive real axis). There are two possible maximum extensions for $\alpha(z)$, $\alpha_+(z)$ and $\alpha_-(z)$, onto the upper and the lower half-plane, respectively. To define $\alpha_+(z)$ in the lower half-plane we must cross the cut. The extension that takes into account such a discontinuity is

$$\alpha_{II}(z) = \alpha(z) + 2\pi i \lambda^2 |g(z)|^2, \tag{4.3}$$

where II stands for the second Riemann-sheet.

In this way $\alpha_{II}(z)$ has a zero in $z = z_0$. For small values of λ, z_0 can be estimated as

$$z_o \approx [\omega_0 + P \int_0^\infty d\omega \frac{\lambda^2 |g(\omega)|^2}{\omega_0 - \omega}] + i[-\pi \lambda^2 |g(\omega_0)|^2], \tag{4.4}$$

where P denotes principal part. The usual expression for z_0 is

$$z_0 = \hat{\omega}_0 - i\frac{\gamma}{2}, \tag{4.5}$$

where $\hat{\omega}_0 - \omega_0$ is the level shift and $1/\gamma$ is the mean-life time of the unstable level.

Performing complex integration using the residues theorem, a complex contribution coming from the pole at z_0 gives rise to a new set of states that fulfill the spectral resolution of H. These eigenvectors are

$$|1^-\rangle = [\alpha'_{II}(z_0)]^{-\frac{1}{2}} (|1\rangle + \int_0^\infty d\omega \frac{\lambda g(\omega)}{[z_0 - \omega]_+} |\omega\rangle), \tag{4.6}$$

$$\langle 1^+| = [\alpha'_{II}(z_0)]^{-\frac{1}{2}} (\langle 1| + \int_0^\infty d\omega \frac{\lambda g(\omega)}{[z_0 - \omega]_+} \langle\omega|), \tag{4.7}$$

corresponding to the discrete part (where $[z_0 - \omega]_+$ must be understood as an indication that the curve of integration is the positive real axis with an extension that passes nearly below the pole z_0), and

$$|\omega^-\rangle = |\omega\rangle + \frac{\lambda g(\omega)}{\alpha_{II}(\omega)} [|1\rangle + \int_0^\infty d\omega' \frac{\lambda g(\omega')}{\omega - \omega' + i\epsilon} |\omega'\rangle],$$

$$\langle\omega^+| = \langle\omega| + \frac{\lambda g(\omega)}{\alpha_-(\omega)} [\langle 1| + \int_0^\infty d\omega' \frac{\lambda g(\omega')}{\omega - \omega' - i\epsilon} \langle\omega'|], \tag{4.8}$$

corresponding to the continuous solutions.

The new orthonormality relations read

$$\langle 1^+|1^-\rangle = 1, \quad \langle 1^+|\omega^-\rangle = \langle\omega^+|1^-\rangle = 0, \quad \langle\omega^+|\omega'^-\rangle = \delta(\omega - \omega'). \tag{4.9}$$

The closure relation is

$$|1^-\rangle\langle 1^+| + \int_0^\infty d\omega |\omega^-\rangle\langle\omega^+| = 1. \tag{4.10}$$

Finally, the spectral decomposition of H is

$$H = z_0|1^-\rangle\langle 1^+| + \int_0^\infty d\omega\, \omega |\omega^-\rangle\langle\omega^+|. \tag{4.11}$$

We have obtained the last relations only in a formal way. We must give an appropriate mathematical structure where all the new elements be rigorously defined. Anyhow the complex eigenvalue z_0, satisfying

$$\langle 1^+|H = z_0\langle 1^+|$$

$$\tag{4.12}$$

$$H|1^-\rangle = z_0|1^-\rangle$$

for right [Eq. (4.6)] and left [Eq. (4.7)] eigenvectors, has no sense in the standard Hilbert space formulation of Quantum Mechanics. However, H admits a self-adjoint extension to *rigged Hilbert spaces*, where the states corresponding to complex eigenvalues acquire a rigorous meaning [24, 25, 26, 27]. In order to continue the exposition a brief comment about rigged Hilbert spaces is needed.

4.1. RIGGED HILBERT SPACES IN QUANTUM MECHANICS

If we restrict the topology of the Hilbert space $\mathcal{H}$ in an appropriate way we find a smaller space which turns out to be a nuclear space Υ such that $\Upsilon \subset \mathcal{H}$. Then, the linear functionals on Υ belong to a bigger space $\Upsilon^\times$, the topological dual of Υ. Then we have defined a Gel'fand triplet [7]: $\Upsilon \subset \mathcal{H} \subset \Upsilon^\times$.

Let A be a linear operator on Υ, then F is a generalized eigenvector of A if

$$F(A\varphi) = \alpha F(\varphi), \quad \forall\varphi \in \Upsilon.$$

We can introduce the action of the operator A, namely $A^\times$, acting on $\Upsilon^\times$ through

$$\langle\varphi A^\times|F\rangle = \langle A\varphi|F\rangle = \alpha\langle\varphi|F\rangle,$$

and then

$$A^\times|F\rangle = \alpha|F\rangle, \quad \forall F \in \Upsilon^\times.$$

In this way we can define the eigenvalues and eigenvectors of any operator A in $\Upsilon^\times$. (The $^\times$ in operators will be implicit from now on.)

The nuclear spaces Υ must be chosen in such a way that all expressions like $\langle\varphi|F\rangle$ be convergent.

5. Discussion

Going back to our case the last condition in the preceeding section is achieved using Hardy class functions $\mathcal{H}_\pm$ [28, 29, 25]. The corresponding Gel'fand triplets now read $\mathcal{H}_\pm \subset \mathcal{H} \subset \mathcal{H}_\pm^\times$. Then, from the Paley-Wiener theorem [28] we know that the time evolution splits into two semigroups, i.e.

$$e^{-iH\tau}|1^-\rangle = e^{-i\tilde{\omega}_0\tau} e^{-\frac{\gamma}{2}\tau}|1^-\rangle, \quad \tau > 0$$
$$e^{-iH\tau}|1^+\rangle = e^{-i\tilde{\omega}_0\tau} e^{\frac{\gamma}{2}\tau}|1^+\rangle, \quad \tau < 0 \tag{5.1}$$

Then, the state $|1^-\rangle$ decays towards the past and the state $|1^+\rangle$ decays towards the future. There are two rigged Hilbert spaces associated with these two states, namely $\mathcal{H}_-^\times$ and $\mathcal{H}_+^\times$ respectively. The physical meaning of the $\mathcal{H}_\pm^\times$-states is the following: The states of $\mathcal{H}_-^\times$ correspond to the evolution from an unstable state towards a stable one (the equilibrium). The states of $\mathcal{H}_+^\times$ correspond to the formation process (that for systems with a large mean-life can be neglected).

We can also find a Lyapunov function and an entropy [30] to characterize the irreversible evolution, using a new representation by means of a non-unitary transformation originally proposed by Prigogine and co-workers [31, 32]. But it is sufficient to give just one "superselection rule" in order to find irreversibility:

The state of the Universe belongs to $\mathcal{H}_-^\times$ (or $\mathcal{H}_+^\times$).

In this way we find a substantial difference between past and future because we call future the direction where the states decay. The choice of $\mathcal{H}_-^\times$ or $\mathcal{H}_+^\times$ is a conventional one, since essentially the two spaces are identical. However in each space past is substantially different from future. (We have neglected the formation process of the unstable initial state of the Universe. In our case this fact is a very reasonable hypothesis and arises as a natural condition because we actually ignore the formation process of this unstable state.)

There are two ways to recover classical Einstein equation [cf. Eq. (2.5)]: a) to go to the "on-shell" theory where the state of the Universe will be the corresponding to the eigenvalue 0, but in this case we lost the unstable state, b) choose as state of the Universe an unstable one like $|1^+\rangle$ or $|1^-\rangle$ (as it is usually considered, e.g. a tunneling state [33]), and in this case the classical Einstein equation is recovered as a mean value $\langle H \rangle = 0$ [cf. Ref. [34]].

The survival amplitude of the state $|1\rangle$ can be shown to have a dominant contribution coming from the exponential term (related to the complex eigenvalue)

$$\langle 1|e^{-iH\tau}|1\rangle = \frac{1}{\alpha'_{II}(z_0)} e^{-iz_0\tau} + \lambda^2 \int \frac{e^{-i\omega\tau} g^2(\omega)}{\alpha_+(\omega)\alpha_-(\omega)} d\omega. \tag{5.2}$$

(This result is valid under the approximations done above, for a short time but greater than the *Zeno* period [35] which is very little.) We see that the state $|1\rangle$ is unstable in favour of transitions to states with greater number of particles. Simultaneously, the radius of the Universe (the scale factor a), coordinated with the "time" τ, grows.

References

[1] E. Calzetta and C. El Hasi, *Chaotic Friedmann-Robertson-Walker Cosmology*. Class. Quantum Grav. **10**, 1825-1841 (1993).

[2] S. Blanco, M. Castagnino, G. Domenech, C. El Hasi, O. Rosso, and N. Umérez, *Chaos in Classical Cosmology*, (unpublished).

[3] E. Calzetta, M. Castagnino, and R. Scoccimarro, *Coarse-graining approach to Quantum Cosmology*. Phys. Rev. **D45**, 2806-2813 (1992).

[4] T.D. Lee, *Some special examples in renormalizable Field Theory*, Phys. Rev. **95**, 1329-1334 (1954); V. Glaser and G. Källén, *A model of an unstable particle*, Nucl. Phys. **2**, 706-722 (1956); N. Nakanishi, *A theory of clothed unstable particles*, Prog. Theor. Phys. **19**, 607-621 (1958); L.A. Khalfin, *Contribution to the decay theory of a quasi-stationary state*, Soviet Physics JETP. **6**, 1053-1063 (1958) (translated from Exptl. Theoret. Phys. (USSR) **33**, 1371-1382 (1957)); M. Lévy, *On the description of unstable particles in Quantum Field Theory*, Nuovo Cimento **13**, 115-143 (1959); M. Lévy, *On the validity of the exponential law for the decay of an unstable particle*, Nuovo Cimento **14**, 612-624 (1959); J. Schwinger, *Field Theory of unstable particles*, Ann. Phys. (NY) **9**, 169-193 (1960); M.L. Goldberger and K.M. Watson, *Lifetime and decay of unstable particles in S-matrix theory*, Phys. Rev. **136**, B1472-B1480 (1964); A. Messiah, *Mécanique Quantique*, (Dunod, Paris, 1964), chap. XXI; J.L. Pietenpol. *Simple soluble model for the decay of an unstable state*, Phys. Rev **162**, 1301-1305 (1967); N. Bleistein, H. Neumann, R. Handelsman, and L.P. Horwitz, *Analysis of a nondegenerate decay system*, Nuovo Cimento **41A**, 389-405 (1977); T.K. Bailey and W.C. Schieve, *Complex energy eigenstates in quantum decay models*, Nuovo Cimento **47A**, 231-250 (1978); I. Fonda, G.C Ghirardi, and A. Rimini, *Decay theory of unstable quantum systems*, Rep. Prog. Phys. **41**, 587-631 (1978); H. Baumgärtel, *Mathematical remarks to resonances and their eigenfunctionals in decay-scattering systems, demonstrated by means of Friedrichs' model*, in "Lectures Notes in Physics," vol. 94, eds. A. Böhm et al., (Springer-Verlag, Berlin, 1979); A. Martin, *unstable particles*, (to be published).

[5] E.C.G. Sudarshan, C.B. Chiu, and V. Gorini, *Decaying states as complex energy eigenvectors in generalized Quantum Mechanics*, Phys. Rev. **D18**, 2914-2929 (1978); G. Parravicini V. Gorini, and E.C.G. Sudarshan, *Resonances, scattering theory, and rigged Hilbert spaces*, J. Math Phys. **21**, 2208-2226 (1980); E.C.G. Sudarshan, *Quantum dynamics, metastable states, and contractive semigroups*, Phys. Rev. **A46**, 37-48 (1992); E.C.G. Sudarshan and C.B. Chiu, *Analytic continuation of quantum systems and their temporal evolution*, Phys. Rev. **D47**, 2602-2614 (1993).

[6] I.E. Antoniou and I. Prigogine, *Intrinsic irreversibility and integrability of dynamics*, Physica **A192**, 443-464 (1993). This work comes from a long series of papers of the Brussels group where they intend to unify Dynamics and Thermodynamics. See, for example, I. Prigogine, *From being to becoming*, (W.H. Freeman & Co., San Francisco, 1980); T. Petrosky and I. Prigogine, *Alternative formulation of classical and quantum dynamics for non-integrable systems*, Physica **A175**, 146-209 (1991). In the line of research of the first work, see also, I. Antoniou, *Classical and Quantum systems: Foundations and symmetries*, in "Proceedings of the 2nd International Wigner Symposium," eds. H. Doebner and F. Schroeck, (World Scientific, Singapore, 1992); I. Antoniou and I. Prigogine, *Dynamics and intrinsic irreversibility*, Nuovo Cimento **219**, 93 (1992); I Antoniou and S. Tasaki, *Generalized spectral decompositions of mixing dynamical systems*, Int. J. Quantum Chem. **46**, 425-474 (1993)

[7] I.M. Gel'fand and N.Y. Vilenkin, *Generalized Functions*, vol. 4: "Application of harmonic analysis," (Academic Press, New York, 1964).

[8] A. Böhm, *The Rigged Hilbert Space and Quantum Mechanics*, "Lectures Notes in Physics," vol. 78, eds. A. Böhm and J.D. Dollard, (Springer-Verlag, Berlin, 1978).

[9] An extensive study of the interpretation of complex eigenvalues in Quantum Mechanics was made by A.J. Kálnay and B.P. Toledo, *A reinterpretation of the notion of Localization*, Nuovo Cimento **48**, 997-1007 (1967); A.J. Kálnay, *Complex physical quantities and space-like states*, "Tachions, monopoles, and related topics," ed. E. Recami, (North-Holland, Amsterdam, 1978), p. 53-59. See also, J.M. Lévy-Leblond, *Who is afraid of nonhermitian operators? A quantum description of angle and phase*, Ann. Phys. (NY) **101**, 319-341 (1976).

[10] K.O. Friedrichs, *On the perturbation of continuous spectra*, Comm. Pure Appl. Math **1**, 361-406 (1948).

[11] A great part of this section is based on the work of J.J Halliwell, *Correlations in the wave function of the Universe*, Phys. Rev. **D36**, 3626-3640 (1987).

[12] R. Arnowitt, S. Desser, and C. Misner, *The dynamics of General Relativity*, in "Gravitation: An introduction to current research," ed. L. Witten, (John Wiley & Sons, New York, 1962), p. 227-265.

[13] B.S. DeWitt, *Quantum Theory of Gravity. I. The Canonical Theory*, Phys. Rev. **160**, 1113-1148 (1967).

[14] K.V. Kuchař, *Time and interpretations of Quantum Gravity*, in "Proceedings of the 4th Canadian Conference on General Relativity and Relativistic Astrophysics," eds. G. Kunstatter, D. Vincent, and J. Williams, (World Scientific, Singapore, 1992), p. 1-104.

[15] W.G. Unruh and R.M. Wald, *Time and the interpretation of Canonical Quantum Gravity*, Phys. Rev. **D40**, 2598-2614 (1989).

[16] It has been shown [see, M. Castagnino and F. Lombardo, *Origin and measurement of time in Quantum Cosmology*, Phys. Rev. **D48**, 1722-1735 (1993)] that if the wave function is peaked on a value of τ equal to a function $\theta(a)$ of the scale factor a, then a serves as a good clock variable and a possible candidate to this function can be the "probabilistic time" previously introduced by one of us [see, M. Castagnino, *Probabilistic time in Quantum Gravity*, Phys. Rev. **D39**, 2216-2228 (1989); M.A. Castagnino and F.D. Mazzitelli, *Notion of time and the semiclassical regime of Quantum Gravity*, Phys. Rev. **D42**, 482-487 (1990)].

[17] F.H. Gaioli and E.T. Garcia-Alvarez, *The problem of time in parametrized theories*, (submitted to Gen. Rel. Grav.).

[18] J.J. Halliwell, *Introductory Lectures on Quantum Cosmology*, in "Proceedings of the Jerusalem Winter School for Theoretical Physics: Quantum Cosmology and Baby Universe," vol. 7, eds. S. Coleman, J.B. Hartle, T. Piran, and S. Weinberg, (World Scientific, Singapore, 1991), p. 159-243.

[19] A work similar to ours but with a different choice of the relevant and irrelevant variables, where the concept of time used is not clearly defined and with a very unrealistic linearized interaction, was recently written by S. Tasaki, I. Prigogine, and P. Nardone, *Negative energy instability in a toy model of Quantum Gravity*, Preprint ULB (1992).

[20] B.L. Hu and D. Pavon, *Intrinsic measures of field entropy in cosmological particle creation*, Phys. Lett. **B180**, 329-334 (1986).

[21] A similar short time approximation was made in the study of collective motion in Nuclear Physics by Tamm and Dancoff. See, for example, A.M. Lane, *Nuclear Theory: pairing force, correlations, and collective motion*, (W.A. Benjamin, New York, 1964), chap. 6.

[22] M. Castagnino, F. Gaioli, and E. Gunzig, *Cosmological features of the time-asymmetry*, Preprint IAFE (1993).

[23] See, for example, the first paper of Ref. 5. See also, T. Petrosky, I. Prigogine, and S. Tasaki, *Quantum Theory of non-integrable systems*, Physica **A173**, 175-242 (1991).

[24] See the second work of Ref. 5.

[25] See the first paper of Ref. 6.

[26] A. Böhm, *Resonance poles and Gamow vectors in the rigged Hilbert space formulation of Quantum Mechanics*, J. Math. Phys. **22**, 2813-2823 (1981); A. Böhm, M. Gadella, and G.B. Mainland, *Gamow vectors and decaying states*, Am. J. Phys. **57**, 1103-1107 (1989); A. Böhm and M. Gadella, *Dirac kets, Gamow vectors, and Gel'fand triplets*, "Lectures Notes in Physics," vol. 348, eds. A. Böhm and J.D. Dollard, (Springer-Verlag, Berlin, 1989).

[27] M. Castagnino and F. Gaioli, *Autovalores complejos en Mecánica Cuántica*, Anales AFA **4** (1993), (in press).

[28] See, for example, P. Duren, *Theory of H^p-spaces*, (Academic Press, New York, 1970).

[29] M. Gadella, *A rigged Hilbert space of Hardy-class functions: Applications to resonances*, J. Math. Phys. **24**, 1462-1469 (1983); *A description of virtual scattering states in the rigged Hilbert space formulation of Quantum Mechanics*, ibid **24**, 2142-2145 (1983); *On the RHS description of resonances and virtual states*, ibid **25**, 2481-2485 (1984).

[30] M. Castagnino, F. Gaioli, and D. Sforza, *Evolución irreversible en Cosmología Cuántica*, (submitted to Anales AFA).

[31] I. Prigogine, C. George, F. Henin, and L. Rosenfeld, *A unified formulation of Dynamics and Thermodynamics*, Chem. Scripta **4**, 5-32 (1973).

[32] H. Hasegawa, T. Petrosky, I. Prigogine, and S. Tasaki, *Quantum Mechanics and the direction of time*, Found. Phys. **21**, 263-281 (1991).

[33] A. Vilenkin, *Creation of Universes from nothing*, Phys. Lett. **117B**, 25-28 (1982); *Birth of inflationary Universes*, Phys. Rev. **D27**, 2848-2855 (1983).

[34] M. Castagnino and R. Laura, *The cosmological essence of time asymmetry*, in "Proceedings SILARG VIII," ed. W. Rodrigues, (World Scientific, Singapore, 1994).

[35] B. Misra and E.C.G. Sudarshan, *The Zeno's paradox in Quantum Theory*, J. Math. Phys. **18**, 756-763 (1977); C.B. Chiu, E.C.G. Sudarshan, and B. Misra, *Time evolution of unstable quantum states and a resolution of Zeno's paradox*, Phys. Rev. D**16**, 520-529 (1977).

GENERALIZED SPECTRAL DECOMPOSITIONS FOR A CLASS OF ONE-DIMENSIONAL CHAOTIC MAPS

D. J. DRIEBE and H. H. HASEGAWA
Center for Studies in Statistical Mechanics and Complex Systems
University of Texas at Austin, Austin TX 78712
and
International Solvay Institutes for Physics and Chemistry
1050 Brussels, Belgium

ABSTRACT. We discuss the generalized spectral representation of the Frobenius-Perron operator for a class of piecewise-linear one-dimensional maps with non-uniform stretching factors. The class includes maps with a fractal repellor.

1. Introduction

One-dimensional chaotic maps furnish examples of unstable (semi-)dynamical systems that may be investigated in full detail due to their relative simplicity [1]. Recently, new spectral decompositions of the time evolution operator (Frobenius–Perron operator) for densities of some simple chaotic maps have been constructed [2-4]. These spectral decompositions are constructed in generalized functional spaces. The motivation for the construction of the generalized spectral decompositions was to understand how a system with unitary time evolution may nevertheless display approach to equilibrium for a class of observables. In the generalized representation, the operator which is unitary in a Hilbert space setting has eigenvalues inside the unit circle corresponding to decay modes. For the unitary, two-dimensional, baker and multibaker maps this construction has been performed and the class of observables and densities for which it is valid has been characterized [2]. The main features of this analysis may be gleaned from the spectral decompositions of the one-dimensional projections of these systems governing the stretching dynamics.

In this paper we present some results on the generalized spectral decomposition for a class of piecewise-linear one-dimensional maps of the interval. After reviewing in Section 2 the generalized spectral decomposition of the dyadic Bernoulli map we present in Section

247

E. Tirapegui and W. Zeller (eds.), Instabilities and Nonequilibrium Structures V, 247–257.
© *1996 Kluwer Academic Publishers.*

3 the general class of maps whose spectral decomposition we have constructed. Section 4 presents the results for the special case of the asymmetric Bernoulli map. Finally, in Section 5, as another example of our general case, we give the generalized spectral decomposition for a map of the interval with a gap so that a fractal repellor appears.

2. The Bernoulli Basis

We will construct the generalized spectral decomposition of the Frobenius–Perron operator for the class of maps we wish to consider by starting with a basis of generalized eigenstates of the simple dyadic Bernoulli map. We will see that this basis is very convenient starting point for maps whose Frobenius–Perron operator (or elements of a decomposition of it) satisfies a simple intertwining relation with the derivative operator.

The dyadic Bernoulli map [1] is a transformation on the unit interval given by the rule $x_{n+1} = S_B(x_n) = 2x_n$ (mod 1). The Lyapunov exponent for this map is $\log 2$. The positivity of the Lyapunov exponent for this system means that it is chaotic and that trajectories are unstable. Due to the instability of the trajectories it is more meaningful to consider the dynamical evolution of the system from the point of view of ensemble theory. The ensemble is described by a probability density.

The evolution of a probability density, $\rho(x,t)$, is given essentially by superposing trajectories. For a map, $S(x)$, the probability density will evolve as

$$\rho(x,t+1) = U\rho(x,t) \equiv \int d\tilde{x} \; \delta(x - S(\tilde{x})) \; \rho(\tilde{x},t), \tag{1}$$

where U is the Frobenius–Perron operator [1]. For a map on the interval we may express this operator as

$$U\rho(x,t) = \frac{d}{dx} \int_{S^{-1}([0,x])} \rho(\tilde{x},t)\,d\tilde{x}, \tag{2}$$

where the integration is done over the inverse branches of the map. The Frobenius–Perron operator, $\bar{U}_B$, for the Bernoulli map is then given by

$$\rho(x,t+1) = \bar{U}_B\rho(x,t) = \frac{1}{2}\left[\rho(\frac{x}{2},t) + \rho(\frac{x+1}{2},t)\right]. \tag{3}$$

The uniform density, $\rho(x,t) = 1$, is easily seen to be the invariant density (or equilibrium state) of $\bar{U}_B$. We are interested in the evolution of a nonequilibrium density. We will be considering the complete spectral representation of the Frobenius–Perron operator so we will need both its right and left eigenstates. The left eigenstates are eigenstates of the adjoint of the Frobenius–Perron operator, which is known as the Koopman operator. It acts on an observable, $A(x)$, as $U^\dagger A(x) \equiv A(S(x))$.

A spectral decomposition of $\bar{U}_B$ may not be performed in the Hilbert space of square integrable functions on the unit interval, $L^2(0,1)$, because the Koopman operator does not have any eigenstates in that space (except for the trivial constant state). The Frobenius–Perron operator admits eigenstates in $L^2(0,1)$ with any eigenvalue inside the unit circle. Recently, a complete spectral representation of $\bar{U}_B$ has been constructed [2]. It is a generalized spectral representation because the eigenstates of the Koopman operator are distributions and so belong to a generalized functional space.

The generalized spectral representation of $\bar{U}_B$ is given by

$$\bar{U}_B \cdot = \sum_{m=0}^{\infty} e^{-\gamma^{(m)}} \beta^{(m)}(x) \langle \tilde{\beta}^{(m)} | \cdot \rangle, \tag{4}$$

where $e^{-\gamma^{(m)}} = 1/2^m$ are the discrete eigenvalues of $\bar{U}_B$ associated with the right eigenstates, $\beta^{(m)}(x)$, and the left eigendistributions, $\langle \tilde{\beta}^{(m)} | \cdot \rangle$. (We are using the bra-ket notation $\langle f | g \rangle \equiv \int dx\, f^*(x)\, g(x)$ for an inner product.) Note that the decay rates, $\gamma^{(m)} = m \log 2$, are multiples of the Lyapunov exponent of the map. The right eigenstates are the Bernoulli polynomials of degree m in x. The left eigenstate $\tilde{\beta}^{(m)}(x)$ is a Schwartz distribution (i.e., functional) on the space of smooth test functions expandable in terms of the right eigenstates. The left eigenstate acts on a function as

$$\langle \tilde{\beta}^{(m)} | f \rangle = \int_0^1 dx\, \frac{d^m}{dx^m} f(x) = f^{(m-1)}(1) - f^{(m-1)}(0), \tag{5}$$

so that $\tilde{\beta}^{(m)}(x) = d^m/dx^m$ and the derivative operator is considered in the sense of a generalized function [5].

The representation (4) is valid for a distribution ρ that is a C^∞ function. (More precise characterizations of the test space are given in the references cited.) In particular, the generalized representation is not valid for a singular density corresponding to a trajectory. In this sense it is an irreducible representation. It is interesting that even though the Frobenius–Perron operator is obtained by superposing trajectories, we may obtain a representaion of the operator that is not decomposable into trajectories.

For a valid density it is sometimes convenient to do a partial spectral decomposition where we include a few decaying terms and a background term decaying more quickly than the rest. For the Bernoulli map this type of partial decomposition takes the form of an Euler-Maclaurin expansion [2,3].

Using a bra-ket notation we can write the Euler–Maclaurin expansion [6] of $f(x)$ up to Mth order as

$$\begin{aligned} f(x) &= \sum_{m=0}^{M} \beta_m(x) \int_0^1 dx'\, \frac{d^m}{dx'^m} f(x') - \mathcal{B}_M(x) \cdot \int_0^1 dx'\, \mathbf{e}^*(x') \frac{d^M}{dx'^M} f(x') \\ &= \sum_{m=0}^{M} \beta_m(x) \langle \tilde{\beta}_m | f \rangle - \mathcal{B}_M(x) \langle \mathbf{e}\tilde{\beta}_M | f \rangle, \end{aligned} \tag{6}$$

where $\mathcal{B}_M(x) \cdot \mathbf{e}(x') = \sum_k \beta_{M,k}(x)e_k(x')$ and $\beta_{M,k}(x) = -e^{2\pi i k x}/(2\pi i k)^M$ for $(k \neq 0)$, $\beta_{M,0}(x) = 0$ and $e_k(x) = e^{2\pi i k x}$. In the space of functions whose Mth derivative belongs to L_2

$$\sum_{m=0}^{M} |\beta_m\rangle\langle\tilde{\beta}_m| - |\mathcal{B}_M\rangle\langle e\tilde{\beta}_M| = I_M, \tag{7}$$

is a unit operator. We refer to the basis of this expansion as the Bernoulli basis (with remainder).

The time evolution of an initial probability density evolving under the Bernoulli map is expressed in terms of the Euler–Maclaurin expansion as

$$\begin{aligned}
\rho(x,t) &= \bar{U}_\mathrm{B}^t \rho(x,0) \\
&= \sum_{m=0}^{M} \beta_m(x)\langle\tilde{\beta}_m|\bar{U}_\mathrm{B}^t|\rho\rangle - \mathcal{B}_M(x)\langle e\tilde{\beta}_M|\bar{U}_\mathrm{B}^t|\rho\rangle \\
&= \sum_{m=0}^{M} \beta_m(x)\langle\tilde{\beta}_0|\frac{d^m}{dx^m}\bar{U}_\mathrm{B}^t|\rho\rangle - \mathcal{B}_M(x)\langle e\tilde{\beta}_0|\frac{d^M}{dx^M}\bar{U}_\mathrm{B}^t|\rho\rangle,
\end{aligned} \tag{8}$$

where we assumed that $(d^M/dx^M)\rho(x,0) \in L_2$.

The Bernoulli map satisfies the intertwining relation between the Frobenius–Perron operator and the derivative operator of $d/dx\ \bar{U}_\mathrm{B} f(x) = (1/2)\bar{U}_\mathrm{B}\ d/dx f(x)$. We may iterate this relation to obtain

$$\frac{d^m}{dx^m}\ \bar{U}_\mathrm{B}^n f(x) = (\frac{1}{2})^{mn}\bar{U}_\mathrm{B}^n\frac{d^m}{dx^m}f(x), \tag{9}$$

if $f(x)$ is at least m-times differentiable. Using the intertwining relation and that $\bar{U}_\mathrm{B}^t 1 = 1$, we have from (8) then that

$$\rho(x,t) = \sum_{m=0}^{M} e^{-\gamma^{(m)}t}\beta_m(x)\langle\tilde{\beta}_m|\rho\rangle - e^{-\gamma^{(M)}t}\mathcal{B}_M(x)\langle e\tilde{\beta}_0|\bar{U}_\mathrm{B}^t\frac{d^M}{dx^M}|\rho\rangle, \tag{10}$$

expressing the evolution of $\rho(x,t)$ in terms of the first M discrete decay modes (along with the invariant $m = 0$ mode) and a background term with more rapid decay.

3. The Piecewise-Linear Onto Map

The Bernoulli basis is convenient for constructing the spectral representation for a quite general class of maps whose Frobenius–Perron operator (or elements of a decomposition of it) satisfies an intertwining relation with the derivative operator like (9) [3,8]. A class of maps of this form are those which are made up of a succession of linear parts which map onto the interval [0,1]. Here the stretching factor may not be uniform over the interval, as it is for the dyadic Bernoulli and r-adic maps, but in general may be different in each interval of the partition. The asymmetric Bernoulli map is a special example from this class that will be considered in the next section.

Consider a map $S(x)$ on $[0, 1]$ such that there is a partition $0 = a_0 < a_1 < \cdots < a_r = 1$ of $[0, 1]$ and for each integer $i = 1, 2, \ldots, r$ the restriction of S to the interval $[a_{i-1}, a_i]$, $S_i(x)$ is linear and onto. $S_i(x)$ will then have the form

$$S_i(x) = \begin{cases} \frac{1}{a_i - a_{i-1}} x - \frac{a_{i-1}}{a_i - a_{i-1}} & a_{i-1} \le x < a_i; \\ 0 & x \notin [a_{i-1}, a_i), \end{cases} \tag{11}$$

and the full map $S(x)$ is given by $\sum_{i=1}^{r} S_i(x)$. The inverse image on each interval is

$$S_i^{-1}(x) = (a_i - a_{i-1})x + a_{i-1}. \tag{12}$$

Using (2) we obtain the Frobenius–Perron operator for this map as

$$\begin{aligned} U\rho(x) &= \frac{d}{dx} \sum_{i=1}^{r} \int_{a_{i-1}}^{(a_i - a_{i-1})x + a_{i-1}} \rho(u)\, du \\ &= \sum_{i=1}^{r} \Delta_i\, \rho(\Delta_i x + a_{i-1}), \end{aligned} \tag{13}$$

where $\Delta_i \equiv a_i - a_{i-1}$. We may thus decompose the Frobenius–Perron operator as

$$U\rho(x) = \left(\sum_{i=1}^{r} U_i \right) \rho(x), \tag{14a}$$

where

$$U_i\rho(x) = \Delta_i\rho(\Delta_i x + a_{i-1}). \tag{14b}$$

The operators U_i each satisfy their own intertwining relation with the derivative operator of

$$\frac{d^m}{dx^m}\, U_i^n \rho(x) = (\Delta_i)^{mn} U_i^n \frac{d^m}{dx^m}\rho(x). \tag{15}$$

The adjoint operators, $U_i^\dagger$, are given by

$$U_i^\dagger\rho(x) = \rho(S_{(i)}(x)) = \begin{cases} \rho\left(\frac{x - a_{i-1}}{\Delta_i}\right) & a_{i-1} \le x < a_i; \\ 0 & x \notin [a_{i-1}, a_i]. \end{cases} \tag{16}$$

Using the intertwining relation it is easy to determine the matrix elements of U in the Bernoulli basis as

$$\begin{aligned} \langle \tilde{\beta}_j | U | \beta_{j'} \rangle &= \sum_{i=1}^{r} \langle \tilde{\beta}_j | U_i | \beta_{j'} \rangle \\ &= \begin{cases} \sum_{i=1}^{r} (\Delta_i)^j \langle \tilde{\beta}_0 | U_i | \beta_{j'-j} \rangle & j \le j'; \\ 0 & j > j', \end{cases} \end{aligned} \tag{17}$$

so that U is upper triangular. We may thus decompose U into diagonal, U_0, and off-diagonal, δU, parts. Now, for $j' > j$,

$$
\begin{aligned}
\langle \tilde{\beta}_0 | U_i | \beta_{j'-j} \rangle &= \int_0^1 dx \beta_{j'-j}(x)\, U_i^\dagger 1 \\
&= \int_{a_{i-1}}^{a_i} dx\, \beta_{j'-j}(x) \\
&= \beta_{j'-j+1}(a_i) - \beta_{j'-j+1}(a_{i-1}).
\end{aligned}
\tag{18}
$$

Thus we have that,

$$
\langle \tilde{\beta}_j | U_0 | \beta_{j'} \rangle = e^{-\gamma^{(j)}} \delta_{j,j'},
\tag{19}
$$

where

$$
e^{-\gamma^{(j)}} = \sum_{i=1}^r (\Delta_i)^{j+1},
\tag{20}
$$

and

$$
\langle \tilde{\beta}_j | \delta U | \beta_{j'} \rangle = \begin{cases} \sum_{i=1}^r (\Delta_i)^j [\beta_{j'-j+1}(a_i) - \beta_{j'-j+1}(a_{i-1})] & j < j'; \\ 0 & j \geq j'. \end{cases}
\tag{21}
$$

The construction of the generalized spectral representation for Frobenius–Perron operators which are upper-triangular in some basis have been given now in several publications [2-4] so we will not reproduce the details here. The time evolution of $\rho(x,t)$ is written in terms of the resolvent of the Frobenius–Perron operator and then decaying eigenstates are obtained by picking up contributions from poles of the resolvent inside the unit circle. The poles are determined by employing the intertwining relation in the context of the Bernoulli basis where the left basis states are derivative operators [3]. For a density (i.e., the domain of the Frobenius–Perron operator) which is M-times differentiable we may explicitly determine the analytic structure of the resolvent as poles down to a radius of $e^{-\gamma^{(M)}}$.

The time evolution of the density can be written as a sum of contributions from each pole and a background integral whose contour is just outside of a circle whose radius is $e^{-\gamma^{(M)}}$ as

$$
\rho(x,t) = \sum_{m=0}^{M-1} \rho^{(m)}(x) e^{-\gamma^{(m)} t} + R^{(M)}(x,t),
\tag{22}
$$

where the pole contributions are

$$
\rho^{(m)}(x) = \frac{1}{2\pi i} \oint_{z=e^{-\gamma^{(m)}}} dz\, z^t\, \frac{1}{z - \bar{U}} \rho(x,0),
\tag{23}
$$

and the background term decaying more rapidly than the pole contributions is

$$
R^{(M)}(x,t) = \frac{1}{2\pi i} \oint_{|z|=e^{-\gamma^{(m)}}+\epsilon} dz\, z^t\, \frac{1}{z - \bar{U}} \rho(x,0),
\tag{24}
$$

with ϵ an infinitesimal positive number.

In order to evaluate $\rho_s^{(m)}(x,t)$ we assume that $(d^{m+1}/dx^{m+1})\rho(x,0) \in L_1$ and use the Euler–Maclaurin expansion up to the m-th order since the pole $1/(z - e^{-\gamma^{(m)}})$ is associated with $|\beta_m\rangle\langle\tilde{\beta}_m|$. Because of the simple pole, $\rho^{(m)}(x,t)$ can be written in terms of right and left decaying eigenstates as

$$\rho^{(m)}(x,t) = \gamma^{(m)}(x)\langle\tilde{\gamma}^{(m)}|\rho\rangle, \tag{25}$$

where

$$\gamma^{(m)}(x) = \sum_{j=0}^{m} \beta_j(x)\langle\tilde{\beta}_j|1 + \frac{1}{e^{-\gamma^{(m)}} - \bar{U}}\delta\bar{U}|\beta_m\rangle, \tag{26}$$

and

$$\langle\tilde{\gamma}^{(m)}|\rho\rangle = \langle\tilde{\beta}_m|\rho_s\rangle - \langle\tilde{\beta}_m|\delta\bar{U}|\beta_m\rangle\langle e\tilde{\beta}_m|\frac{1}{e^{-\gamma^{(m)}} - \bar{U}}|\rho_s\rangle. \tag{27}$$

4. The Asymmetric Bernoulli Map

As an example of the piecewise-linear onto map we consider here the analysis of the asymmetric Bernoulli map [7,8]

$$S(x) = \begin{cases} px & 0 \le x < 1/p; \\ q(x-1) + 1 & 1 - 1/q \le x \le 1, \end{cases} \tag{28}$$

where we assume that $1/p + 1/q = 1$. If $1/p + 1/q < 1$ a gap appears in the definition (28) and we will take the map as zero in that gap. The map is then not measure preserving with respect to Lebesgue measure. In the next section we will consider such a case.

Comparing with the previous section we see that $a_0 = 0$, $a_1 = 1/p$, and $a_2 = 1$; so that $\Delta_1 = 1/p$ and $\Delta_2 = 1/q$. The Frobenius–Perron operator may thus be decomposed into two parts following (14) as

$$U\rho(x,t) = U_p\rho(x,t) + U_q\rho(x,t), \tag{29a}$$

where

$$U_p\rho(x,t) = \frac{1}{p}\rho(\frac{x}{p}), \tag{29b}$$

and

$$U_q\rho(x,t) = \frac{1}{q}\rho(1 - \frac{1-x}{q}). \tag{29c}$$

The operators U_p and U_q satisfy the intertwining relations

$$\frac{d^m}{dx^m} U_p\rho(x) = \frac{1}{p^m}U_p\frac{d^m}{dx^m}\rho(x), \tag{30a}$$

$$\frac{d^m}{dx^m}\,U_q\rho(x) = \frac{1}{q^m}U_q\frac{d^m}{dx^m}\rho(x). \tag{30b}$$

The matrix elements of the Frobenius–Perron operator are

$$\langle\tilde\beta_j|U|\beta_{j'}\rangle = \begin{cases} \left(\frac{1}{p^j}-\frac{1}{q^j}\right)[\beta_{j'-j+1}(\frac{1}{p})-\beta_{j'-j+1}(0)] & j < j'; \\ e^{-\gamma^{(j)}} & j = j'; \\ 0 & j > j', \end{cases} \tag{31}$$

where

$$e^{-\gamma_j} \equiv \frac{1}{p^{j+1}} + \frac{1}{q^{j+1}}. \tag{32}$$

From (26) we may write the following recursion for the right eigenstates

$$\gamma^{(m)}(x) = \beta_m(x) + \sum_{j=0}^{m-1}\beta_j(x)\frac{1}{e^{-\gamma^{(m)}} - e^{-\gamma^{(j)}}}\langle\tilde\beta_j|\delta\bar U|\gamma^{(m)}\rangle, \tag{33}$$

The explicit evaluation of (33) gives for the first three right eigenstates

$$\gamma^{(0)}(x) = \beta_0(x) = 1, \tag{34a}$$

$$\gamma^{(1)}(x) = \beta_1(x) = x - \frac{1}{2}, \tag{34b}$$

$$\begin{aligned}
\gamma^{(2)}(x) &= \beta_2(x) + \frac{1}{2}\left(\frac{1}{p}-\frac{1}{q}\right)\beta_1(x) \\
&= \frac{1}{2}x^2 - \frac{1}{q}x + \frac{1}{4}\left(\frac{1}{3}-\frac{1}{p}+\frac{1}{q}\right).
\end{aligned} \tag{34c}$$

By extending the Euler–Maclaurin expansion in (27) up one order we may express the left eigenstate as

$$\langle\tilde\gamma^{(m)}|\rho\rangle = \langle\beta_0|\frac{d^m}{dx^m}\rho\rangle + \langle\xi_m|\frac{d^{m+1}}{dx^{m+1}}\rho\rangle, \tag{35}$$

where

$$\xi_m(x) = \frac{1}{e^{-\gamma^{(m)}} - \frac{U_p^\dagger}{p^{m+1}} - \frac{U_q^\dagger}{q^{m+1}}}\eta_m(x), \tag{36}$$

and

$$\begin{aligned}
\eta_m(x) &\equiv \langle\beta_0|\frac{U_p}{p^m}+\frac{U_q}{q^m}|\beta_1\rangle - \langle\beta_0|\frac{U_p}{p^m}+\frac{U_q}{q^m}|\beta_1\rangle e(x) \\
&= \begin{cases} \left(\frac{1}{p^{m+1}}-\frac{1}{p^m}+\frac{1}{q^{m+1}}\right)x & 0 \le x < 1/p; \\ \frac{1}{p^{m+1}}(x-1)+\frac{1}{q^{m+1}}x & 1/p < x < 1-1/q; \\ \left(\frac{1}{p^{m+1}}-\frac{1}{q^m}+\frac{1}{q^{m+1}}\right)(x-1) & 1-1/q < x \le 1, \end{cases}
\end{aligned} \tag{37}$$

and we have included here the case of $1/p+1/q < 1$. Since $\left|\frac{1}{p^{m+1}} - \frac{1}{q^{m+1}}\right| < e^{-\gamma^{(m)}}$ we may rewrite $\xi_m(x)$ as the convergent series

$$\xi_m(x) = e^{\gamma^{(m)}}\sum_{n=0}^{\infty}\left(\frac{U_p^\dagger}{1+(\frac{p}{q})^{m+1}}+\frac{U_q^\dagger}{1+(\frac{q}{p})^{m+1}}\right)^n\eta_m(x). \tag{38}$$

We may express the left eigenstate in terms of its Riesz representation by integrating by parts expression (35). We then obtain the eigenfunctional in terms of a Riemann-Stieltjes integration as

$$\langle \tilde{\gamma}^{(m)} | \rho \rangle = \int_0^1 dG_m(x) \frac{d^m}{dx^m} \rho(x), \qquad (39a)$$

where

$$G_m(x) = x + \xi_m(x). \qquad (39b)$$

5. A Map with a Fractal Repellor

We may easily generalize the map considered in Section 3 to include maps with gaps in a countable number of intervals. Points falling in these gaps will thus "escape" and asymptotically only points lying on a fractal set will survive. More precisely, the invariant measure for the map will be some singular fractal measure.

For illustration, we will not consider here the most general case of the piecewise-linear map with gaps but will consider the example of what we call the "Cantor map".

The Cantor map is defined by

$$S_{\mathrm{C}}(x) = \begin{cases} 3x & 0 \le x < 1/3; \\ 0 & 1/3 \le x < 2/3; \\ 3x - 2 & 2/3 \le x \le 1. \end{cases} \qquad (40)$$

All points on the "middle thirds" Cantor set escape from this map. The Frobenius–Perron operator (with respect to Lebesgue measure) for this map is

$$\rho(x, t+1) = U_{\mathrm{C}}\rho(x, t) = \frac{1}{3}\left[\rho(\frac{x}{3}, t) + \rho(\frac{x+2}{3}, t) \right]. \qquad (41)$$

The matrix elements with respect to the Bernoulli basis are

$$\langle \tilde{\beta}_j | U_{\mathrm{C}} | \beta_{j'} \rangle = \begin{cases} (1/3^j) \left[\beta_{j'-j+1}(\frac{1}{3}) - \beta_{j'+j+1}(\frac{2}{3}) \right] & j \le j'; \\ (2/3^{j+1}) & j = j'. \\ 0 & j > j', \end{cases} \qquad (42)$$

The diagonal elements are the eigenvalues

$$e^{-\gamma^{(j)}} = \frac{2}{3^{j+1}} = \frac{2}{3} e^{-j \log 3} = e^{\log 2 - (j+1) \log 3}. \qquad (43)$$

The first eigenvalue, $e^{-\gamma^{(0)}} = (2/3)$, so that any density which is absolutely continuous with respect to Lebesgue measure will decay to zero. (Alternatively, we may say that the Cantor map is a dissipative system since it does not preserve probability on its phase space. This

256

is due to the fact that the transformation is not a measure-preserving transformation with respect to Lebesgue measure.)

From (37) we obtain here ($p = q = 3$)

$$\eta_m(x) = -\frac{1}{3^{m+1}} \begin{cases} x & 0 \le x < 1/3; \\ 1 - 2x & 1/3 < x < 2/3; \\ x - 1 & 2/3 < x1. \end{cases} \tag{44}$$

Then from (38) and (39) we may write down the left eigenfunctionals.

An invariant fractal measure for this map is the Cantor function, $F_C(x)$, defined by

$$F_C(x) = \begin{cases} 1/2 F_C(3x) & 0 \le x < 1/3; \\ 1/2 & 1/3 \le x < 2/3; \\ 1/2 F_C(3x - 2) + 1/2 & 2/3 \le x \le 1. \end{cases} \tag{45}$$

This corresponds to the integral measure $G_m(x)$ so that the Riesz representation of the left eigendistributions are given by the Riemann–Stieltjes integral

$$\langle \tilde{\gamma}_C^{(m)} | f \rangle = \int_0^1 dF_C \frac{d^m}{dx^m} f(x). \tag{46}$$

6. Conclusions

The generalized spectral representation of the Frobenius–Perron operator extracts exponentially decaying modes of the probability density for chaotic systems. We have given the generalized representation for a class of maps of the interval made up of piecewise-linear parts which may have different slopes.

When there are gaps the map admits a fractal repellor. The generalized representation may also then be constructed and the left eigendistributions are seen to reflect the properties of the singular invariant fractal measure of the map.

ACKNOWLEDGEMENT. This research was supported by the Robert Welch fundation Grant No. F-0365, U.S. Department of Energy Grant No. DE-FG05-88ER13897, and the European Community Contract No. PSS*0661.

References

1. H.G. Schuster, *Deterministic Chaos* (VGH publishers, Berlin 1989);
   A. Lasota and M. Mackey, *Probabilistic Properties of Deterministic Systems*, (Cambridge University Press, Cambridge, 1985).
2. H.H. Hasegawa and W.C. Saphir, Phys. Rev. A 46, 7401 (1992);

  P. Gaspard, J. Phys. A <u>25</u>, L483 (1992);
  I. Antoniou and S. Tasaki, J. Phys. A <u>26</u>, 73 (1992);
  H.H. Hasegawa and D.J. Driebe, Phys. Rev. E , *to appear* (1994).

3. H.H. Hasegawa and D.J. Driebe, Phys. Lett. A <u>176</u>, 193 (1993).

4. P. Gaspard, Phys. Lett. A <u>168</u>, 13 (1992);
  H.H. Hasegawa and D.J. Driebe, Phys. Lett. A <u>168</u>, 18 (1992);
  S. Tasaki, I. Antoniou and Z. Suchanecki, Phys. Lett. A <u>179</u>, 97 (1993).

5. J. Mikusiński and T. Boehme, *Operational Calculus*, (Pergamon Press, Oxford, 1987).

6. G.H. Hardy, *Divergent Series*, (Oxford, 1949).

7. H.H. Hasegawa and W.C. Saphir, Phys. Lett. A <u>161</u>, 471 (1992).

8. D.J. Driebe, *Ph.D. Dissertation*, The University of Texas at Austin, (1993).

9. S. Tasaki, Z. Suchanecki and I. Antoniou, Phys. Lett. A <u>179</u>, 103 (1993).

TWO-PHOTON MICROMASER:
FROM TRAPPING TO IDEAL SQUEEZED STATES

MIGUEL ORSZAG, LUIS ROA, AND RICARDO RAMIREZ

Facultad de Física

Pontificia Universidad Católica de Chile,

Casilla 306 Santiago 22, Chile

ABSTRACT. In this paper, we study the generation of pure states in a two-photon micromaser with finite detuning of the intermediate atomic level. This study was done via the temporal evolution of the discrete master equation. We show that it is possible to generate the ideal squeezed vacuum for a broad range of detunings in a lossless cavity and away from the trapping condition. We also consider the effect of cavity losses.

I. INTRODUCTION

Micromasers have been in the Quantum Optics literature for the last decade. In a micromaser, two a three-level Rydberg atoms, prepared initially in one state or superposition state, are injected into a high Q quality factor microwave cavity. It has been shown experimentally that maser oscilations can be sustained even when the flux is so low that only one atom is in the cavity at a time.

The theoretical description of a micromaser [1] shows that the photon statistics is essentially determined by the atomic flight-time of the atom through the cavity as well as the rate of incoming atoms and the cavity decay constant.

Recently, pure states have been found for various micromaser systems, such as the one-photon [1,2] and twophoton [3,4] micromaser as well as in the λ system [5], under idealized conditions such as having a lossless cavity, infinite atomic lifetimes, and a large detuning of the intermediate atomic level, in the case of the two-photon micromaser [3]. Some of these pure states present a strong quadrature noise reduction [2,6]. Nonideal effects have also been studied in the past in a one-photon micromaser [7,8]. Examples of the above-mentioned states are the tangent and cotangent states for the one-photon micromaser and the even and odd states for the two-photon micromaser. The even and odd states [3] for the twophoton micromaser were derived under the condition of high detuning of the intermediate atomic level (threelevel system) with respect to the midpoint between the upper and lower levels and for special values of the interaction time (trapping condition). In addition, the initial state of the field, say for an even state, has to be a coherent superposition of even number states within the given trap. Alternatively, it could also be an incoherent field with nonzero density matrix elements corresponding only to even indices, again within the trap. These initial conditions are, in general, difficult to realize experimentally, with the exception of the vacuum state, which we use here. In this work, we will study how the even states are modified by the fact that the detuning is finite

259

E. Tirapegui and W. Zeller (eds.), Instabilities and Nonequilibrium Structures V, 259–271.
© 1996 *Kluwer Academic Publishers.*

and also by the effect of cavity losses. We further find in this work that pure states can be generated in the two-photon micromaser without satisfying the trapping conditions. These states turn out to be the perfectly squeezed vacuum [9].

This paper is organized as follows: In Sec. II we obtain the pure states for the field, for an arbitrary finite detuning, in a lossless cavity. In Sec. III we derive the discrete master equation for the reduced field density matrix. This equation enables us to verify dynamically the existence of those pure states. In Sec. IV we include the cavity losses in the temporal evolution and compare them with the previous results. Finally, Sec. V is a summary and conclusion.

II. TRAPPING STATES FOR ARBITRARY DETUNING

We consider the three-level atom, shown in Fig. 1, interacting with a single mode of the electromagnetic field in a lossless microwave cavity. The well-known temporal evolution operator for this system [10] is given by

$$U(\tau) = \begin{vmatrix} 1-a\hat{G}a^{\dagger} & -ia\hat{H}e^{-i\epsilon\phi} & -a\hat{G}a \\ -ie^{i\epsilon\phi}\hat{H}a^{\dagger} & e^{i\epsilon\phi}\hat{L}^{\dagger} & -ie^{i\epsilon\phi}\hat{H}a \\ -a^{\dagger}\hat{G}a^{\dagger} & -ia^{\dagger}\hat{H}e^{-i\epsilon\phi} & 1-a^{\dagger}\hat{G}a \end{vmatrix} , \tag{1}$$

where we have defined

$$\hat{H} = \frac{\sin(\phi\sqrt{2a^{\dagger}a+1+\epsilon^{2}})}{\sqrt{2a^{\dagger}a+1+\epsilon^{2}}} , \tag{2}$$

$$\hat{L} = \cos(\phi\sqrt{2a^{\dagger}a+1+\epsilon^{2}}) - i\epsilon\hat{H} , \tag{3}$$

and

$$\hat{G} = \frac{1-e^{-i\epsilon\phi}\hat{L}}{2a^{\dagger}a+1} \tag{4}$$

with $\epsilon = \Delta/2g$ and $\phi = g\tau$, Δ being the detuning (Fig. 1), g the coupling constant, and

τ the interaction time between the atoms and the cavity mode.

Here we assume that all the atoms enter the cavity in the same coherent superposition:

$$|\Phi\rangle = \alpha|a\rangle + \gamma|c\rangle \qquad (5)$$

The evolution of a pure state of the atom-field system is given by

$$
\begin{aligned}
\sum_n S_n|n\rangle(\alpha|a\rangle+\gamma|c\rangle) \rightarrow \sum_n S_n \Bigg[&\left\{\alpha\left[1-\frac{n+1}{2n+3}R(n+1)\right]|n\rangle - \gamma\frac{\sqrt{n(n-1)}}{2n-1}R(n-1)|n-2\rangle\right\}|a\rangle \\
&-ie^{i\epsilon\phi}\left\{\alpha\left[\frac{n+1}{2n+3+\epsilon^2}\right]^{1/2}\sin(\phi\sqrt{2n+3+\epsilon^2})|n+1\rangle\right. \\
&\left.+\gamma\left[\frac{n}{2n-1+\epsilon^2}\right]^{1/2}\sin(\phi\sqrt{2n-1+\epsilon^2})|n-1\rangle\right\}|b\rangle \\
&+\left\{-\alpha\frac{\sqrt{(n+1)(n+2)}}{2n+3}R(n+1)|n+2\rangle\right. \\
&\left.+\gamma\left[1-\frac{n}{2n-1}R(n-1)\right]|n\rangle\right\}|c\rangle\Bigg],
\end{aligned}
\qquad (6)
$$

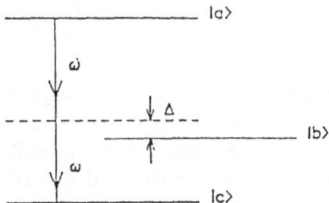

FIG. 1. Energy levels relevant to the two-photon micromaser.

where we have defined

$$R(n) = 1 - e^{-i\epsilon\phi}\left[\cos(\phi\sqrt{2n+1+\epsilon^2})\right.$$
$$\left. + i\epsilon\frac{\sin(\phi\sqrt{2n+1+\epsilon^2})}{\sqrt{2n+1+\epsilon^2}}\right]. \qquad (7)$$

From Eq. (6) we see that the upward trapping conditions are

$$\phi\sqrt{2n_u + 3 + \epsilon^2} = 2p\pi, \qquad (8a)$$

where p is an integer, and

$$\epsilon\phi = 2r\pi, \qquad (8b)$$

where r is an integer, which remove the transition from the $|n_u\rangle$ state to the $|n_u + 2\rangle$ and $|n_u + 1\rangle$ states. The following additional condition is necessary in order to remove the two-photon transition from $|n_u - 1\rangle$ to $|n_u + 1\rangle$:

$$\phi\sqrt{2n_u + 1 + \epsilon^2} = 2q\pi, \qquad (8c)$$

where q is an integer. Conditions (8a) and (8c) are inconsistent because there is no fi such that will satisfy both conditions.

For $n_u \gg 1$ or $\epsilon \gg 1$ the conditions (8a) and (8c) are approximately the same. When $\epsilon \gg 1$ all the one-photon transitions are removed, and we recover the two-photon case [3]. There is a third case with no trapping condition that will be considered next. Let us assume that the field reached a pure steady state. If one additional atom crosses the cavity, it only modifies the state of the system by a global phase factor and in general yields a different atomic state, that is,

$$\sum_{n=0} S_n |n\rangle(\alpha|a\rangle + \gamma|c\rangle)$$
$$\rightarrow \sum_{n=0} S_n |n\rangle(\alpha'|a\rangle + \beta'|b\rangle + \gamma'|c\rangle). \qquad (9)$$

Comparing the right-hand side of Eqs. (6) and (9), we obtain

$$\alpha' S_n = \alpha \left\{ 1 - \frac{n+1}{2n+3} R(n+1) \right\} S_n$$
$$- \gamma \frac{\sqrt{(n+1)(n+2)}}{2n+3} R(n+1) S_{n+2} , \tag{10a}$$

$$\beta' S_n = -ie^{i\epsilon\phi} \frac{\sin(\phi\sqrt{2n+1+\epsilon^2})}{\sqrt{2n+1+\epsilon^2}}$$
$$\times \{ \alpha\sqrt{n} S_{n-1} + \gamma\sqrt{n+1} S_{n+1} \} , \tag{10b}$$

and

$$\gamma' S_n = -\alpha \frac{\sqrt{n(n-1)}}{2n-1} R(n-1) S_{n-2}$$
$$+ \gamma \left\{ 1 - \frac{n}{2n-1} R(n-1) \right\} S_n . \tag{10c}$$

For these equations to be consistent it is necessary that

$$\alpha' = \alpha , \quad \beta' = 0 , \quad \gamma' = \gamma . \tag{11}$$

With the conditions given above, Eqs. (10) become to single equation

$$S_{n+2} = -\frac{\alpha}{\gamma} \left[\frac{n+1}{n+2} \right]^{1/2} S_n \tag{12}$$

This is the well-known recursion relation for the even an, odd states found in the two-photon micromaser model [3]. It is interesting to set $n = 0$ in Eqs. (10). Equation (10a) and (10c) are consistent with Eq. (11); however from Eq. (10b) we get $S_l = 0$. Therefore, in the case of finite detuning the steady state of the field contains only superposition of even photon-number states since all the odd components vanish. Hence, when we have no trapping condition, the solution for the recursion relation (12) gives us the squeezed vacuum state (see the Appendix). Comparing the recursion relation (12) with Eq. (A3) we find that the squeezing parameter is given by

$$\xi = e^{i\theta_{at}} \ln \left[\left[\frac{1 + \left| \frac{\alpha}{\gamma} \right|}{1 - \left| \frac{\alpha}{\gamma} \right|} \right]^{1/2} \right] \tag{13}$$

where θ_{at} is the relative atomic phase between α and γ. For $\theta_{at} = \pi$ we get noise reduction in the a_2 quadrature [9]. The fluctuations of a_2 are given by (A4b),

$$\langle (\Delta a_2)^2 \rangle = \frac{1}{4} \left\{ \frac{1 - \left| \dfrac{\alpha}{\gamma} \right|}{1 + \left| \dfrac{\alpha}{\gamma} \right|} \right\}. \tag{14}$$

From this last equation we see that in the neighbourhood of $|\alpha/\gamma| = 1.$, slightly to the left, the quantum noise can be made nearly zero. A valid question is, how can we generate one of these states? This question is answered in the following section.

III. GENERATION OF PURE STATES VIA THE TEMPORAL EVOLUTION

In this section we consider the dynamic behaviour of the reduced density matrix of the field including arbitrary detuning. We have a special interest in the steady state reached by the cavity field. We calculate the reduced field density matrix by using the time evolution operator given by Eq. (1), after the ($k + 1$)th atom has crossed the cavity, that is,

$$\rho^{k+1} = \text{Tr}_{\text{at}}[U(\tau)\rho_{\text{at}}\rho^k U^\dagger(\tau)], \tag{15}$$

where ρ_{at} is the initial atomic density matrix and ρ_k is the reduced density matrix of the field after interacting with the kth atom. Using the photon-number basis, after a straightforward calculation, we obtain the following discrete master equation:

$$
\begin{aligned}
\rho_{n,m}^{k+1} =& \left\{ |\alpha|^2 \left[1 - \frac{n+1}{2n+3} R(n+1) \right] \left[1 - \frac{m+1}{2m+3} R^*(m+1) \right] \right. \\
&\left. + |\gamma|^2 \left[1 - \frac{n}{2n-1} R(n-1) \right] \left[1 - \frac{m}{2m-1} R^*(m-1) \right] \right\} \rho_{n,m}^k \\
&+ |\alpha|^2 \frac{\sqrt{n(n-1)m(m-1)}}{(2n-1)(2m-1)} R(n-1)R^*(m-1)\rho_{n-2,m-2}^k \\
&+ |\gamma|^2 \frac{\sqrt{(n+1)(n+2)(m+1)(m+2)}}{(2n+3)(2m+3)} R(n+1)R^*(m+1)\rho_{n+2,m+2}^k \\
&- \gamma\alpha^* \left[1 - \frac{m+1}{2m+3} R^*(m+1) \right] \frac{\sqrt{(n+1)(n+2)}}{2n+3} R(n+1)\rho_{n+2,m}^k \\
&- \gamma\alpha^* \left[1 - \frac{n}{2n-1} R(n-1) \right] \frac{\sqrt{m(m-1)}}{2m-1} R^*(m-1)\rho_{n,m-2}^k \\
&- \alpha\gamma^* \left[1 - \frac{n+1}{2n+3} R(n+1) \right] \frac{\sqrt{(m+1)(m+2)}}{2m+3} R^*(m+1)\rho_{n,m+2}^k \\
&- \alpha\gamma^* \left[1 - \frac{m}{2m-1} R^*(m-1) \right] \frac{\sqrt{n(n-1)}}{2n-1} R(n-1)\rho_{n-2,m}^k \\
&+ \frac{\sin(\phi\sqrt{2n+1+\epsilon^2})\sin(\phi\sqrt{2m+1+\epsilon^2})}{\sqrt{(2n+1+\epsilon^2)(2m+1+\epsilon^2)}} \\
&\times [|\alpha|^2\sqrt{nm}\, \rho_{n-1,m-1}^k + |\gamma|^2\sqrt{(n+1)(m+1)}\rho_{n+1,m+1}^k \\
&+ \gamma\alpha^*\sqrt{(n+1)m}\, \rho_{n+1,m-1}^k + \alpha\gamma^*\sqrt{n(m+1)}\rho_{n-1,m+1}^k].
\end{aligned} \tag{16}
$$

The last four terms account for the one-photon transitions and they depend on a factor $1/(2n+1+\epsilon^2) < 1$. These terms are very small in the regime $|\epsilon| \gg 1$. In the following, we present and discuss some numerical calculations of the temporal evolution of this discrete master equation. Figure 2(a) shows the quadrature fluctuations $\langle(\Delta a_2)^2\rangle$ for three times, as a function of e, the detuning parameter, in logarithmic scale. The initial condition is the vacuum state, $\alpha/\gamma = -\sqrt{0.7}$ and the reduced interaction time is $\phi = 4\pi\epsilon/23$, which corresponds to $n_u = 10$ for high detuning [3]. We observe that for $\epsilon > 10^3$ the value of $\langle(\Delta a_2)^2\rangle$ is in agreement with the value corresponding to the trapped even states. In the region $\epsilon < 10$ the fluctuations are constant and smaller than in the previous region. The value $\langle(\Delta a_2)^2\rangle = 0.022$ is in agreement with the corresponding number obtained from Eq. (14) for $|\alpha/\gamma| = \sqrt{0.7}$. In Fig. 2(b) we show the corresponding entropy for the same cases as in Fig. 2(a). Here we observe that the entropy vanishes in the two s regions discussed above. Hence, in those regions we have pure states. Figure 2(c) shows the ratio (versus detuning)

$$r_1 = \frac{\sum_{n=1,3,5\ldots} \rho_{n,n}}{\sum_{n=0,2,4\ldots} \rho_{n,n}}, \tag{17}$$

which gives us an idea about the behaviour of the onephoton transitions and how these are removed in time. We see that in the ϵ prameter ranges specified above only the even photon-number components survive. Hence, in the steady state, the one-photon transitions have been totally suppressed. Figure 2(d) shows the ratio (versus detuning)

$$r_2 = \frac{\sum_{n=11,12,13\ldots} \rho_{n,n}}{\sum_{n=0,1,2,\ldots,10} \rho_{n,n}}, \tag{18}$$

which gives us information about the probability leakage from the even trap corresponding to $n_u = 10$ Here we see that $n_u = 10$ is upward trapping state for $\epsilon > 10^3$. On the other hand, for $\epsilon > 10$, this ratio is a constant different from zero. This number agrees with the r_2 value as obtained from the ideal squeezed vacuum [9].

Figure 3(a) shows the quadrature fluctuations as function of l, which is number of atoms crossing the cavity. Here the interaction time $\phi = 4\pi\epsilon/23$, $|\alpha/\gamma| = \sqrt{0.7}$, $\epsilon = 1$, and the initial condition corresponds to a coherent state with $\langle n \rangle = 16$. We observe that in steady state, the quadrature fluctuations reduce to the same value as obtained with the squeezed vacuum. Figure 3(b) shows the entropy for the previous case. It goes to zero when approaching steady state. Figure 3(c) shows the temporal evolution of the ratio r_1 which also vanishes in steady state; thus only the coherent superposition of even photon-number states survive. The same is true with any other initial condition, therefore, we have numerical evidence to state that the two-photon micromaser with arbitrary detuning generates a perfectly squeezed vacuum, with the complex squeezing parameter given by Eq. (13).

IV. THE EFFECT OFF CAVITY LOSSES

In this section we study the effects of the cavity losses on the pure states, through the dynamic behaviour of the discrete master equation for the field. Here we consider atoms being injected at a constant rate $R < \tau^{-1}$ We also assume that the cavity damping time tcav is much larger than $t_{at} = R^{-1}$ and τ. With these assumptions, the field

266

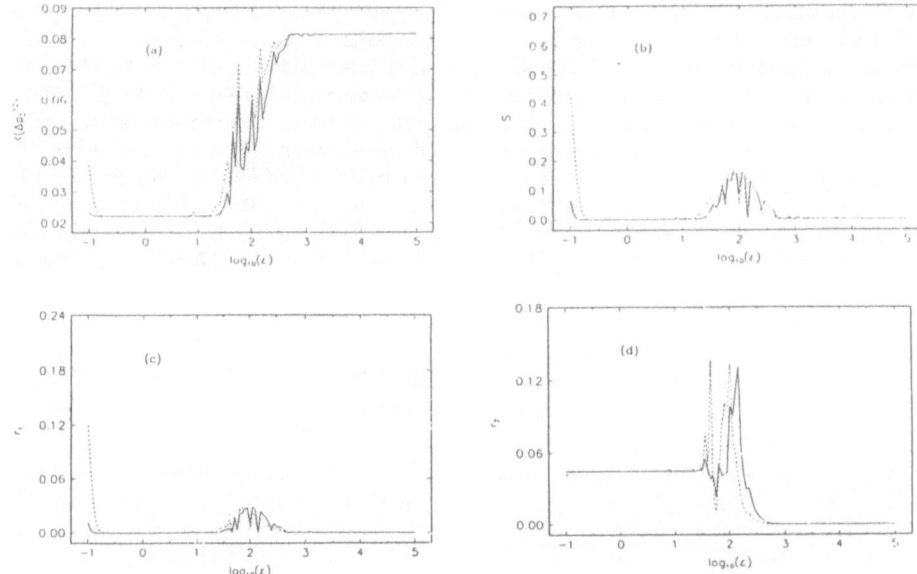

FIG. 2. (a) $\langle(\Delta a_2)^2\rangle$, (b) entropy S. (c) r_1 and (d) r_2 as a function of $\log_{10}(\epsilon)$ for various numbers of atoms crossing the cavity. 3000 atoms (dotted line), 5000 atoms (dashed line), and 10000 atoms (solid line). Here $\alpha/\gamma = -\sqrt{0.7} = 4\pi\epsilon/23$, and the initial condition of the field is $|0\rangle$.

density matrix after the $(k + 1)$th atom has gone through the cavity is given by [8]

$$\rho^{k+1} = e^{R^{-1}L} \cdot M\rho^k, \qquad (19)$$

where L is the well-known cavity loss operator and M the one-atom gain operator, that is, it represents the gain for the field due to the effect of one-atom crossing the cavity. Consistent with the assumptions stated above, we neglect the losses while the atom interacts with the field. We define the parameter $N_{ex} = t_{cav}/t_{at}$ which represents the number of atoms crossing the cavity during the time t_{cav}. In the following numerical calculations we will neglect the thermal photons (low-temperature regime). In Fig.4(a) we show the steady state of the quadrature fluctuations as a function of N_{ex} in logarithmic scale for $\epsilon = 10^4$ and 1. Here $\alpha/\gamma = -\sqrt{0.7} = 4\pi\epsilon/23$, and we took the initial condition to be $|0\rangle$.

We observe that, for $N_{ex} > 10^7$ and $\epsilon = 10^4$, $\langle(\Delta a_2)^2\rangle$ settles at about the same value as the one obtained with the trapped even state; however, for $N_{ex} > 10^5$ and $\epsilon = 1$, this fluctuation stabilizes around the value for the squeezed vacuum. Figure 4(b) shows the entropy as a function of N_{ex} with the same parametes as in Fig.4(a). We obtain quasipure states in the same N_{ex}

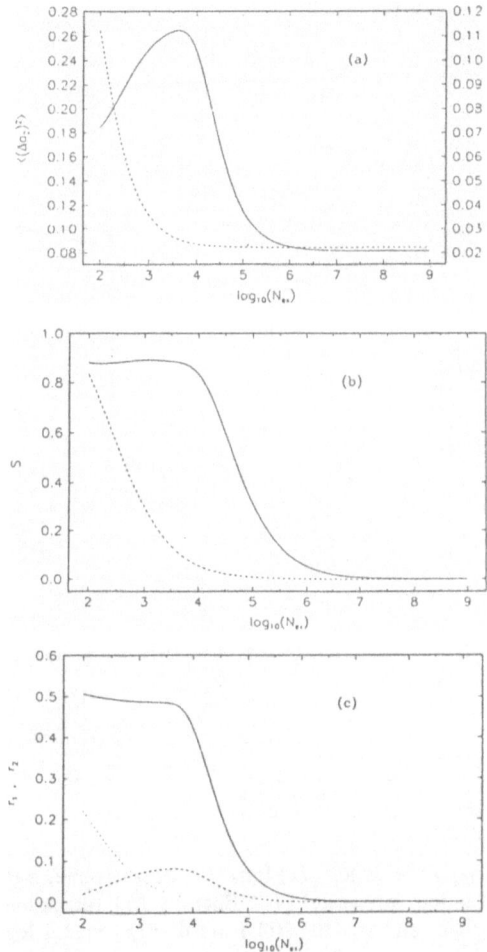

FIG. 3. (a) $\langle(\Delta a_2)^2\rangle$, (b) entropy S. and (c) r_1 versus the number of atoms l crossing the cavity. Here $\alpha/\gamma = -\sqrt{0.7} = 4\pi\epsilon/23$, $\epsilon = 1$ and the initial condition is a coherent state with $\langle n \rangle = 16$.

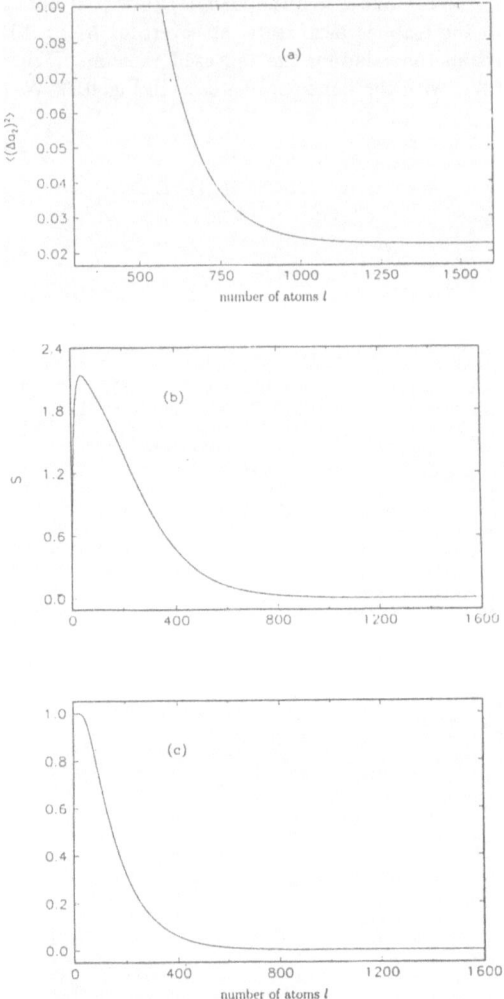

FIG. 4. (a) $\langle(\Delta a_2)^2\rangle$ versus $\log_{10}(\epsilon)$ for $\epsilon = 1$ (dashed line, right scale) and $\epsilon = 10^4$ (solid line, left scale). (b) Entropy (S) versus $\log_{10}(N_{\mathrm{ex}})$ for $\epsilon = 1$ (dashed line) and $\epsilon = 10^4$ (solid line). (c) r_1 versus $\log_{10}(N_{\mathrm{ex}})$ for $\epsilon = 1$ (dotted line), $\epsilon = 10^4$ (solid line), and r_2 versus $\log_{10}(N_{\mathrm{ex}})$ (dashed line) for $\epsilon = 10^4$. Here the initial condition for the field is $|0\rangle$, $\phi = 4\pi\epsilon/23$, and $\alpha/\gamma = -\sqrt{0.7}$

range as in the previous cases. Figure (4c) shows the steady state of the r_1 and r_2 ratios as a function of N_{ex} for $\epsilon = 10^4$ and the r_1 ratio for $\epsilon = 1$. We see that for $\epsilon = 10^4$ and $N_{\mathrm{ex}} > 10^7$ the trap in $|0\rangle$ does not leak, and only even number states are important. Also for $\epsilon = 1$ and $N_{ex} > 10^5$ only even number states are important.

V. CONCLUSION

In this paper we have studied the generation of pure states in a cascade three-level system with finite detuning by means of the temporal behaviour. We also considered the cavity losses taking realistic values [11-15] of the physical parameters e and Nex.

As a first conclusion, we can say that for α/γ fixed and a reduced interaction time Sb corresponding to an upward trapping condition for the high detuning [3], the steady state for the field will have a greater quadrature noise reduction for smaller ϵ values. Furthermore, this steady state is insensitive to initial conditions. The experimental range for this parameter ϵ is between -10^3 and 10^4 for actual micromasers with Rydberg atoms [11 - 15]. Therefore, one could, in principle, generate an ideal squeezed vacuum state, for an arbitrary initial condition, provided that the detuning is small enough so as to have both two and one-photon transitions. Finally, in the case of lossy cavities we have concluded that within an experimentally reachable range of N_{ex} values and realistic values of ϵ we get quasipure states for the electromagnetic field, with properties not very different from the even or ideal squeezed vacuum states. We emphasize that the large detuning case is much more sensitive to the cavity losses than the finite-detuning case. This can be seen, for example, from Fig. 4(b), where the entropy vanishes for a N_{ex} value at least two orders of magnitude smaller for the small detuning case $\epsilon = 1$ ideal squeezing) as compared to the large detuning case ($\epsilon = 10^4$, even states). This is another advantage for the ideal squeezed vacuum state over the even states.

ACKNOWLEDGMENTS

This work is supported by the FONDECYT Grants Nos. 1930568 and 2930032. One of us acknowledges CONICYT support.

APPENDIX

In this appendix we calculate the recursion relation for the expansion coefficients in terms of number states. A squeezed state [9] can be written as

$$|\bar{a}, \xi\rangle = \sum_{n=0}^{\infty} [n! \cosh(s)]^{-1/2} [\tfrac{1}{2} e^{i\theta} \tanh(s)]^{n/2}$$
$$\times e^{-(1/2)[|\bar{a}|^2 + \bar{a}^{*2} e^{i\theta} \tanh(s)]}$$
$$\times H_n \left[\frac{\bar{a} + \bar{a}^* e^{i\theta} \tanh(s)}{\sqrt{2 e^{i\theta} \tanh(s)}} \right] |n\rangle, \qquad \text{(A1)}$$

where H_n is the Hermite polynomial of nth degree, $\epsilon = s e^{i\theta}$, $\theta/2$ is a geometrical phase, and a is the displacement of the error ellipse with respect to the origin [9]. Now, if we set $\bar{a} = 0$ in Eq. (A1), then the probability amplitude for the n-photon number state is given by

$$S_n = \begin{cases} [\cosh(s)]^{-1/2}[e^{i\theta}\tanh(s)]^{n/2}\frac{\sqrt{n!}}{2^{n/2}[\frac{n}{2}]!} & \text{for } n \text{ even} \\ 0 & \text{for } n \text{ odd} \end{cases} \qquad (A2)$$

It is simple to verify that the S_n satisfy the recursion relation

$$S_{n+2} = e^{i\theta}\tanh(s)[\frac{n+1}{n+2}]^{1/2}S_n \qquad (A2)$$

From Ref.[9], for an ideal squeezed vacuum state, the quadrature uncertainties are given by

$$\langle(\Delta a_1)^2\rangle = \frac{1}{4}\left\{e^{-2s}\cos^2\left[\frac{\theta}{2}\right] + e^{2s}\sin^2\left[\frac{\theta}{2}\right]\right\}, \qquad (A4a)$$

$$\langle(\Delta a_2)^2\rangle = \frac{1}{4}\left\{e^{-2s}\sin^2\left[\frac{\theta}{2}\right] + e^{2s}\cos^2\left[\frac{\theta}{2}\right]\right\}. \qquad (A4b)$$

References

[1] P. Filipowicz, J. Javanainen and P. Meystre, J. Opt. Soc. Am. B 3, 906 (1986); Phys. Rev. A 34, 3077 (1986).

[2] J. J. Slosser, P. Meystre, and S. L. Braunstein, Phys. Rev. Lett. 63, 934 (1989); J. J. Slosser and P. Meystre, Phys. Rev. A 41, 3867 (1990).

[3] M. Orszag, R. Ramírez, J. C. Retamal, and L. Roa, Phys. Rev. A A45, 6717 (1992).

[4] P. Meystre and M. Wilkens (unpublished).

[5] J. Retamal, L. Roa, and C. Saavedra, Phys. Rev. A 45, 1876 (1992).

[6] P. Meystre, J. Slosser, and M. Wilkens, Phys. Rev. A 43, 4959 (1991).

[7] Shi-Yao Zhu and L. Z. Wang, Phys. Rev. A 42, 5798 (1990); Shi-Yao Zhu, L. Z. Wang, and Heidi Fearn, ibid. 44, 737 (1991).

[8] L. Roa, J. C Retamal, and C. Saavedra, Phys. Rev. A 47, 620 (1993).

[9] R. Loudon and P. L. Knight, J. Mod. Opt. 34, 709 (1987).

[10] M. Orszag and J. C. Retamal, Opt. Commun. 79, 455 (1990)

[11] D. Merschede, H. Walther, and G. Muller, Phys. Rev. Lett. 54, 551 (1985).

[12] G. Rempe, H. Walther, and N. Klein, Phys. Rev. Lett. 58, 353 (1987).

[13] M. Brune, J. M. Raimond, and S. Haroche, Phys. Rev. A 35,154 (1987).

[14] M. Brune, J. Raimond, P. Goy, L. Davidovich, and S. Haroche, Phys. Rev. Lett. 56, 1899 (1987).

[15] G. Rempe, F. Schmidt-Kaler, and H. Walther, Phys. Rev. Lett. 64, 2783 (1990).

THERMODYNAMIC INSTABILITIES IN QUANTUM LIQUIDS

E.S. Hernández

Departamento de Física - Facultad de Ciencias Exactas y Naturales,
Universidad de Buenos Aires, 1428 Buenos Aires, Argentina

We propose a simple generalization of the criteria for thermodynamic stability that the Landau theory of Fermi liquids provides for zero temperatures. If the latter is nonvanishing, loss in the definition of the Fermi surface inhibits a well - established relationship between each thermodynamic response strength and one selected Landau parameter. We show that an extension of the Landau parameter to a temperature dependent field allows one to formulate the stability criteria in a similar fashion as in the zero temperature case. We discuss the application of these concepts to channels other than the spin symmetric - isospin symmetric one where the response to a scalar compression field takes place.

1. Introduction

The purpose of this work is to discuss a possible generalization to finite temperatures of some aspects of Landau theory of Fermi liquids. Indeed, the latter has proven to be highly satisfactory to describe elementary excitations, dynamic and static response, collective spectrum, scattering properties of single particles in the bulk and transport coefficients[1,2] and, more recenly, the second order phase transition of liquid ^3He and finite droplets[3]. The validity of the theory is restricted to long wavelength perturbations and extremely low temperatures. In such a context, the sharp Fermi surface plays a relevant role as the locus of every dynamical microscopic phenomenon that happens in the liquid. As a consequence, the static thermal response to external fields measured, for example, by the isothermal compressibility or the magnetic susceptibility is associated to the strength of one internal effective field acting on the Fermi surface, i. e., a Landau parameter. In particular, one may locate the onset of thermodynamic instabilities like the exponential growth of density fluctuations[4]; for spin or isospin unsaturated liquids, a similar approach may apply to fluctuations in the magnetization or in the neutron - proton excess.

As a thermal difussivity smoothes the sharp surface, the relation between the microscopic and the macroscopic worlds, respectively represented by the effective in-

E. Tirapegui and W. Zeller (eds.), Instabilities and Nonequilibrium Structures V, 273–280.

teraction between Landau quasiparticles (qp's) and by thermal response coefficients is no longer obvious; a legitimate inquiry concerns the survival of the concept of Landau parameters at finite temperatures. A definite step in this direction has been advanced by Heiselberg, Pethick and Ravenhall[5] in an investigation of the bulk modulus in thermally excited nuclear matter driven by a density dependent, zero range effective interaction. In a recent work carried jointly with J. Ventura, A. Polls and X. Viñas from the University of Barcelona[6], we have generalized the previous authors' approach to a finite range interaction - the so called Gogny D1 force[7] - and have stated a clear prescription to relate the nuclear matter incompressibility to the first order expansion of an averaged internal field that plays the role of the zero temperature Landau coefficient F_0^{ss}. In what follows, I will briefly expose the main ideas presented in the above generalization and indicate why and how a similar treatment could be applied to compute, i. e., the magnetic susceptibility and the symmetry energy.

2. A brief on Landau's theory of Fermi liquids

Excellent treatises on the Landau theory of Fermi liquids are available (see, for example, refs. [1,2]), then I shall just enumerate the necessary tools for the computation of static response functions and the determination of the onset of instabilities. Let us consider a Fermi liquid in thermodynamic equilibrium at temperature T and chemical potential μ. The equilibrium distribution can be established looking for the minimum of the grand potential

$$A = E - TS - \mu N, \tag{2.1}$$

where E is the total energy, S the entropy and N the number of particles. One further assumes that in terms of the occupation numbers n_i, the entropy takes the same form as in a free Fermi gas and writes,

$$A = E - \mu N - \sum_i [n_i \, ln(n_i) + (1 - n_i) \, ln(1 - n_i)]. \tag{2.2}$$

Setting the variations $\delta A/\delta n_i$ equal to zero one reaches the Fermi Dirac distribution

$$n_i^0 = \frac{1}{1 + e^{(\varepsilon_i - \mu)/T}}, \tag{2.3}$$

with the single particle energies

$$\varepsilon_i = \frac{\delta E}{\delta n_i}. \tag{2.4}$$

The qp with energy ε_i possesses an effective masses m^* related to the qp velocity $\vec{v}_i = \nabla_{\vec{k}_i} \varepsilon_i$ by

$$\vec{v}_i = \frac{\vec{k}_i}{m^*(k_i)}. \tag{2.5}$$

Furthermore, the qp's interact with an effective interaction given by

$$f_{ij} = \frac{\delta^2 E}{\delta n_i \, \delta n_j}, \tag{2.6}$$

and the qp energy (2.4) can be decomposed as

$$\varepsilon_i = \varepsilon_i^0 + \sum_i f_{ij} \delta n_j, \tag{2.7}$$

where ε_i^0 is the qp energy in the liquid when only the occupation number of the qp level $|i>$ is varied and the second term in (2.7) represents the effect of the interaction with excited qp's whose occupations are changing as $n_i = n_i^0 + \delta n_i$.

For a two qp event where initial momenta labelled as $(\vec{k} + \vec{q}/2, \vec{k}' - \vec{q}/2)$ scatter into $(\vec{k} - \vec{q}/2, \vec{k}' + \vec{q}/2)$, thus transferring momentum $\vec{q}$, the effective interaction is most conveniently denoted as $f_{\vec{k}\vec{k}'}(\vec{q})$. Assuming rotational invariance in space, spin and isospin variables, one writes,

$$f_{\vec{k}\vec{k}'}(\vec{q}) = f_{\vec{k}\vec{k}'}^{ss}(\vec{q}) + f_{\vec{k}\vec{k}'}^{as}(\vec{q})\,\sigma \cdot \sigma' + f_{\vec{k}\vec{k}'}^{sa}(\vec{q})\,\tau \cdot \tau' + f_{\vec{k}\vec{k}'}^{aa}(\vec{q})\,\sigma \cdot \sigma'\,\tau \cdot \tau', \tag{2.8}$$

where $f_{\vec{k}\vec{k}'}^{\sigma\tau}(\vec{q})$ is the effective interaction in the spin - isospin channel $\sigma\tau$ and the labels s, a respectively indicate the symmetric or asymmetric channel in the given angular momentum variable. Moreover, carrying the expansion in Legendre polynomials

$$f_{\vec{k}\vec{k}'}^{\sigma\tau}(\vec{q}) = \sum_l f_l^{\sigma\tau}(k, k', q) P_l(\hat{k} \cdot \hat{k}'), \tag{2.9}$$

one may identify the multipole amplitudes or fields $f_l^{\sigma\tau}$ whose corresponding limits for vanishing wave number q, evaluated on the Fermi surface, give rise to the so called (dimensionless) Landau parameters $F_l^{\sigma\tau} = N(0)f_l^{\sigma\tau}(k_F, k_F, 0)$, with $N(0)$ being the density of qp states at the Fermi level at zero temperature.

3. Thermal instabilities: the zero temperature case

The Landau theory of Fermi liquids furnishes strict relationships between each Landau parameter and a macroscopic magnitude measuring the static response of the liquid to a given perturbation at zero temperature. More specifically, these coefficients represent the relative deviation of the corresponding response with respect to its value for a free Fermi gas of particles with effective mass $m^*(k_F)$. In general, this deviation is represented by the quantity $F_l^{\sigma\tau}/(2l+1)$. The most familiar relationships involve the compressibility, specific heat and magnetic susceptibility of the liquid, and the symmetry energy in the nuclear matter case, the corresponding Landau parameters being F_0^{ss}, F_1^{ss}, F_0^{as} and F_0^{sa}.

The general criterion for thermodynamic stability against fluctuations in the $l-th$ harmonic of the $(\sigma\tau)$ density reads

$$1 + \frac{F_l^{\sigma\tau}}{2l+1} > 0, \tag{3.1}$$

for each channel $\sigma\tau$ and multipolarity l. In particular, from the isothermal incompressibility

$$\frac{1}{\kappa} = \frac{1}{\kappa}\Big)_{free}(1 + F_0^{ss}), \tag{3.2}$$

we realize that the spinodal points $1/\kappa = 0$ correspond to $F_0^{ss} = -1$. For $F_0^{ss} > -1$ the liquid is stable against density fluctuations while for $F_0^{ss} < -1$, these fluctuations can grow leading to spinodal decomposition, i. e., liquid - gas separation.

4. Thermal instabilities: the finite temperature case

If the temperature is nonvanishing, however much smaller than the chemical potential μ, the Fermi surface is well defined within an uncertainty width proportional to T itself. The chemical potential, rather than the Fermi energy, is the leading energy parameter of the system and the Fermi momentum is just a scaling of the total density ρ. One may consider a thermal Fermi momentum k_μ as the momentum value at which the occupation number n_i is one half its maximum value, in other words, we define k_μ by

$$\mu = \frac{k_\mu^2}{2m^*(k_\mu)} + U(k_\mu),$$ (4.1)

with $U(k)$ the momentum dependent mean field. Since microscopic quantities such as the qp spectrum (2.4) and the effective mass (2.5), together with the statistical distribution n_i and the macroscopic chemical potential μ are determined as usually, it becomes an important issue to see to what an extent the typical Fermi liquid relationships among Landau parameters and strengths of thermodynamic responses are valid. An illustrative situation can be provided by the isothermal incompressibility[5,6], that from the thermodynamic viewpoint can be written as

$$\frac{1}{\kappa} = \rho \frac{\partial P}{\partial \rho})_T$$

$$= \rho^2 \frac{\partial \mu}{\partial \rho})_T,$$ (4.2)

where the density is $\rho = g \sum_i n_i$, g being the spin - isospin degeneracy. Using (2.7), the partial derivatives entering $\partial \rho / \partial \mu$ are the solutions of an integral equation, namely,

$$\frac{\partial n_i}{\partial \mu} = -\frac{\partial n_i^0}{\partial \varepsilon_i} [1 - g \sum_j f_0^{ss}(k, k_j, 0) \frac{\partial n_j}{\partial \mu}].$$ (4.3)

It is now clear that if f_0^{ss} is a constant independent of the momentum labels, a summation over the labels i easily leads to the already known relationship (3.2), with $N(0)$ replaced by $N(T) = g \sum_i (-\partial n_i^0 / \partial \varepsilon_i)$. An exact solution of Eq. (4.3) can also be encountered if f_0^s is a quadratic function of either momenta[5]. For other momentum dependence the integral equation is not analytically solvable; however, for sufficiently smooth fields, having in mind that Eq. (4.3) shows that the unknown $\partial n_i / \partial \mu$ is proportional to $-\partial n_i^0/\partial \varepsilon_i$, one might expect that the sharp peak in the latter distribution at low temperatures allows the replacement of $f_0^{ss}(k, k_j, 0)$ by $f_0^{ss}(k_\mu, k_\mu, 0)$. In ref. [6] we have shown that such an assumption does not hold for a D1 - interaction in nuclear matter; however, replacement by

$$\bar{f}_l^{\sigma\tau}(k_\mu, k_\mu, 0) = f_l^{\sigma\tau}(k_\mu, k_\mu, 0) + d\, f_l^{\sigma\tau}(k_\mu, k_\mu, 0), \tag{4.4}$$

with

$$d\, f = 2\frac{\partial f_0^{ss}}{\partial k}\Big)_{k_\mu} m_1(T), \tag{4.5}$$

where $m_1 + k_\mu$ locates the centroid of the distribution $-\partial n_i^0/\partial \varepsilon_i$, gives rise to an excellent fit to the exact compressibility.

The point to be stressed here is the fact that the integral equation (4.3) also holds for perturbation in asymmetric spin or isospin channels, with the proper identification of the field $f_0^{\sigma\tau}$ that defines the kernel. Let us consider for example the symmetry energy in nuclear matter, defined as

$$E_{sym} = \frac{\partial^2 E}{\partial \rho_r^2}\Big)_{\rho_r = 0}, \tag{4.6}$$

where

$$\rho_r = \rho_n - \rho_p, \tag{4.7}$$

is the relative neutron - proton density. In the lack of asymmetry, either density is one half the total one and the neutron and proton chemical potentials are equal, i. e., one has $\rho_n = \rho_p = \rho/2$ and $\mu_n = \mu_p = \mu$. Consequently, Eq. (4.3) is just

$$E_{sym} = \frac{\partial \mu}{\partial \rho}\Big)_{\rho_r = 0}. \tag{4.8}$$

Since in this case variations in the chemical potential take place in the neighbourhood of $\rho_r = 0$, the operator responsible of the appearance of fluctuations $\delta \rho_r$ is the isospin flip one, $\tau \cdot \tau'$. This condition selects the field f_0^{sa}, rather than f_0^{ss}, in the integral kernel of (4.3). One can easily convince oneself that a similar argument may be invoked for the computation of the spin susceptibility (replacing the neutron and proton labels by spin up and spin down, respectively, and the isospin projector by $\sigma \cdot \sigma'$. In other words, it appears that there is one integral equation for $\partial n_i / \partial \mu$ leading to thermodynamic response, whose kernel $f_0^{\sigma\tau}$ has to be chosen according to the channel where the processes that give rise to the fluctuations $\delta\mu$ takes place.

5. Summary

The Landau theory of Fermi liquids provides thermodynamic stability criteria at zero temperature, which can be cast in terms of the dimensionless strengths $F_l^{\sigma\tau} = N(0) f_l^{\sigma\tau}(k_F, k_F, 0)$ of the internal fields. At nonzero temperatures, the Fermi momentum k_F is no longer the most relevant momentum scale, however a related quantity is the momentum k_μ defined in equation (4.1). In general, it is not true that the relations (3.2) hold with the replacement of k_F by k_μ. Instead, the finite temperature generalization of the Landau parameters appears to be

$$ F_l^{\sigma\tau}(k,T) = g \sum_j (-\partial n_j / \partial\varepsilon_j) f_l^{\sigma\tau}(k, k_j, 0). \qquad (5.1) $$

A good approximation to the exact formula (5.1) can be seen to be

$$ \bar{F}_l^{\sigma\tau} = N(T)\, \bar{f}_l^{\sigma\tau}, \qquad (5.2) $$

with $\bar{f}_l^{\sigma\tau}$ given by (4.4). Although this approximation has only been tested for nuclear matter incompressibility with the Gogny $D1$ force as the effective interaction between bare fermions, it is apparent that the same reasoning applies to the symmetry energy and the magnetic susceptibility. We are then in a position to advance that the general stability criterion (3.1) holds at finite temperatures for the strengths (5.2); in particular, it is possible to show that positiveness of the specific heat at constant volume is ruled by the modified (3.1) with $\bar{F}_1^{ss8}$, the latter given by the corresponding relation (5.2).

It is a pleasure to acknowledge J. Ventura, A. Polls and X. Viñas the collaboration in whose frame the results here quoted[6] were performed.

References

1. D. Pines and P. Nozières, The Theory of Quantum Liquids, vol. 1 (Benjamin, New York, 1966)

2. G. Baym and C.J. Pethick, Landau Fermi-Liquid Theory, (Wiley, New York, 1991)

3. E. S. Hernández, M. Barranco and A. Polls, Phys. Lett. A171 (1992) 119; M. Barranco, D. M. Jezek, E. S. Hernández, J. Navarro y Ll. Serra, Zeit. Phys. D28 (1993) 257.

4. H. Heiselberg, C.J. Pethick and D.G. Ravenhall, Phys. Rev. Lett. 61 (1988); H. Heiselberg, C.J. Pethick and D.G. Ravenhall, Nucl. Phys. A519 (1990) 279c.

5. H. Heiselberg, C.J. Pethick and D.G. Ravenhall, Ann. Phys. 223 (1993) 37.

6. J. Ventura, A. Polls, X. Viñas y E. S. Hernández, Nucl. Phys. A (1993), in press.

7. J. Dechargé and D. Gogny, Phys. Rev. C21 (1980) 1568.

8. J. Ventura, private communication.

Construction of Hamiltonian Structures for Dynamical Systems from Scratch

Sergio A. Hojman
Departamento de Física, Facultad de Ciencias, Universidad de Chile
Casilla 653, Santiago, Chile

ABSTRACT. Hamiltonian structures for dynamical systems are constructed starting from the equations of motion only. Neither previous knowledge of a Lagrangian nor even its existence are required for the construction. Two different approaches to the problem are discussed. The first one is based on constants of motion for the system, while the second one uses just one constant of the motion and one symmetry transformation of the equations of motion to construct Hamiltonian structures for systems with either finite or infinite number of degrees of freedom. Examples are presented.

1. Introduction

The standard construction of Hamiltonian theories starts from a Lagrangian. The procedure is well known and it is the subject of very many textbooks which describe it in detail. The case of regular Lagrangians is the one which is most often treated [1,2], but there are also books which describe what to do in the case of singular Lagrangians, see for instance [3,4].

Hamiltonian structures have been used successfully in several branches of physics. Besides, one of the standard methods of quantization is based on the construction of a Hamiltonian theory for the system at hand, using the Hamiltonian as the generator of the dynamics and the symplectic matrix to define the commutation relations of the basic operators. Group theoretical approaches to the study of symmetries of dynamical systems rely also on the knowledge of the Hamiltonian description of the system considered. There are, in consequence, very many good reasons to try to get a better understanding of Hamiltonian structures on their own right. It is interesting to investigate whether it is possible to quantize or to construct Hamiltonian structures for classical systems of differential equations, without using a Lagrangian [5,6,7,8]. No general procedure seems to be available at present to create a Hamiltonian theory starting from the equations of motion only, in spite of several successful examples already provided in fluid dynamics and in field theory by different authors [9,10], who seem to have obtained them by extremely inspired guesswork.

We present in this note two different approaches to this problem based on the construction of constants of the motion of the system at hand, in the first case [11,12,13,14], while one constant of the motion and one symmetry transformation are the basic ingredients for the second procedure [15].

E. Tirapegui and W. Zeller (eds.), Instabilities and Nonequilibrium Structures V, 281–288.
© 1996 Kluwer Academic Publishers.

2. Basic Definitions

We start with the basic ideas which define a Hamiltonian structure for a dynamical system. Consider the equations of motion for an autonomous first order differential system,

$$\frac{dx^a}{dt} = f^a(x^b) \qquad a, b = 1, \ldots\ldots, N \ . \tag{1}$$

A differential system of any order can always be cast in first order form using standard textbook techniques. We may define a Hamiltonian structure for (1) in terms of an anti-symmetric symplectic matrix $J^{ab}(x^c)$ and a Hamiltonian $H(x^c)$ satisfying

$$J^{ab} = -J^{ba} \qquad a, b, c, \ldots.. = 1, \ldots\ldots, N \ , \tag{2}$$

$$J^{ab}{}_{,d} J^{dc} + J^{bc}{}_{,d} J^{da} + J^{ca}{}_{,d} J^{db} \equiv 0 \ , \tag{3}$$

and,

$$J^{ab} \frac{\partial H}{\partial x^b} = f^a \ . \tag{4}$$

The Poisson bracket for two dynamical variables $A(x^a)$ and $B(x^b)$ is defined by

$$[A, B] = \frac{\partial A}{\partial x^a} J^{ab} \frac{\partial B}{\partial x^b} \ . \tag{5}$$

It is an straightforward exercise to prove that the Poisson bracket structure (5) satisfies all the usual algebraic and differential properties because of the antisymmetry condition (2) and the Jacobi identity.
If we require

$$\det J^{ab} \neq 0 \ , \tag{6}$$

then the symplectic matrix is regular. Condition (6) cannot always be met. For Euler's equations, N is odd, therefore, (6) cannot be satisfied because of (2), as it will be seen in the sequel [11]. For singular systems as well as in many examples in fluid dynamics condition (6) is not satisfied. For details, see [3,9,15,16]. We define a Hamiltonian theory in terms of a symplectic matrix J^{ab} and a Hamiltonian H which satisfy (2), (3), and (4), without requiring that condition (6) be fulfilled. Of course, the usual textbook Hamiltonian structures satisfy all of them.
It can be easily seen that $H(x^a)$ is a time independent constant of the motion for system (1)by using (2) and (4). In fact, H satisfies

$$\frac{\partial H}{\partial x^a} f^a = \mathcal{L}_f H = 0 \ , \tag{7}$$

which also means that the Lie derivative of H along f vanishes. To get acquainted with Lie derivatives, see [17].

3. The Constants of Motion Approach

One way to construct Hamiltonian structures is based on the knowledge of constants of the motion for the dynamical systems under consideration. Some examples have already been discussed in connection with quantum cosmology [12,13,14], so we will deal here with just one example involving a singular, non–linear system: the Euler top [11].
Consider the equations of motion

$$\frac{dL^i}{dt} = \epsilon^{ijk}\Omega_j L_k \qquad i,j,k = 1,2,3 \quad , \tag{8}$$

with

$$\Omega_j = \frac{L_j}{I_j} \quad , \tag{9}$$

where L_i, Ω_i, and I_i are the components of the angular momentum vector, the angular velocity vector and the eigenvalue of the tensor of inertia for a top along the ith direction, respectively.
It is well known that C_1 and C_2 given by

$$C_1 = \left(L^1\right)^2 + \left(L^2\right)^2 + \left(L^3\right)^2 \quad , \tag{10}$$

and

$$C_2 = \frac{\left(L^1\right)^2}{2I_1} + \frac{\left(L^2\right)^2}{2I_2} + \frac{\left(L^3\right)^2}{2I_3} \quad , \tag{11}$$

are constants of the motion for the dynamics generated by (8). We have already seen that the Hamiltonian for any system must be a constant of the motion. Therefore, C_1 and C_2 are, in principle, possible Hamiltonians for the top. It is a straightforward matter to prove that for any μ and D arbitrary functions of C_1 and C_2

$$J^{ij} = \mu\epsilon^{ijk}\frac{\partial D}{\partial L^k} \quad , \tag{12}$$

satisfies (2) and (3). Equation (4) may be fulfilled provided the Hamiltonian H is defined as any function $H = H(C_1, C_2)$ which is independent of D, i.e., if

$$\Delta \equiv \frac{\partial H}{\partial C_1}\frac{\partial D}{\partial C_2} - \frac{\partial H}{\partial C_2}\frac{\partial D}{\partial C_1} \neq 0 \quad . \tag{13}$$

In this case, the choice

$$\mu = \frac{1}{2\Delta} \quad , \tag{14}$$

insures that (4) be satisfied. We have thus constructed infinitely many Hamiltonian structures for the top depending on the different choices of the arbitrary functions D and H.
The vector V_i

$$V_i = \epsilon_{ijk}J^{jk} \quad , \tag{15}$$

is a null eigenvector for any three–dimensional antisymmetrical matrix J^{ij}. Therefore, our symplectic matrices (12) are always singular. Note that the vector V_i is proportional to the gradient of the constant D with respect to L^i. In other words, the Poisson bracket of D

with any dynamical variable vanishes. One migth have foreseen the fact that the Poisson bracket of D and H should vanish, because D is a constant of motion, but the vanishing of all of the Poisson brackets in which D is involved is somewhat surprising. The function with vanishing Poisson brackets may be chosen at will provided it be a constant of the motion, because, given H, the choice of D determines the symplectic matrix J^{ij} uniquely. This result is similar to the one obtained for second class constraints in Dirac's treatment of singular systems and to the appearance of the so called Casimir functions in fluid dynamics, for details, see [3,4,11,15,16]. We have showed how to construct Hamiltonian structures for dynamical systems using the constants of motion as building blocks. These findings may be generalized for higher dimensional phase spaces as it has been done in [12,13,14].

4. The Symmetry Approach

We are now going to display a different scheme to construct Hamiltonian structures from scratch, based mainly on the symmetry properties of the system at hand. For this purpose it is useful to write down the symmetry (perturbation) equation of (1), for details, see [18]. Consider the transformation

$$\tilde{x}^a = x^a + \epsilon \, \eta^a(x^b, t) \ , \tag{16}$$

where $\epsilon \, \eta^a(x^b, t)$ is a small perturbation which maps solutions of (1) in solutions of the same equation, up to first order in ϵ. The equation that the perturbation vector η satisfies is,

$$\partial_t \, \eta^a + \eta^a{}_{,b} \, f^b - f^a{}_{,b} \, \eta^b = 0 \ , \tag{17}$$

or,

$$(\partial_t + \mathcal{L}_f) \, \eta^a = 0 . \tag{18}$$

It is not difficult to prove that K, defined by

$$K \equiv \frac{\partial H}{\partial x^a} \, \eta^a = \mathcal{L}_\eta H \ , \tag{19}$$

is also a constant of the motion for the same system, if η satisfies Eq. (18). By the same token, a new symmetry transformation $\bar{\eta}$ which satisfies Eq. (18) can be constructed using a symmetry transformation η and a constant of motion K by

$$\bar{\eta}^a = \frac{\eta^a}{K} \ . \tag{20}$$

A detailed account of these results may be found in [19].
Consider the following *ansatz* for the symplectic matrix J^{ab}

$$J^{ab} = f^a \, \eta^b - f^b \, \eta^a \ , \tag{21}$$

where η satisfies (17) and has been normalized using (20) in such a way that J^{ab} fulfill (4) identically. Of course, condition (2) is trivially met. The Jacobi identity (3) imposes the following condition

$$J^{bc} \, \mathcal{L}_f \, \eta^a + J^{ca} \, \mathcal{L}_f \, \eta^b + J^{ab} \, \mathcal{L}_f \, \eta^c = 0 \ , \tag{22}$$

which is satisfied by a particular, time independent symmetry vector η_0 which solves (18) defined by

$$\mathcal{L}_f \, \eta_0{}^a \;=\; 0 \;. \tag{23}$$

A more interesting solution η_1 is given by the condition

$$\mathcal{L}_f \, \eta_1{}^a \;=\; \lambda \, f^a \;, \tag{24}$$

which will be most useful in many instances.

We have have thus contructed a Hamiltonian structure for (1) based on the knowledge of just one symmetry vector (either η_0 or η_1) and only one constant of the motion, H, of the system under consideration (assuming a non vanishing K, which can be easily achieved as it will be seen in the examples).

Let us now make a few comments. The rank of the symplectic matrix J^{ab} given by (21) is two. Therefore, it will be, in most cases, singular. A procedure to enlarge its rank is described after the examples. The method we have presented will, in general, yield a Hamil-tonian structure written in terms of non–canonical coordinates. It is worth mentioning that the natural coordinates of non–Lagrangian systems are always non–canonical. This problem is dealt with in some detail in [8]. Even though this procedure differs from the usual one for the case of Lagrangian systems, it may sometimes reproduce the well known results in terms of canonical coordinates, as it is shown in one of the examples below.

Let us now consider some examples. The systems are completely described by the evolution vector f, or the equations of motion written in its first order version. The Hamiltonian structure is completely specified by one constant of motion, the Hamiltonian H, and one symmetry vector η_0 or η_1, which we will quote before the normalization done in (20).

Example 1.– One dimensional monomial force.
This example is defined by the equations of motion

$$f^1 \;=\; x^2 \;, \; f^2 \;=\; - \, c \, (n+1) \, (x^1)^n \;, \tag{25}$$

while a Hamiltonian

$$H \;=\; \frac{(x^2)^2}{2} + c \, (x^1)^{n+1} \;, \tag{26}$$

and one symmetry transformation are given by

$$\eta^1 \;=\; x^1 + \frac{n-1}{2} \, t \, f^1 \;, \; \eta^2 \;=\; \frac{n+1}{2} \, x^2 + \frac{n-1}{2} \, t \, f^2 \;. \tag{27}$$

This is, of course, a very trivial example, which, nonetheless, shows how the scheme presented here can reproduce the usual results. In this case, the symplectic matrix is regular. The harmonic oscillator and the free particle are special cases in this example. Note that this treatment can be extended to any number of dimensions provided the force be a homogeneous function of degree n in the coordinates.

Example 2.– Two dimensional non–conservative force.
This system is defined by

$$f^i \;=\; x^{i+2} \;, \; f^{i+2} \;=\; R \, x^i \;, \; R \equiv (x^1)^3 + (x^2)^3 \;, \; i \;=\; 1, 2 \;. \tag{28}$$

and one possible Hamiltonian structure can be obtained from

$$H = x^1 x^4 - x^2 x^3 \, , \; \eta^i = 2 \, x^i + 3 \, t \, f^i \, , \; \eta^{i+2} = 5 \, x^{i+2} + 3 \, t \, f^{i+2} \, . \tag{29}$$

note that this Hamiltonian *is not equal to the energy*, which, in this case, is not even defined.

Example 3.– Korteweg–de Vries equation.
The equation of motion is

$$u_t + u \, u_x + u_{xxx} = 0. \tag{30}$$

One possible Hamiltonian density is u^2 while one symmetry vector is

$$\eta = -2 \, u - x \, u_x + 3 \, t \, (\, u \, u_x + u_{xxx} \,) \, . \tag{31}$$

As far as we know this is a new Hamiltonian structure for the KdV equation. Note that other Hamiltonian densities can be used as well, in conjunction with the same symmetry vector.

Example 4.– Non–linear Schrödinger equations.
The equations of motion are

$$i \, \psi_t + \psi_{xx} + \psi^2 \, \psi^* = 0 \, , \tag{32}$$

and its complex conjugate. One possible non–standard Hamiltonian density is $\psi \, \psi^*$, and the symmetry vectors are

$$\eta = -\psi - x \, \psi_x + 2 \, t \, (\, \psi_{xx} + \psi^2 \, \psi^* \,) \, , \tag{33}$$

and its complex conjugate. This Hamiltonian structure also appears to be new.

Let us now describe one way of increasing the rank of the symplectic matrix. Consider two new time independent symmetry vectors η_2 and η_3 such that the Lie derivatives of the Hamiltonian H along them vanish, i.e.,

$$\frac{\partial H}{\partial x^a} \, \eta_2{}^a = \frac{\partial H}{\partial x^a} \, \eta_3{}^a = 0 \, , \tag{34}$$

and that the Lie derivatives of η_2 along η_3 as well as those of η_2 and η_3 along η_1 (or η_0) vanish. Then, the new symplectic matrix $J_1{}^{ab}$ defined by

$$J_1{}^{ab} = J^{ab} + \eta_2{}^a \eta_3{}^b - \eta_3{}^a \eta_2{}^b \, , \tag{35}$$

satisfies all of the requirements which define a symplectic matrix (2), (4), and even the non–linear Jacobi identity (3), and its rank is equal to four. This procedure can be repeated at will, producing an increase of two units in the rank of the symplectic matrix each time that it is performed. Note that this construction clearly shows that the Poisson bracket structure is not uniquely determined by the dynamics. If, eventually, one gets a regular symplectic matrix, the method presented here may constitute an alternative to construct a Lagrangian description of the system (1), yielding a novel, symmetry based, approach to the classical Inverse Problem of the Calculus of Variations [20,21,22,23].

We remark that singular symplectic matrices are present in Dirac's construction of Hamiltonian structures, as we have already mentioned above. We are currently investigating the possibility of applying this method, which naturally leads to singular symplectic matrices, to deal with gauge and constrained systems, as an alternative to Dirac's method when no Lagrangian is available. We are also studying whether it is possible to obtain constants of the motion of the system at hand as Casimir functions of the singular symplectic matrix constructed here.

Acknowledgements

The author would like to express his gratitude to P. Ripa and J. Sheinbaum for inspiration which lead him to undertake the study of Hamiltonian systems. It is a pleasure to thank P.J. Morrison for enlightning discussions. He has also enjoyed interesting conversations with O. Castaños, A.M. Cetto, A. Frank, A. Gomberoff, L. de la Peña, M.P. Ryan, Jr., E.C.G. Sudarshan, L.C. Shepley, L.F. Urrutia, at different times and places. This work has been supported in part by Fondo Nacional de Ciencia y Tecnología (Chile) grant 93–0883, and a bi–national grant funded by Comisión Nacional de Investigación Científica y Tecnológica–Fundación Andes (Chile) and Consejo Nacional de Ciencia y Tecnología (México).

References

[1] H. Goldstein, *Classical Mechanics* (Addison–Wesley, Reading, 1980).

[2] L.D. Landau and E.M. Lifschitz, *Mechanics* (Pergamon, Oxford, 1976).

[3] P.A.M. Dirac, *Lectures on Quantum Mechanics* (Yeshiva University, New York, 1964).

[4] E.C.G. Sudarshan and N. Mukunda, *Classical Dynamics: A Modern Perspective* (Wiley, New York, 1974).

[5] E.P. Wigner, Phys. Rev. 77, 711 (1950).

[6] R. Hojman and S. Hojman, Il Nuovo Cimento B, 90B, 143, (1985).

[7] F.J. Dyson, Am. J. Phys. 58, 209 (1990).

[8] S.A. Hojman and L.C. Shepley, J. Math. Phys. 32, 142 (1991).

[9] P.J. Morrison, "Hamiltonian Description of the Ideal Fluid", (to appear in the Proceedings of the 1993 Summer Study Program in Geophysical Fluid Dynamics, Woods Hole Oceanographic Institution Report, WHOI 94, (1994)).

[10] C. Rebbi and G. Soliani, *Solitons and Particles* (World Scientific, Singapore, 1984).

[11] S.A. Hojman, J.Phys. A: Math. Gen. (Letters) 24, L249 (1991).

[12] S.A. Hojman, D. Núñez, and M.P. Ryan, Jr., Phys. Rev. D 45, 3523 (1992).

[13] S.A. Hojman, D. Núñez, and M.P. Ryan, Jr., Proceedings of the Sixth Marcel Grossmann Meeting (Part A), Kyoto, Japan, H. Sato and T. Nakamura Editors, page 718 (World Scientific Publishing Co., Singapur, 1992).

[14] M.P. Ryan, Jr. and S.A. Hojman, Directions in General Relativity, Proceedings of the 1993 International Simposium, Maryland, Papers in Honor of Charles Misner, Cambridge University Press, edited by B.L. Hu, M.P. Ryan Jr. and C.V. Vishveshwara, page 300, Vol. 1 (1993).

[15] S.A. Hojman, "Non–Lagrangian Construction of Hamiltonian Structures", (submitted for publication).

[16] R.G. Littlejohn, A.I.P. Conf. Proc. 28, 47 (1982).

[17] S. Weinberg, *Gravitation and Cosmology* (Wiley, New York, 1975).

[18] S. Hojman, J. Phys. A: Math. Gen. 17, 2399 (1984).

[19] S. Hojman, L. Núñez, A. Patiño, and H. Rago, J.Math. Phys. 27, 281 (1986).

[20] H. Helmholtz, Journal für die Reine und Angewandte Mathematik (Berlin) 100, 137 (1887).

[21] G. Darboux, *Leçons sur la Théorie Générale des Surfaces* IIIe partie, (Gauthier-Villars, Paris, 1894).

[22] J. Douglas, Trans. Am. Math. Soc. 50, 71 (1941).

[23] S. Hojman and L. F. Urrutia, J. Math. Phys. 22, 1896 (1981).

PART V
GRANULAR MATTER

A VISIT OF GRANULAR MATTER

Etienne Guyon

Ecole Normale Supérieure, 45 rue d'Ulm, 75230 Paris Cedex 05

Jean-Paul Troadec

Groupe Matière Condensée et Matériaux, URA CNRS 804 Université de Rennes 1,

35042 Rennes Cedex

ABSTRACT

Granular materials have been used for ages in many domains of technology and have served as models in describing the differents states of matter for many centuries. However, their study knows presently new spectacular developments thanks to the use of methods and concepts of statistical physics introduced to describe the disorder of matter. We describe a few properties and experiments wich often display surprising behaviors and begin to be understood. This note is a reduced version of a recently written book "*du sac de billes au tas de sable*" [1] which deals with the coordinated activity on the subject , in particular in France.

1 - INTRODUCTION

Granular matter is present throughout the history of humanity. It has inspired the atomism of Democrite & Lucrece. The names of Kepler, Buffon & Hauÿ are associated with the first crystallographic models made of piles of identical objects, a subject which has known modern developments with the works of Bernal & his school in Britanny in their understanding of liquid and glassy state [2]. Molecular dynamic studies on the same problems initiated by Alder [3], as well as analog experimental simulations on microscopically disordered systems are presently used to study granular materials, where disorder takes place at a mesoscopic scale implying a revisit of past descriptions and experiments (this is the case of the dynamical structure induced in a vertically vibrating layer of sand presented by F. Melo during the conference, which has a long history). The lasting interest of the "bag of marbles" (or "hard sphere") model comes from its simplicity as there

E. Tirapegui and W. Zeller (eds.), Instabilities and Nonequilibrium Structures V, 291–298.

are no length or energy scales unit. This is in fact the simplest model one can think of to introduce the notion of multiplicity of scales. The ancient concept of "sorite" already dealt with the idea that a pile can never be reduced to the sum of its grains. In fact some problems of granular matter, such as the recently introduced concept of *self organized criticality* dealing initially with the angle of a sand cone and with the avalanches of all scales taking place on it, do not possess any characteristic length scale between the individual grain one and the whole pile.

The initial hard sphere model can be enriched by considering distributions of sizes, by introducing soft sphere interactions related to deformation and penetrability of grains. Parametrisation also deals with intergranular friction and geometrical parameters such as flatness and roughness of grains. However a compromise must be found between the accuracy in the description and the simplicity and universality found in the initial model.

2 - VARIOUS PACKINGS

Although it has been posed long ago, the simple problem of maximum compactness of monodisperse hard sphere packings is not fully understood. One has not yet been able to show rigorously that the crystalline face centered cubic packing provides the maximum packing fraction (74%). Local structures constructed from tetrahedrons of spheres are more probable and locally more compact. However the resulting local structure is an isosahedron with five fold symmetry incompatible with long range crystalline order. Disordered structures which result from random packings have a packing fraction wich varies between 60 and 64 % depending on the detailed conditions of filling and shaking. These packings correspond to very different organizations, and applying pressure is not enough to go continuously from one state to the other one. The reorganization needed to increase packing fraction is a direct illustration of the simulated annealing technique used to solve similar complex (non polynomial) problems. An attempt has been made by Sam Edwards [4] to create a "thermodynamics" of disordered packings by using packing fraction and compactivity (rate of change of packing fraction) as coupled variables, but is not fully satisfactory.

It is of pratical interest to know how to increase or reduce the packing fraction C appreciably. It is well known that binary mixtures of spheres have a maximum of C for around 25 % of small spheres. One can iterate the procedure for many classes of spheres leading eventually to Apollonian packings where the voids of one class of spheres are filled by spheres fitting into it and so on. High performance concrete having strength up to 10 times larger than classical one makes use of the reduction of porosity, $P = 1 - C$, by using a distribution of grains over a large range of sizes from d_1 to d_2 ($d_1 << d_2$). Caquot [5] has proposed an empirical formula $P = (d_1/d_2)^{1/5}$ which can be understood using the fractal dimension d_f in the case of the apollonian case and a Caquot exponent $d - d_f$, where d is the space dimension.

Sintering implies heating a granular medium at temperatures well below the melting point. Soldering takes place at the points of contact and involves displacement of atoms along the surface or in the bulk of the grains. The process can be pushed up to the point of zero interconnected porosity. But here as in other material problems (ceramics in particular) the homogeneity of the resulting material depends on the initial setting. This justifies in part the recent interest for granular flow processus.

Weak compactness is indeed obtained in an uncontroled way in poor packings. The formation of arches can leave open voids and be very detrimental for mechanical or permeability properties. We have yet no good general understanding of it although it is clear that the surface state of grains plays a major role in it.

Colloidal interactions between brownian grains in suspension lead to weakly connected structures. Fractal cluster-cluster aggregation describes the irreversible growth and sticking of fractal aggregates forming a weak interconnected gel. The resulting material can be dried naturally forming the well known dry ponds. In the case of the very light aerogels, supercritical conditions are used to avoid capillary forces during drying.

Loose packings are also obtained with non spherical objects. For packings of spheres with radius R, packing fraction is described through a single invariant $NR^3 \cong 1$ where N is the number of spheres per unit volume. On the other hand, for rods of length L and diameter d one can obtain three limit behaviors :

$NL^3 \cong 1$ is the dilute limit; the spheres of influence of rods just touch.

$NLd^2 \cong 1$ is a condition of compactness obtained for fully aligned rods in contact.

$NL^2d \cong 1$ is the interesting limit dealing with disordered jammed rods forming a weak solid (like in the case of thermally insulating matresses of glass fibers) .

3 - MECHANICS OF GRANULAR MATERIALS

At first sight, on can think of characterizing the mechanical properties of an assembly of grains by sums and averages over pairs of grains. Such a method usually called *homogeneization* is often satisfactory to describe composite material behavior. We use the expression *"weak disorder"* to characterize such systems. Homogeneization methods do not apply well in the present case due to the strong heterogeneities in the intergrain contacts, even in dense random packings. This is well revealed in photoelastic experiments with photoelastic aligned cylinders (modeling 2D packing) or grains. When pressure is applied on such an assembly, the photoelastic experiment shows lattices of stressed elements joining the limiting plates between which stress is applied. The set of lines becomes richer and more interconnected when the packing is improved by increasing pressure. In such a case, a knowlege of the full packing is needed to describe the mechanical response. We talk in this case of *"large disorder"*. In a previous report of the same series [6], we have described this behavior in connection with the fragmentation of grains which takes place along the lines of stresses. More detailed information on this paragraph will be found in this reference.

4 - FLOW OF GRANULAR MATERIALS

The observation of a granular flow on an inclined surface can be used to show different modes of particulate transport. In fast flows, surface grains are strongly shaken; those near the bottom have restricted relative motion with durable contacts between grains, leading to frequent reorganization of the packing. It is reasonable to apply to the surface grain flow a kinetic theory model, introducing in particular an *effective temperature* to quantify agitation, although there is no microscopic justification for it and equipartition does not apply well. This model has been widely used by mechanicians in correspondence with molecular models and simulations. Losses in collisions are introduced through a *restitution coefficient* e (o < e < 1), energy being fed back by the overall flow to compensate for energy losses. One expects kinetic models to work only if e is not far form unity. In the concentrated phase the concept of *dilatancy* (figure 1) introduced by Reynolds [7] and developped by Bagnold [8] is a basic ingredient : if one shears a compact packing of grains, the volume occupied by the assembly must increase to allow for the local reorganizations of grains.

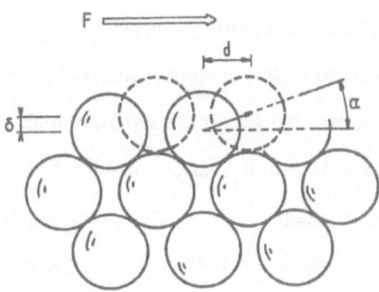

Figure 1. In order they can move under a shear stress imposed in the horizontal direction, the grains must rise a little. The result of this is that there is an upward motion δ per layer and a small angle α of order of δ/d where d is the mean distance between two neighbors grains.

Many experiments and numerical works have been published recently, often in a joint effort between numericians and experimentalists. The comparison are often good and lead to major improvements in the description of the flow mechanisms dealing with the full range of grain concentrations. But an adequate continuum theory is still missing. The experiments can be classified starting from a few simple model ones :

- The *flow of grains in a silo* shows a very heterogeneous distribution of stresses which manifests itself by the formation of density waves which propagate within the medium. An extreme case is the formation of a static arch (already mentionned in §2) which blocks the flow. Recent experiments have shown that a simple hour glass with two closed interconnected volumes shows a periodic flow of grains if the diameter of the grains is not too small with respect to that of the orifice [9]. This relaxation phenomenon is due to the necessary upflow of air through the grains in the upper part to compensate for the filling of the lower half of the hour glass by grains. Quite generally, even for "dry flows" of grains, it is of importance to include the effect of the flow of air between the grains, in particular if the grains are small.

- The *sandpile* problem has been made fashionable to theoreticians since the proposal by Bak et al [10] that it is the archetype of *self organized criticality* (SOC), a concept they have introduced. If one tries to modify the sand pack angle by adding grains one after one at the top, it will tend to keep its cone angle constant by forming avalanches. In the initial model, avalanches of all possible scales are found which represent the fluctuations associated with the critical state. This critical state is attractive rather than repulsive as it is the case for usual critical phase transitions (except just at the critical point, renormalization procedures push the system away from criticality). The model is still an object of active controversy in its various applications (distribution of seismic events, kraches in economy) as well as in the basic model itself [11]. In particular two limit cone angles appear. If one increases the cone angle regularly (for exemple by rotating an horizontal cylinder half filled with grains), these is a maximum cone angle for stability, θ_M. The avalanche develops at θ_M, and stops at a lower angle θ_m. The difference $\theta_M - \theta_m = \alpha$ is related to the dilatancy angle introduced in the figure 1. The hysteretic behavior is clearly incompatible with the critical character of SOC. But experiments with a cylinder small enough so that α is larger than the ratio of grain size to cylinder size appear to behave critically as expected. SOC can also been observed (at least numerically) by imposing certain (automata) rules for initiation, propagation, stops of avalanches.

- *Collective motions* of vibrated grains are also an active subject described by Faraday long ago and illustrated in this conference by the experiments of F. Melo. If one submits a layer of grains to a periodic vibration (amplitude a, frequency ω), one observes that the flat free surface becomes unstable over a characteristic vibration wich can be measured naturally by the ratio $a\omega^2/g$ (where g is the gravity constant). A typical hump is formed due to a continuous upward motion of grains in the central part balanced by downward flow on the side slopes. The behavior becomes more complex when the dimensionless ratio increases. But ordered structures made of a periodic alternance of bumps and valleys can be found. Note that the process is different from the classical

Chladni figures obtained from motion of particles on a vibrating plate, mapping the eigenstructures of the plate.

- Possibly the most spectacular effects are obtained with *polydisperse granular assemblies* [12]. By shaking vertically a mixture of spheres of two different diameters one gets segregation with larger spheres at the upper free surface. Several mechanisms are at play. On one hand, the vertical upward displacement of one grain leaves lower space for smaller spheres to move under it and limits its fall. But there also exists a convective motion which carries the big spheres to the top surface and is induced by the presence of the vertical side walls. When they reach the top surface, they are trapped there and cannot move down. Another experiment involves pouring a mixture of grains on an inclined plane. The possibility for small grains to fall vertically within large gaps of the structure leads to the accumulation of the large spheres on the upper layers during the flow process. Such a phenomenon occurs during avalanches and land slides. However the presence of particules plus fluid suggests a continuum description. The problem becomes analogous to a density flow problem which occurs when a dense liquid poured on an horizontal or tilted plate spreads on it.

We have presented here some properties of granular materials to which physicists have recently contributed. They involve multiscale descriptions ranging from mechanics of grains to quasi continuum models. Non linear effects enter due to local properties. Collective behaviors are found, sometimes in striking analogy with fluid mechanical situations, but a general frame of description of the flow is still missing. Finally we stress the practical impact of such studies in engineering, as well as the possible extensive use of such models in teaching natural sciences in a more pratical way .

REFERENCES

[1] E. Guyon and J.P. Troadec, "Du sac de billes au tas de sable", Editions Odile Jacob (France), 1994.

[2] J. Bernal, Proc. Roy. Soc. of London, A280, 299, 1964.

[3] B.J. Alder and T.E. Wainwright, Phys. Rev. Lett., 18, 988, 1967.

[4] S. Edwards and R.B.S. Oakeshott, Physica A, 157, 1080, 1989.

[5] M.A. Caquot, Mém. Soc. Ingén. Civils de France, 562, 1937.

[6] E. Guyon, in Instabilities and Non equilibrium Structures III, p.347, E. Tirapegui & W. Zeller Eds, Klewer 1991.

[7] O. Reynolds, Phil. Mag., 20, 469, 1885.

[8] R.A. Bagnold, Proc. Roy. Soc. of London, A295, 219, 1966.

[9] X.I. Wu, K.J. Måløy, A. Hansen, M. Ammi and D. Bideau, Phys. Rev. Lett., 71, 1363, 1993.

[10] P. Bak, C. Tang, and K. Wiesenfeld, Phys. Rev. Lett., 59, 381, 1987.

[11] D. Sornette, Les phénomènes critiques autoorganisés (édition du CNRS 1993) p.9.

[12] A. Rosato, K.J. Strandburg, F. Prinz and R.H. Swendsen, Phys. Rev. Lett., 58, 1038, 1987.

A Dissipative Model for Parametric Waves in Granular Materials

Enrique Cerda
Departamento de Física de la Universidad de Chile
Beaucheff 850 Santiago-Chile
and
Francisco Melo H.
Departamento de Física de la Universidad de Santiago de Chile
Av Ecuador 3493, Casilla 307 Correo 2 Santiago-Chile

Abstract.

A simple dissipative model for describing the dynamics of parametric waves in granular materials is proposed. The instability mechanism in the model is due to two elementary processes in competition: a focusing effect that concentrates particles in space, and a diffusion effect that relaxes large thickness gradients in the system. The same mechanism are predicted to occur in a very viscous fluid layer submitted to impulsive forcing.

Introduction.

Recent experiments examining thin layers of vertically vibrated granular material have revealed that parametric surface waves occur when the dimensionless acceleration $\Gamma = (2\pi f)^2 a/g$, exceeds a critical value [1], where g is the acceleration of gravity, 'a' is the amplitude of the vibrating surface, and f is the frequency of the driving force.

In the experimental system the cell is being driven sinusoidally, however the effective forcing experienced by the layer is more complicated since it is in freefall for part of the cycle and not in contact with the cell. We therefore consider the effective driving to be composed of a series of impulsive accelerations that occur when the layer collides with the plate. Two primary wave regimes are observed depending on the ratio between the particle diameter 'd' and the amplitude of the plate. The wavelength of the pattern λ is proportional to $1/f^2$ when a/d is large, but is proportional to $1/f$ when a/d is small.

In both regimes, a lateral motion of grains exists that is similar to the motion of a fluid particle in a standing surface wave. In fact, in the granular case, the lateral transfer of mass is

299

E. Tirapegui and W. Zeller (eds.), Instabilities and Nonequilibrium Structures V, 299–311.
© 1996 *Kluwer Academic Publishers.*

induced just after the collision between the layer and the bottom of the container. Figure 1 shows the time evolution of the waves close to the collision point. Right after the collision, a sand bump splits into two packets which travel laterally and carry mass. This is one of the key experimental points on which our model is based. Later, packets collide and new bumps shifted in space by half the wavelength of the pattern are formed. After two plate collisions the bumps return to their original positions. The pattern is thus parametric, meaning that the pattern oscillates at half the drive frequency.

This article is divided into two parts; in the first we introduce a dissipative model for explaining the instability threshold of a granular layer submitted to impulsive forcing. In the second part, we show that the same behavior is expected to occur in a thin layer of viscous fluid. This analogy allow us to assign physical meaning to the quantities introduced in the analysis of the granular layer.

I. A dissipative model for parametric granular waves.

In this model the granular dynamics will be described in the continuum limit using two variables: $h(x,t)$, the local layer thickness, and $v_\perp(x,t)$, the horizontal component of the grain velocity. (x is the horizontal coordinate and t is the time). We consider $v_\perp(x,t)$ to be indepent of the vertical coordinate. With these variables, the continuity equation that accounts for mass conservation reads,

$$h(x,t_1) = h(x,0) + \int_{-\infty}^{+\infty} dr \left\{ \delta\left[x - \left(r + t_1 v_\perp(r,0)\right)\right] h(r,0) - \delta\left[r - \left(x + t_1 v_\perp(x,0)\right)\right] h(x,0) \right\}. \quad (1)$$

This equation reflects the fact that some of the grains that were at position r at t=0 move a distance $v_\perp(r,0)$t₁. t₁ can be interpreted as a characteristic time during which the inertial mode of the problem relaxes. This point will be illustrated later; for the moment let us consider t₁ as a parameter of the problem. Equation (1) is an approximation, in reality there is a distribution of grain speeds right after the collision and we consider here only the mean lateral velocity of such a distribution. This approximation is valid when the characteristic horizontal length, the wavelength of the pattern λ, is appreciably larger than the thickness of the layer $h(x,t)$. This limit is analogous to the shallow water approximation for a fluid.

A formal derivation of the velocity distribution would solve the entire dynamic problem of motion in the granular layer but appears to be a difficult problem to solve analytically. For this reason we adopt a phenomenological point of view in which the mean

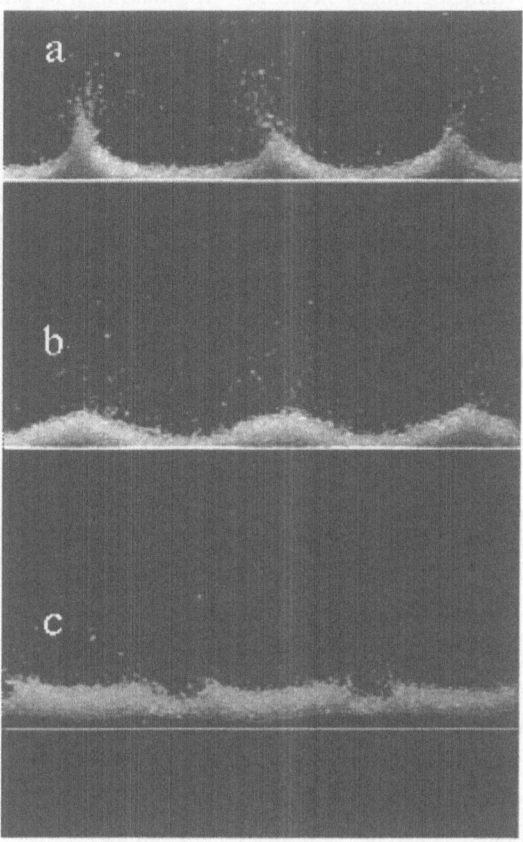

Fig. 1. Side view of a parametric wave. The horizontal white line indicates the moving plate. a) and b) Wavy pattern before the collision. c) After the collision, a lateral transfer of mass occurs.

lateral speed is determined during the collision. It should then be proportional to both the collision velocity v_o and the slope of the free surface of the layer at the time of collision,

$$v_\perp(x,\tau) = -v_o \nabla_\perp h(x,0), \qquad (2)$$

where τ is the collision duration and $\nabla_\perp = \frac{\partial}{\partial x}$.

Our next step is to introduce a mechanism to prevent the accumulation of sand in regions where two opposite fluxes of sand meet. Notice that such fluxes are required to sustain a standing wave. A simple way to implement this idea is through a diffusion equation which automatically conserves mass,

$$\frac{\partial}{\partial t} h(x,t) = D \nabla_\perp^2 h(x,t), \qquad (3)$$

where t is in the interval t1<t<T, T is the time between successive collisions, and D is a diffusion coefficient. The diffusion mechanism that we introduce here acts similarly to a relaxation process invoked previously by Nishimori and Ouchi [2] to capture the principal features of the ripple pattern on dunes created by wind-blown sand.

In our case, equation (3) reflects processes in which height relaxations of a grain hill take place on the surface and are induced by gravity. An idea of the dimensional dependence of D can be obtained by considering the flowing layer to be a fluid with constant viscosity η. In this situation, the velocity of the grains is obtained by equilibrating the viscous force and the gravity force tangent to the interface. The total flux of particles "j" flowing in the moving layer is then,

$$j = \frac{g\rho}{3\eta} [n(e)d]^3 \frac{\partial h}{\partial x}. \qquad (4)$$

Here ρ is the particle density, d is the particle diameter, e the restitution coefficient, and n(e) a function that take into account the penetration depth of the motion. Introducing this expression into the mass conservation equation for the grains, $\frac{\partial h}{\partial t} + \frac{\partial j}{\partial x} = 0$, we obtain equation (3) and determine that the diffusion coefficient is $D = \frac{g\rho}{3\eta}[n(e)d]^3$. It is well known that viscosity in a granular material is not clearly defined. Moreover, for high shear rates the tangential stress is not proportional to the shear rate. However, for simplicity, we will keep this description. Implications of using other laws for thickness diffusion processes will be discussed elsewhere.

In the following, we study the stability of a vibrated flat layer by performing a linear stability analysis. We define $h = h_0 + \xi(x,t)$, where h_0 is the thickness of the flat layer and $\xi(x,t)$ is a disturbance in the height. Using equations 1, 2 and 3 and retaining first order terms, we find,

$$\xi(x,t_1) = \xi(x,0) + h_o v_o t_1 \nabla^2 \bot \xi(x,0) \tag{5}$$

and,

$$\xi(x,T) = \frac{1}{2\sqrt{\pi D(T - t_1)}} \int_{-\infty}^{+\infty} dx' \xi(x',t_1) \exp\left[-(x - x')^2 \Big/ 4D(T - t_1) \right]. \tag{6}$$

Equation (6) is the solution of diffusion equation (3) including the initial condition established by relation (5). Writing $\xi(x,t) = \Phi_k(t)e^{ikx} + c.c.$ in terms of a perturbation of wave vector k and amplitude Φ_k, equations (5) and (6) become,

$$\Phi_k(t_1) = (1 - h_o v_o t_1 k^2) \Phi_k(0) \tag{7}$$

and

$$\Phi_k(T) = e^{-D(T - t_1)k^2} \Phi_k(t_1). \tag{8}$$

Combining both relations we obtain,

$$\Phi_k(T) = f(k) \Phi_k(0) \tag{9}$$

with

$$f(k) = (1 - h_o v_o t_1 k^2)e^{-D(T - t_1)k^2}. \tag{10}$$

304

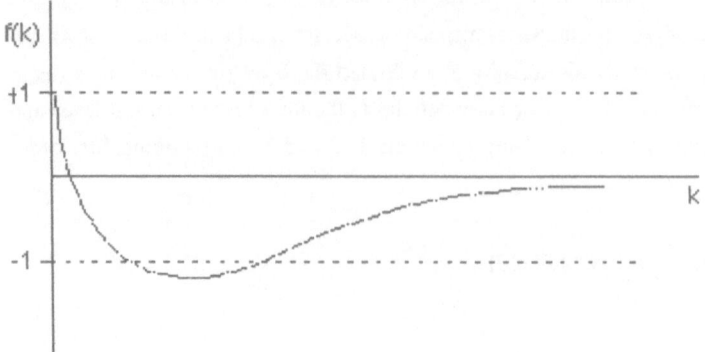

Fig. 2. f(k) as a function of k. f(k) has a minimum for $k_*^2 = \frac{1}{d} + \frac{1}{D}$, corresponding to the most unstable mode of the parametric instability.

Thus, equation (9) tells us that after each collision the amplitude of mode k, Φ_k, is amplified by $f(k)$ and grows by $f(k)^n$ after n collisions. Thus, if $|f(k)|>1$, the system is unstable. It can be seen in Fig.2, where $f(k)$ is plotted as a function of k, that the only way to obtain amplification of disturbances is for $f(k)<-1$. The meaning of this result is clear; at each collision the phase of the unstable mode must change by π, changing the sign of the disturbance. Such a phase change is characteristic of a parametric standing wave since the oscillation is at half the excitation frequency; after two collision the perturbation returns to its original phase.

We now focus on the most unstable mode at the instability onset. f(k) can be written as,

$$f(k) = (1 - \hat{d}k^2)e^{-\hat{D}k^2} ,$$ (11)

where $\hat{d} = h_o v_o t_1$ and $\hat{D} = D(T - t_1)$.

f(k) is a minimum at $k_*^2 = \frac{1}{\hat{d}} + \frac{1}{\hat{D}}$, which is the most unstable mode and a minimum value of

$f(k_*) = -\frac{\hat{d}}{\hat{D}}\exp\left(-1 - \frac{\hat{D}}{\hat{d}}\right)$. $f(k_*)$ is smaller than -1 when the natural control parameter of the instability $\tilde{\Gamma} \equiv \hat{d}/_{\hat{D}}$, exceeds a critical value $\tilde{\Gamma}_c \approx 3.6$. Thus, using the original variables, the most unstable mode at the instability onset becomes

$$k = \frac{1}{\sqrt{D(T-t_1)}},\qquad(12)$$

and the control parameter $\tilde{\Gamma}$ can be written

$$\tilde{\Gamma} \equiv \frac{h_o v_o t_1}{D(T-t_1)} .\qquad(13)$$

We now discuss the meaning of equations (12) and (13). At the moment, we have little insight about t_1, however numerical simulations [3] show that it should be a function of the restitution coefficient and the number of collisions necessary to dissipate the injected kinetic energy, this is $t_1 = f(e)v_o/N^{3/2}g$, where $N=h_o/d$. Furthermore, t_1 should also be related to the viscosity of the granular layer through a diffusion law, $t_1 = [f(e)d]^2 \rho/\eta$. This relation actually defines the viscosity η in our case. The thickness diffusion constant D can also be calculated from the same arguments. We obtain $D=dv_o f(e)^2/3N^{3/2}$.

Finally, $\tilde{\Gamma}$ can be written as, $\tilde{\Gamma} = \frac{3h_o}{f(e)dg} v_o f$, which is proportional to $v_o f$ and fits nicely with the control parameter that is shown to be pertinent in recent experiments [4]. The critical value of $\tilde{\Gamma}$ predicted here differs from experimental measurements as should be expected when numerical coefficients have been disgarded.

On the other hand, the selected wavelength can be obtained from relation (12) by substituting the value of D. λ varies like $1/f$, and is not in agreement with experimental results in the low frequency regime. One possible reason is that in the low frequency regime the instability is subcritical and the wave vector calculated here is not necessarily the one selected in the experiment. Moreover, in the experiment, immediately after the collision a bump splits into two bumps that travel freely when the layer takes off. The dissipated energy should be small when the layer is in free flight and the wavelength should then be selected by the speed of the bumps and by the time allowed for lateral motion. This argument gives the correct dispersion relation for low frequency waves [4].

In the high frequency regime a comparison between this model and experimental results seems more appropriate since the model predicts the correct dependence of λ on f [4]. Although a clear time scale separation between relaxation and thickness diffusion mechanism has not been shown experimentally, this model should be valuable in raising interesting question about these phenomena.

In conclusion, even though the model proposed here does not explain all the features of parametric waves, it gives valuable insight into the instability mechanism. When a disturbance appears in the free surface of the layer during a collision, a lateral flux is immediately induced. Depending on the shape of the disturbance, regions where grains accumulate can exits. For instance, if we focus on particles that are located at position r (the disturbance region) that go to position $x = r - t_1 v_o \frac{\partial h}{\partial r}$ (flat region) after a collision, and in addition impose mass conservation $n(x) = n(r) \frac{\partial r}{\partial x}$, we find that a self-focusing effect occurs when $\frac{\partial r}{\partial x} > 1$. The disturbance will thus be larger at x than it was at r. It can be shown that this self-focusing mechanism always leads to spatial instability with no spatial scale selection. In our case, processes of thickness diffusion allow us to select a wavelength for the wave pattern.

II.   Parametric instability in a viscous fluid layer submitted to impulsive periodic forcing.

In this section we give a derivation of the parametric instability in a viscous fluid [5]. We consider a thin layer of viscous fluid submitted to strong (and short in time) vertical accelerations. The peaks in acceleration repeat in time with period T, and are intended to capture the effects of the collision of the layer with the cell bottom in the granular material. This derivation allow us to give natural interpretations of the parameters introduced in the previous model. In particular, we show that D is a diffusion coefficient for disturbances in the layer height, and t_1 is the relaxation time of an inertial mode of the system. In the viscous fluid case, $\tilde{\Gamma}$ has its usual meaning as the ratio between the characteristic acceleration of the plate and the acceleration of gravity.

For a thin horizontal layer of viscous fluid, in the lubrication approximation, the hydrodynamic mass and momentum conservation equations are [5-6],

$$\frac{\partial}{\partial t} \xi(x,t) = -h_o \nabla_\perp \cdot v_\perp \tag{14}$$

and

$$\frac{\partial}{\partial t} v_\perp(x,t) = -(g + \ddot{z}_p) \nabla_\perp \xi(x,t) - 3 \frac{v}{h_o^2} v_\perp(x,t), \tag{15}$$

where $\ddot{z}_p$ is the acceleration of the moving plate, h_0 is the thickness of the flat layer, $\perp$ represents coordinates in the horizontal direction, and v is the kinematics viscosity.

To perform a linear stability analysis we write

$$\xi(x,t) = \phi_k e^{ikx} + c.c.$$

$$v_\perp(x,t) = v_k e^{ikx} + c.c.$$

(16)

where ϕ_k and v_k are the fourier amplitudes of ξ and $v_\perp$ respectively. Introducing equation (16) in equations (14) and (15) we obtain the following equation for a forced damped oscillator,

$$\ddot{\phi}_k + \frac{3v}{h_0^2}\dot{\phi}_k + \frac{v^2}{h_0^4}\varepsilon_k\left(1 + \frac{\ddot{z}_p}{g}\right)\phi_k = 0,$$

(17)

where we have defined a dimensionless number $\varepsilon_k \equiv \frac{h_0^5 g k^2}{v^2}$. The lubrication approximation requires that $\varepsilon_k \ll 1$. Equations (14) and (15) can be easily modified to account for surface tension effects. In our case this effect can be neglected by taking k small enough.

A single collision can be studied by introducing the following quantity,

$$v_0 \equiv \int_0^\tau \ddot{z}_p dt,$$

(18)

where τ is the collision time duration which is small with respect to T. Integrating of equation (17) between 0 and τ shows that the effect of the collision on the oscillator described by (17) is a change in the speed,

$$\dot{\phi}_k(\tau) = \dot{\phi}_k(0) - \frac{v^2 v_0}{h_0^4 g}\varepsilon_k \phi_k(0).$$

(19)

This equation is equivalent to the condition obtained by integrating equation (15) in time. It is also the same as equation (2) for the granular case.

Immediately after the collision the general solution of equation (17) is

$$\phi_k(t) = C_+ e^{w_+ t} + C_- e^{w_- t}, \tag{20}$$

where C_+ and C_- are constants determined by initial condition (19), and

$$w_\pm = \frac{3v}{2h_o^2}\left[-1\pm\left(1-\frac{4}{9}\varepsilon_k\right)^{1/2}\right] \tag{21}$$

are the oscillator eigenfrequencies.

Each collision gives rise to a discontinuity of $\phi_k(t)$, or equivalently of C_+ and C_-. After n collisions, $nT<t<(n+1)T$, $\phi_k(t)$ becomes

$$\phi_k(t) = C_+^n e^{w_+(t-nT)} + C_-^n e^{w_-(t-nT)}. \tag{22}$$

Using the facts that $\dot{\phi}_k$ has a discontinuity at each collision and ϕ_k does not change during a collision the following relations can be established

$$\phi_k\big|_{t=nT+\tau} = \phi_k\big|_{t=nT}$$

$$\dot{\phi}_k\big|_{t=nT+\tau} = \dot{\phi}_k\big|_{t=nT} - \frac{v^2 v_o}{h_o^4 g}\varepsilon_k\phi_k \tag{23}$$

This allow us to write the following recursion relation between $\{C_+^{n-1}, C_-^{n-1}\}$ and $\{C_+^n, C_-^n\}$,

$$\begin{pmatrix} C_+^n \\ C_-^n \end{pmatrix} = \begin{pmatrix} \left[1 - \dfrac{v^2 v_o}{h_o^4 g(w_+ - w_-)}\varepsilon_k\right]e^{w_+ T} & 0 \\ \dfrac{v^2 v_o}{h_o^4 g(w_+ - w_-)}\varepsilon_k e^{w_+ T} & 0 \end{pmatrix}\begin{pmatrix} C_+^{n-1} \\ C_-^{n-1} \end{pmatrix}, \tag{24}$$

where we have assumed that $h_o^2/v \ll T$, and used the following approximate relations for w_+ and w_-:

$$w_+ = -\frac{v}{3h_o^2}\varepsilon_k + \vartheta\left(\varepsilon_k\right)^2$$

$$w_- = -\frac{3v}{h_o^2} + \frac{v}{3h_o^2}\varepsilon_k + \vartheta\left(\varepsilon_k\right)^2$$

(25)

Constants $\{C_+^n, C_-^n\}$ which describe the dynamics of the oscillator submitted to n impacts, can be found from the initial conditions $\{C_+^0, C_-^0\}$, by use of the following relations,

$$C_+^n = (f(k))^n C_+^0$$

$$C_-^n = \frac{v^2 v_o}{h_o^4 g(w_+ - w_-)}\varepsilon_k e^{w_+ T}(f(k))^n C_+^0$$

(26)

where we have defined the quantity f(k) as

$$f(k) \equiv \left[1 - \frac{v^2 v_o}{h_o^4 g(w_+ - w_-)}\varepsilon_k\right]e^{w_+ T}.$$

(27)

In terms of the original variables and at the lower order in ε_k, (27) becomes

$$f(k) = \left[1 - \left(\frac{v_o h_o^3}{3v}\right)k^2\right]\exp\left[-\left(\frac{gh_o^3 T}{3v}\right)k^2\right].$$

(28)

This function dominates the stability of the oscillator and is exactly equation (11) established previously for the granular case if t_1 is identified as h_o^2/v. As pointed out before, the system is unstable to parametric disturbances when $|f(k)| > 1$. Thus the instability occurs when the control parameter, $\tilde{\Gamma} = \frac{v_o}{gT}$, is larger than 3.6. The most unstable mode is

$$k^2 = \frac{3v}{gh_o^3 T}.$$

(29)

The drive frequency selects a mode that has a diffusion time equal to T. In fact, such a time can be derived from equations (14) and (15) (without the inertial term) by introducing a disturbance of scale $1/k$. We found a characteristic diffusion time $t_D = 3v/gh_0^3k^2$ and a characteristic diffusion constant $D = gh_0^3/3v$. t_D should then be exactly the period of the driving force at the instability threshold. The meaning of this result is clear - in order to sustain a parametric disturbance in a dissipative system the energy injection should take place before the potential energy has been completely dissipated. In addition, the energy injection should be large enough to compensate the dissipation that takes place on a time of the order of $t_1 = h_0^2/v$, which is the relaxation time of the inertial mode of the system.

In conclusion, the parametric wave instability also appears in the viscous case and presents the same features already discussed for the granular case. However, for the viscous fluid, quantities like t_1 and D have a clear physical meaning. In addition, the viscous analysis shows that the approximation introduced in equation (1) for handling the relaxation process in the granular case, is valid. The important question that remains to be answered for the granular layer is whether or not diffusion and relaxation processes can be modeled by a viscous analogy. Experiments should bring some light to answer these questions.

Finally it is interesting to point out that the control parameter Γ does not appear to be directly related to the dissipation of energy as is the case in the low dissipation regime.

Conclusion.

We have derived a simple dissipative model for describing the dynamics of parametric waves in a granular material. It has been shown that the instability mechanism arises from two elementary processes in competition: a focusing effect that concentrates particles in space and a diffusion effect that relaxes large thickness gradients. The control parameter is essentially the acceleration of the plate, in agreement with experimental results. Although the model is successful in predicting the wavelength of the pattern in the high frequency regime, we think that more experiments are necessary to better characterize the relaxation and thickness diffusion phenomena in granular materials.

We would like to thank P. Umbanhowar for helpful discussions, and for a critical reading of this manuscript.

References.

[1] F. Melo, P. Umbanhowar and H. L. Swinney; Phys. Rev. Lett. **72**, 172 (1994).
[2] H.Nihimori and N. Ouchi, Phys. Rev. Lett. **71**, 197 (1994).

[3] A numerical simulation allows us to compute t_1 in a one dimensional case. The dependence on N can change with dimensionality, but dependence on other variables should remain the same for the three dimensional case. Experimental evidence supports this assumption [4]. See, S. Luding, E. Clement, A. Blumen, J. Rajchenbach, and J. Duran. Phys. Rev. E. **49**, 1634 (1994).

[4] P. Umbanhowar, F. Melo and H. L. Swinney. To appear in Phys. Rev. E.

[5] For a complete study of the parametric instability in a viscous fluid, see K. Kumar, "Linear theory of Faraday instability in real liquids", (to appear in Proc. of Royal Society of London).

[6] A. Fetter and J. Walecka, Theoretical Mechanics of Particles and Continua, McGraw-Hill, (1980).

Simple Models for Fragmentation

Gonzalo Hernández [*], Hans J. Herrmann [%] and Hisao Nakanishi [†]

HLRZ, Forschungszentrum Jülich,

Postfach 52425, D-5170 Jülich, Germany.

December 5, 1993

Abstract

We study an iterative stochastic process as a model for two dimensional discrete fragmentation. This simple model, which fulfills mass conservation, allows us to introduce different microscopic stress configurations. The introduction of elementary size pieces that can not be broken further imposes a limit to the fragmentation process. Despite their simplicity, our models present complex features that reproduce some of the experimental results that have been obtained previously. Different fragment size distributions are obtained depending on the breaking rules of our models. For some regimes a power law behavior is obtained. For this reason we propose them as basic models that can be substantially refined to describe the fragmentation process of rocks.

1 Introduction

In this work we study simple models for fragmentation processes defined as discrete time stochastic processes on a two dimensional square lattice. At each time step of the process only one piece of the material is broken, chosen according to the rule of the model. We are interested in determining the fragment size distribution.

[*]Permanent address: Departamento de Ingenieria Matemática, Universidad de Chile, Santiago, Chile. e-mail: ghernan@tamarugo.cec.uchile.cl

[%] Permanent address: PMMH, ESPCI (URA CNRS 857), 10 rue Vauquelin, 75234 Paris Cedex 05, France

[†]Permanent address: Physics Department, Purdue University, West Lafayette, IN 47906, USA.

E. Tirapegui and W. Zeller (eds.), Instabilities and Nonequilibrium Structures V, 313–326.
© 1996 *Kluwer Academic Publishers.*

314

Fragmentation processes are common phenomena in nature, that appear in a wide range of scales and situations. In [1] Turcotte gave a long enumeration of natural fragment distributions (asteroids, coal heaps, moraines, etc) for which power-law distributions were measured with exponents ranging from 1.9 to 2.6 concentrating around 2.4.

Fragmentation processes are common phenomena in nature, that appear in a wide range of scales and situations. Turcotte [1] gave a long enumeration of natural fragment distributions (asteroids, coal heaps, moraines, etc) for which power-law distributions were measured with exponents ranging from 1.9 to 2.6 concentrating around 2.4.

In some way, the power law behavior for small fragment masses or sizes seems to be a universal characteristic of the instantaneous breaking of brittle objects.

There are careful experiments in one dimension, see [2], in which long thin glass rods are broken by vertically dropping them. Depending on the height from which the glass rods are dropped, the fragment size distribution varies from a log-normal shape for smaller heights to a power law for increasing height. Stochastic models for one dimensional fragmentation can be found in [3], where randomly fracture points between 0 and 1 are chosen. The fragment size distribution is completely determined by the a priori random distribution of the fracture points. Power-law, exponential and log-normal fragment size distributions can be obtained. By the introduction of a minimal fragment size (a piece that can not be broken further) and a Poisson distribution for the number of fragments into which pieces are broken, the power law behavior is obtained at some stages of the fragmentation process.

In [4], this power law behavior for the fragment size density distribution f is given in the limit of small fragment masses, by the relation:

$$f(x) \propto x^{-(2-\alpha/3)} \tag{1}$$

where x is the mass (or the volume) and the exponent α varies between 0.5 and 1.0 for instantaneous fragmentation and highly energetic breaking. In general, α depends on the material, the kind of fracture process and the external load.

In ref. [5], a model for the fragmentation of gas clouds via gravitational condensation is developed, under the assumptions of mass conservation and no presence of pressure. Initially, clouds split (condense) into $q \geq 2$ equal mass clouds. The process continues in a self-similar way such that the dynamic equations can be solved analytically. Further assumptions on the model allow to prove that the fragment mass distribution follows a power law behavior in the steady state.

The fracture process of a single grain has been studied by many authors . For instance, Gilvarry [6] obtained its results based on the idea of pre-existing flaws (cracks) randomly distributed in the material, that are activated and propagate until they meet another flaw.

A mean-field type approach to describe the fragmentation process can be formulated through the concentration $c(x, t)$ of fragments of mass lower than x in time t by the *rate equations* [4]:

$$\frac{\partial c(x,t)}{\partial t} = -a(x)c(x,t) + \int\limits_{x}^{\infty} c(y,t)a(y)f(x/y)dy \qquad (2)$$

where $a(x)$ is the rate at which fragments of mass x break into smaller ones (this quantity is supposed not to depend on time) and $f(x/y)$ is the conditional probability that a fragment of mass x was produced from a fragment of mass $y \geq x$. Using scaling and homogeneity assumptions, some exact results are obtained in Ref. [4], if some assumptions are made about $f(x/y)$. In general the solutions of the former equations are however very complicated to obtain.

Stochastic processes discrete in space and time have also been studied as models for fragmentation phenomena using cellular automata. In Ref. [7] two and three dimensional cellular automata are proposed to model the power law distribution resulting from shear experiments of a layer of uniformly sized fragments. The fracture probability of a fragment is determined by the relative size of its neighbors. A larger probability is obtained for fragments with a larger number of equal size neighbors. If two blocks have the same fracture probability a random decision is made. The process continues until no blocks of equal size are neighbors. In all cases a power law size distribution is obtained, with an exponent that depends on the former parameters.

In this paper we propose simple discrete stochastic processes on a two dimensional square lattice of size 2^n. At each step only one piece of the material is broken into two equal size fragments (this is the reason for using system sizes of 2^n). The breaking follows specific criteria based on microscopic models of forces, causing either horizontal or vertical breaking. The iterative breaking process stops when a piece of the smallest possible size (unit length) is forced to break. How we choose the piece to break and how the fragmentation process continues will be defined in the next section.

Our objective in this work is to measure the fragment size distribution for our models, and investigate under what circumstances these simple models present power law behavior in the limit of small fragments.

2 The Model

Each realization of our stochastic discrete process for modeling fragmentation considers the breaking of only one piece of the material assuming area (mass) conservation. As initial situation, we consider a two dimensional square lattice of size 2^n. This initial piece of material can be broken either in the x or in the y axis direction (horizontal or vertical fracture) into two new pieces of equal area. In each corner of the piece we put independently and uniformly distributed random numbers between 0 and 1. These numbers should represent the initial (scalar) forces acting on the material (one could also consider them to be local displacements), see figure 1.

For each step of our stochastic breaking process the following ingredients must be defined:

a) A definition and interpretation for the scalar forces as functions of the random numbers in the corners.

For this purpose, we will consider two classes of forces: traction or compression and shear forces, that can be computed from the random numbers in both directions by:

traction or compression:

$$f_x = |a - d| + |b - c| \quad , \quad f_y = |a - b| + |c - d| \tag{3}$$

shear:

$$f_x = |(a - d) - (b - c)| \quad , \quad f_y = |(a - b) - (c - d)| \tag{4}$$

The modulus is considered for both kinds of forces because we are interested in the total forces acting on the material. The forces are supposed to follow a one dimensional Hooke's law.

b) A criterium to choose the piece of material to be broken.

Based on the above definitions for forces at each iteration of the stochastic process the more stressed piece of the material is broken. The piece with maximum stress among both axis directions is selected. Specifically, a piece P_1 with forces given by f_x^1 and f_y^1 is more stressed than a piece P_2 with forces f_x^2 and f_y^2 if:

$$\max(f_x^1, f_y^1) \geq \max(f_x^2, f_y^2) \tag{5}$$

which clearly defines an order relation between the pieces of the material depending on the definition of the forces.

c) A criterium to choose the orientation of the fracture in the selected piece.

Once the piece to be broken is selected by applying the criterium defined in b), the fracture orientation must be defined. In this case the decision is straightforward: an horizontal cut is made if $f_y \geq f_x$, otherwise a vertical cut is made. Two new pieces of the material of equal area are generated.

d) A criterium to define how the breaking process continues.

The idea of our model is to continue the fragmentation process in a self-similar way. In order to do that we have to define the new scalar forces acting on each piece. The external forces of the fractured piece are maintained as it can be see in figure 2, in which the two new pieces are represented after a vertical break. But new forces, say a', b', c', d', must be defined. We suppose Newton's law of action = reaction:

$$\text{horizontal cut} \qquad b' = a' \quad \text{and} \quad d' = c' \tag{6}$$

$$\text{vertical cut} \qquad d' = a' \quad \text{and} \quad c' = b' \tag{7}$$

With this assumption, only two new forces must be defined to continue the fragmentation process. We will consider two possible ways to choose these numbers. The first one is to choose independently random numbers between the corresponding values of the external forces, i.e.:

$$a' = b' = \lambda a + (1 - \lambda)b \qquad \text{and} \qquad c' = d' = \lambda' d + (1 - \lambda')c \tag{8}$$

where λ and λ' are uniformly distributed random numbers between 0 and 1. This kind of model will be called model A.

In the second case, the new forces are independently distributed random numbers between 0 and 1 without any dependence on the previous forces.

This kind of model will be called model B.

e) A stopping criterium for the fragmentation process.

At this point we have to consider the discrete nature of the model. This means that we cannot continue the breaking process for ever. We have to define stopping criteria, that can also be considered as typical for the evolution of the fracture process. It seems natural to impose that it is not possible to break a rectangular piece along a direction into which it has a length of unity.

We will consider two stopping criteria. In the first one we stop the global fragmentation process if the maximally stressed piece is tried to be broken along an elementary side. We call this criterium "fast stopping". The second one will be to continue the fragmentation process with the second most stressed piece and so on until all the pieces of the material have at least one side of length unity. We call this criterium "relaxed stopping".

Clearly, all the former considerations define different stochastic models of fragmentation in the sense that different fragment size distributions will be obtained, and this is exactly the purpose of this work: We will show that even these simple models of fragmentation, based on microscopic strength considerations and well defined (deterministic) breaking criteriums, can give interesting results that agree with the experimentally available data.

3   Results

In this section we present and discuss the numerical results obtained from the study of our models defined in the last section. We average our results over many (between 5.000 and 10.000) independent random initial conditions, characterized by the initial force configuration, i.e., the four random forces. As we said before, the system linear size is 2^n where n, for our simulations, varies between $n = 64$ and $n = 32768$. This wide range for the system size will be explained later.

For each model we computed the fragment size (area) distribution as a function of time. The maximum number of steps the stochastic process can make is $(2^n - 1)$. This gives an exponential bound for the evolution time that was observed for the models with relaxed stopping criterium. For the fast stopping criterium a transient time of order $O(n)$ was observed.

Let us first consider the fast stopping criterium under traction (eq. (3)) for models A (fig. 3) and B (fig. 4). We see from these figures that the process does not have enough time to generate a large quantity of small area fragments. In the case of Model A, (figure 3), the number of big fragments is almost constant up to a certain value that depends on the original size of the system, a fact that can be explained from the definition of the model, because each time that a fragment breaks into two new pieces, the difference between the new forces on the generated fragments becomes smaller. Thus, the probability of generating fragments of small area is small. The time evolution in this case is very fast. Mostly, large area fragments are generated (about 75% of the fragments), which explains the fast convergence. On the other hand, Model B, (see figure 4), shows an almost log-normal behavior, due to the independent choice of the new fragments generated from the fracture process.

Next we consider the relaxed stopping criterium again for models A and B using eq. (3). The fragment size distribution is shown in figures veryd 6, respectively. The computations (made on a Sun 4 Sparc 2) were very time consuming due to the exponential time evolution for the relaxed stopping criterium. Model A, (figure 5), generates roughly a power law behavior for small area fragments. A large fraction of the fragments have the shape of long thin rods, which sugests a kind of shear effect inside the system. This feature of the model can be explained from the definition of the forces on the generated fragments: each time that a piece is broken its forces differences becomes smaller and thus the fragmentation process can be stopped at early stages, generating large area pieces. By the other hand, the relaxed stopping determines the shape of the fragments as thin rods. These two features can produce the shear effect, but a more convincing explanation must be stated from a theoretical study of the model. The fraction and the exponent of the power law depend on the initial size of the material. For the systems studied the exponent for small area fragments is equal to -1.2 ± 0.4. Model B generates a power law fragment size distribution, with an exponent which increases with the size of the studied system. This result can be explained as a finite size effect. Furthermore, due to the high requirements computer time needed for the models with relaxed stopping, it is difficult to predict or extrapolate an exponent for $n \to \infty$. Despite of this fact, we have found power law behavior over the entire range of the area value, which is itself an important result. For the system size of $n = 64$ the power law is of the form:

$$\log F(s) = \alpha \cdot \log(s) + \beta \qquad (9)$$

where $F(s)$ is the fragment size distribution, s is the area of the fragments and $\alpha = -0.85 \pm 0.09$, $\beta = 1.66 \pm 0.03$. The former exponent does not agree with the available three dimensional experimental data, which gives $\alpha = -1.6 \pm 0.2$ since our model is two-dimensional, but represents an important feature of our model, because a power law behavior is produced by a simple stochastic model.

Finally, we studied the relaxed stopping criterium with shear forces defined by eq. (4). In this case, the fragments typically have the shapes of long thin rods. For the system sizes studied, $n = 32, 64$ and 128, the fragments are mostly thin rods of size 1×2^n. The fact that the thin rods have an elementary unit length is not surprising due to the definition of the relaxed stopping criterium. But, the fact that the other side of the long thin rod is substantially larger, needs more attention and simulations.

4 Conclusions

In this work simple models for two dimensional discrete fragmentation were studied. Different behaviors were obtained for the fragment size distribution which includes log-normal and power law behavior, depending on the definition of the parameters of our models, which are: the scalar forces acting on the fragments; the selection of the piece to break; the choice of the fracture orientation (vertical or horizontal); the definition of how the fragmentation process must continue and finally, the definition of a criterium to stop the fragmentation. The quasi-log-normal distribution can be explained from the random processes. The power law distribution is a non-trivial result.

The existence of elementary pieces that cannot be broken anymore introduces an arbitrary assumption. This limitation can be avoided by considering models that are discrete in time but not in space. We are actually developing such continous spatial models taking into account similar assumptions as in the discrete models studied here.

Acknowledgments

This work was partially supported by PG041-92 U. de Chile, C-10003 Fundación Andes and EEC. G. H. wishes to acknowledge financial support of the EEC through a "Marie Curie" fellowship.

References

[1] D.L. Turcotte, Fractals and Chaos in Geology and Geophysics, Cambridge University Press, 1992.

[2] M. Matsushita and T. Ishii, Fragmentation of long thin glass rods, Department of Physics, Chuo University, 1992.

[3] M. Matsushita and K. Sumida, How do thin glass rods break? (Stochastic models for one-dimensional fracture), Chuo University, Vol. 31, pp. 69-79, 1988.

[4] S. Redner, in Statistical Models for the Fracture of Disordered Media, H. Herrmann and S. Roux (eds.), Random Materials and Processes Series, Elsevier Science North Holland Publishers, 1990.

[5] W. Newmann and I. Wasserman, Hierarchical Fragmentation Model for the Evolution of Self-Gravitating Clouds, The Astrophysical Journal, 354, 411-417, 1990.

[6] J.J. Gilvarry, Fracture of Brittle Solids. I. Distribution Function for the Fragment Size in Single Fracture (Theoretical), Journal of Applied Physics, Vol. 32, 3, 391-399, March 1961. J.J. Gilvarry and B.H. Bergstrom, Fracture of Brittle Solids. II. Distribution Function for the Fragment Size in Single Fracture (Experimental), Journal of Applied Physics, Vol. 32, 3, 400-410, March 1961.

[7] S. Steacy and C. Sammis, An Automaton for Fractal Patterns of Fragmentation, Nature, Vol. 353, 250-252, September 1991.

Figures

Figure 1. Initial piece of material, where a, b, c, d are independently and uniformly distributed random numbers

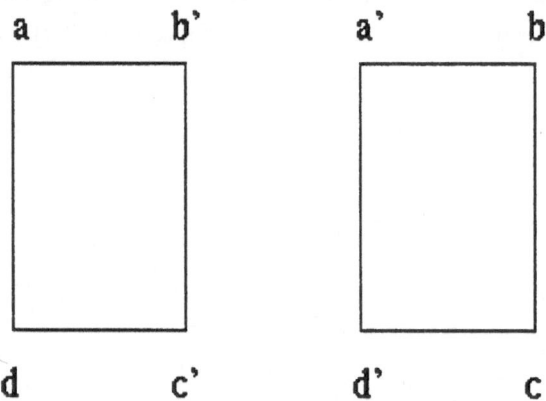

Figure 2. Vertical Breaking the piece of figure 1.

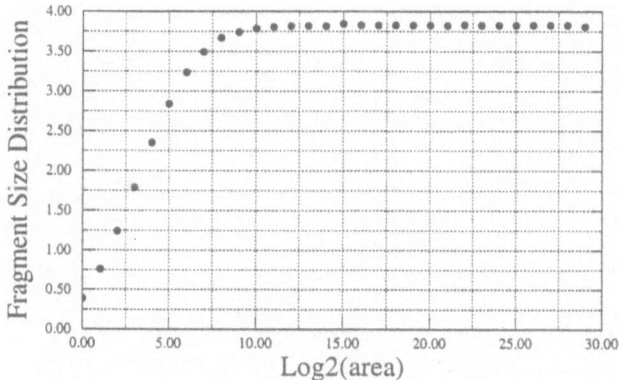

Figure 3. Model A using eq. (3) and fast stopping criterium for $n = 32768$ and averaging over 100.000 samples.

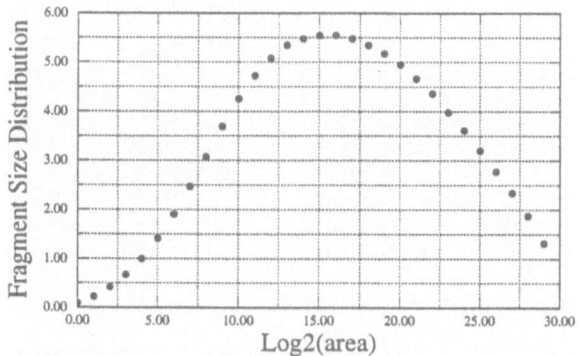

Figure 4. Model B using eq. (3) and fast stopping criterium for $n = 32768$ and averaging over 100.000 samples.

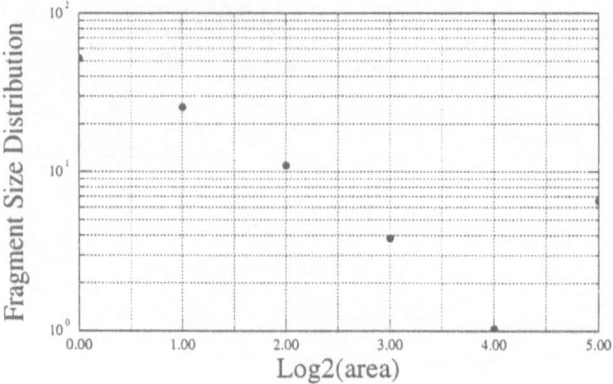

Figure 5. Model A using eq. (3) and relaxed stopping criterium for $n = 64$ and averaging over 100.000 samples.

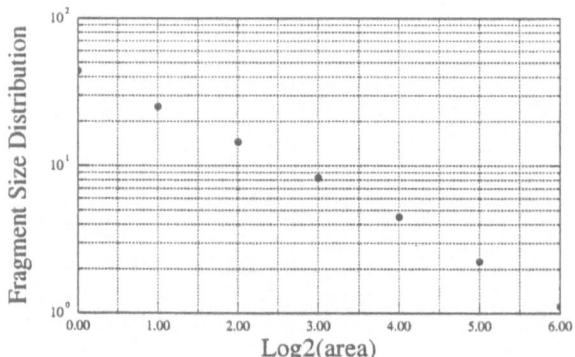

Figure 6. Model B using eq. (3) and relaxed stopping criterium for $n = 64$ and averaging over 100.000 samples.

PART VI
SHORT COMMUNICATIONS

CHAOS AND THREE-FREQUENCY RESONANCES IN A DYNAMICAL SYSTEM

S. BLANCO [1,2], D. GONZALEZ [3], O. A. ROSSO [1,4], F. SPORTOLARI [3]
[1] *CONICET,* [2] *Dpto. de Física - UBA,* [4] *Inst. Cálculo - UBA*
Ciudad Universitaria.
1428 Bs. As. , Argentina.
[3]*Sezione di Cinematografia Scientifica, CNR*
Via dell'Inferno 5
40126 Bologna, Italy

ABSTRACT. The behavior of two parametrically coupled nonlinear oscillators driven by two periodically independent forces is studied. Three-frequency quasiperiodic resonances are observed. An anzat based on a generalized Farey Sum operation between fraction of real numbers is given for the hierarchical arrangement of the stability width of three-frequency resonances.

1. Introduction

There are many kinds of nonlinear oscillators, an usually founded type is the limit cycle oscillator. In difference with the linear oscillators, the frequency, amplitudes and phase of the limit cycle oscyllators are closely related with the intrinsic properties of the system that one try to model.

If we "couple" two nonlinear oscillators in most cases, they will try to "sing in unisone": either the frequency of the stronger one takes over, or they compromise by finding a common frequency, usually at a fixed ratio of the two "free" frequencies. This is the phenomenon of *frequency-locking* or *mode-locking.*

Linear oscillators simply cannot do this, because they are not capable of changing their own frequency. Even coupled, nonlinear oscillators cannot always compromised at a common periodic regime, whence quasiperiodic and chaotic motions come into play.

We report here numerical results of this kind of behavior in a system of two non-linear parametrically coupled oscillators and each one, drives with an external impulsive and periodic force. The locking frequency for an irrational relationship between the frequencies of both external forces are studied using the characteristic frequency of one of the oscillators as tuning parameter.

The model studied here is constructed on the basis of a nonlinear oscillator with exact solution [1], in such a way the overall problem is piece-wise exactly resoluble. This strategy avoids the need to use numerical integration to compute the time evolution of both oscillators, lowering drastically the computation time for the generation of time series.

E. Tirapegui and W. Zeller (eds.), Instabilities and Nonequilibrium Structures V, 329–334.
© *1996 Kluwer Academic Publishers.*

2. The Model

We consider a system of two nonlinear oscillators and each one driven with an impulsive periodic force of amplitude V_{E_i} and frequency $f_i, (i = 1, 2)$.

The oscillators are coupled parametrically changing their proper frequency ω_{0i} any times the coordinates of one of them change sign,

Oscillator I:

$$\omega_{01} = \tilde{\omega}_{01} + sign(X_1) \cdot sign(X_2) \cdot \Delta_1 \tag{1}$$

Oscillator II:

$$\omega_{02} = \tilde{\omega}_{02} - sign(X_1) \cdot sign(X_2) \cdot \Delta_2 \tag{2}$$

where $\tilde{\omega}_{0i}$ and Δ_i are the central frequency and their change (the coupling parameter) for the oscillator i respectively. X_i are the coordinates of oscillator i at time t.

The dynamic of both oscillators is described by the following differential equation:

$$\ddot{x} + (4b\, x^2 - 2a)\, \dot{x} + b^2\, x^5 - 2ab\, x^3 + (\omega_{0i}{}^2 + a^2)\, x = V_{E_i} \sum_n \delta(t - n \cdot \tau_{E_i}) \tag{3}$$

where a , b and ω_{0i} are the parameters of the autonomous oscillator ($a = 1.57079, b = 10\, a$). V_E, τ_E and $f = 2\pi/\tau_E$ are the amplitude, period and frequency of the external forces respectivelly.

The locking frequencies for an irrational relationship between f_1 and f_2 are studied using $\tilde{\omega}_{02}$ as tunning parameter.

3. Results and Discussion

The experiment is carry out in the following way: for a given set of parameters, the time series of the coordenate $X(t)$ (2^{16} data) at sample frequency $\Omega_{sample} = 5$ and corresponding power spectrum for each oscillator are performed. In addition, we determine the frequency responce, f_i^R, that gives the maximun contribution to the total intensity of the power spectrum. The Lyapunov exponents are calculated for some selected points in order to characterize the kind of behavior of the system. The Lyapunov exponents were performed using the algorithm proposed by Eckmann *et al.*[2] using an embedding dimension $M_d = 5$.

In Fig. 1 the frequencies responce of both ocillators as a function of the central frequency of the oscillator one, $\tilde{\omega}_{01}$, is shown. The central frequency of oscillator two is fixed at $\tilde{\omega}_{02} = 1.57079$ and the frequencies of the external forces take the values $f_1 = 11.09553$ and $f_2 = 6.38316$. The amplitudes for both external forces are $V_{E_1} = V_{E_2} = 0.3$ and the change in the central frequencies are $\Delta_1 = \Delta_2 = 0.15$.

Note that three zones in the frequency space can be identified, denoted by I, II, and III in Fig.1. The system behavior fouded in the different zones are:

Zone I:

In this zone, it can distinguish three kinds of behaviors, non synchronized, few and narrow sysnchronized intervals and chaotic.

The intensities associated with the frequencies responce f_1^R and f_2^R are between 30% and 50% of the total intensity of the corresponding pawer spectra for the synchronized

($f_1^R = f_2^R$) and non-synchronized ($f_1^R \neq f_2^R$) behavior respectively. For the chaotic behavior, the intensities are less equal than 8%.

The biggest Lyapunov exponent was $\lambda > 0$ for chaotic motion. $\lambda \sim 10^{-3}$ for synchronized motion and $\lambda \simeq 10^{-2}$ or non convergency is obtained for non-synchronized behaviors.

Note that f^R presents oscillations in its values.

Zone II :

Both oscillators are always synchronized with 1 : 1 locking frequency and present well defined stability intervals.

The frequencies responce $f_1^R = f_2^R$ present increasing values and the average intensity associated is between 60% and 90%.

A similar structure of tipical Devil's Stair of periodically driven nonlinear oscillator is observed. In this case however the stability intervals correspond to mode-locking with three linearly dependent basic frequencies (the frequency plotted plus the two forcing ones) instead of periodic ones. Then we can say that these are quasioeriodic three-frequency resonances.

The range of the biggest Lyapunov exponent are between 10^{-4} and 10^{-5} and the other Lyapunov exponents are less than zero. These results are in complete agreement with the assumption of three-frequency quasiperiodic resonances.

In Fig. 2 a detail of this structure is given; and in Table I we give the values of linear coefficients p_n, q_n and k_n which do the best matching with the correspondig quasiperiodic three-frequency resonance: $f_n^R = (p_n \cdot f_1 + q_n \cdot f_2)/k_n$.

Zone III :

Two cases are observed: the oscillators are not synchronized or both oscillators are synchronized with different locking frequencies to 1 : 1 and present stability width smaller than in Zone II.

The intensity associated with f_1^R and f_2^R are of the same order that in the previous zone, but now they present increasing and decreassing values. The corresponding Lyapunov exponents are in agreement with quasiperiodic motions.

In this communication we focus our attention on the Zone II where three-frequency resonances are observed.

For the description of these resonances we can do the following anzat based in a generalization to fractions of reals numbers of the usual Farey Sum between rational numbers [3, 4].

If p/q and r/s are adjacent rationals, that is, if they are irreducibles and satisfy $|q \cdot r - p \cdot s| = 1$, we can generate the so called mediant with the aid of the Farey Sum, defined as following

$$\frac{p}{q} \oplus \frac{r}{s} = \frac{p+r}{q+s}$$

The *mediant* is the rational with the lowest denominator between p/q and r/s. In periodically driven non-linear oscillators [1] the mediant, as a rotation number, characterizes the most important phase-locked region in parameter's space between those corresponding to p/q and r/s.

To obtain a similar hierarchy in our case, we modify slightly the definition of adjacency: if p_n/q_n is a convergent of f_1/f_2 we define as adjacents any pair of f_i/k_i , f_j/k_j with $f \in R$ and $k \in Z$ that satisfy

$$|f_i \cdot k_j - f_j \cdot k_i| = |f_1 \cdot q_n - f_2 \cdot p_n| \tag{4}$$

Then, inside the frequency interval $(f_2/q_n, f_1/p_n)$ we can define a *generalized Farey Sum* between adjacents as:

$$f_s = \frac{f_i}{k_i} \oplus \frac{f_j}{k_j} = \frac{f_i + f_j}{k_i + k_j} \tag{5}$$

In our numerical model 4/7 is a convergent of $f_2/f_1 = 6.38316/11.09553$. In fact, in the interval $(f_1/7, f_2/4)$, see Fig. 3, we have found solutions up to order $N = 8$ by means of the application in recursive form of the generalized Farey composition. Their stability width is arranged in hierarchical order by means of the generalized mediant operation.

For example, between $f_1/7$ and $f_2/4$ the solution with the gratest stability width corresponds to $f_s^{(N=1)} = (f_1/7) \oplus (f_2/4) = (f_1 + f_2)/(7 + 4) = 1.588971$.

At the second level $(N = 2)$, we have two new intervals: $(f_1/7, f_s^{(N=1)})$ and $(f_s^{(N=1)}, f_2/4,)$, inside of them the solution with gratest stability width corresponds to $f_s^{(N=2)} = (f_1/7) \oplus f_s^{(N=1)} = 1.587456$ and $f_s^{(N=2)} = f_s^{(N=1)} \oplus (f_1/7) = 1.590790$ respectively (see Table I).

Similar results are obtained for other frequencies intervals, i.e., $((f_1 - f_2)/4, f_1/7)$.

In Table I we display the observed resonances values for our numerical model up to $N = 5$ in the frequency response interval $(f_1/7, f_2/4)$ as well as the predicted values by applaying the generalized Farey Sum. The hierarchical order in the Farey tree is denoted by N. We see that the agreement is very good and all visible resonances are well described by the tree structure. That is, their stability width are arranged in hierarchical order by means of the generalized mediant operation, in analogy with the case of phase locked solutions in periodically forced oscillators.

When we change the control parameters Δ and V_E or we choose a different irrational ratio f_1/f_2 similar qualitative results for the stability plots are obtained.

This results lead us to conclude that the behavior described in this communication is structuraly stable and could comprise a whole universaly class of real systems.

Aknowledgment

One of the authors (O.A.R.) is very grateful to the organizers of the workshop for their very kind hospitality during his state in Chile.

Bibliography

1. D. L. Gonzalez and O. Piro, Phys. Rev. Lett. 50, 870 (1983).
2. J. P. Ekmann, S. Oliffson Kamphorst, D. Ruelle and S. Ciliberto, Phys. Rev. A 34, 4871 (1986).
3. A. J. Khinchin, Continued Fractions (University of Chicago, Chicago, 1964).
4. D. L. Gonzalez, O. A. Rosso, F. Sportolari and L. Morettini, to be published.

Table I: The predicted resonances by the generalized Farey Sum (their order in the generalized Farey Tree are denoted by N), their values and the observed resonances for the O.D.E.'s system in the frequency response interval $(f_1/7, f_2/4)$. The external frequencies are fixed at $f_1 = 11.09553$ and $f_2 = 6.38316$. The values of the triplets (p_n, q_n, k_n) which do the best matching of the linear combinations for the resonances are given.

N	Mediant	Theory	Experiment	p_n	q_n	k_n
0	$f_1/7$	1.585075	1.585091	1	0	7
5	$(5f_1 + f_2)/39$	1.586174	1.586182	5	1	39
4	$(4f_1 + f_2)/32$	1.586415	1.586400	4	1	32
5	$(7f_1 + 2f_2)/57$	1.586579	1.586545	7	2	57
3	$(3f_1 + f_2)/25$	1.586790	1.586764	3	1	25
5	$(8f_1 + 3f_2)/68$	1.586966	1.586982	8	3	68
4	$(5f_1 + 2f_2)/43$	1.587069	1.587055	5	2	43
5	$(7f_1 + 3f_2)/61$	1.587183	1.587200	7	3	61
2	$(2f_1 + f_2)/18$	1.587456	1.587491	2	1	18
5	$(7f_1 + 4f_2)/65$	1.587713	1.587709	7	4	65
4	$(5f_1 + 3f_2)/47$	1.587811	1.587855	5	3	47
5	$(8f_1 + 5f_2)/76$	1.587895	1.587927	8	5	76
3	$(3f_1 + 2f_2)/29$	1.588031	1.588000	3	2	29
5	$(7f_1 + 5f_2)/69$	1.588181	1.588146	7	5	69
4	$(4f_1 + 3f_2)/40$	1.588290	1.588291	4	3	40
5	$(5f_1 + 4f_2)/51$	1.588437	1.588437	5	4	51
1	$(f_1 + f_2)/11$	1.588971	1.588946	1	1	11
5	$(4f_1 + 5f_2)/48$	1.589540	1.589528	4	5	48
4	$(3f_1 + 4f_2)/37$	1.589708	1.589746	3	4	37
5	$(5f_1 + 7f_2)/63$	1.589837	1.589891	5	7	63
3	$(2f_1 + 3f_2)/26$	1.590020	1.590037	2	3	26
5	$(5f_1 + 8f_2)/67$	1.590193	1.590182	5	8	67
4	$(3f_1 + 5f_2)/41$	1.590302	1.590328	3	5	41
5	$(4f_1 + 7f_2)/56$	1.590432	1.590473	4	7	56
2	$(f_1 + 2f_2)/15$	1.590790	1.590764	1	2	15
5	$(3f_1 + 7f_2)/49$	1.591198	1.591201	3	7	49
4	$(2f_1 + 5f_2)/34$	1.591378	1.591346	2	5	34
5	$(3f_1 + 8f_2)/53$	1.591544	1.591564	3	8	53
3	$(f_1 + 3f_2)/19$	1.591842	1.591855	1	3	19
5	$(2f_1 + 7f_2)/42$	1.592218	1.592219	2	7	42
4	$(f_1 + 4f_2)/23$	1.592529	1.592510	1	4	23
5	$(f_1 + 5f_2)/27$	1.593012	1.593019	1	5	27
0	$f_2/4$	1.595790	1.595783	0	1	4

Figure 1: Frequency response of both oscillators as a function of the central frequency of the oscillator one.

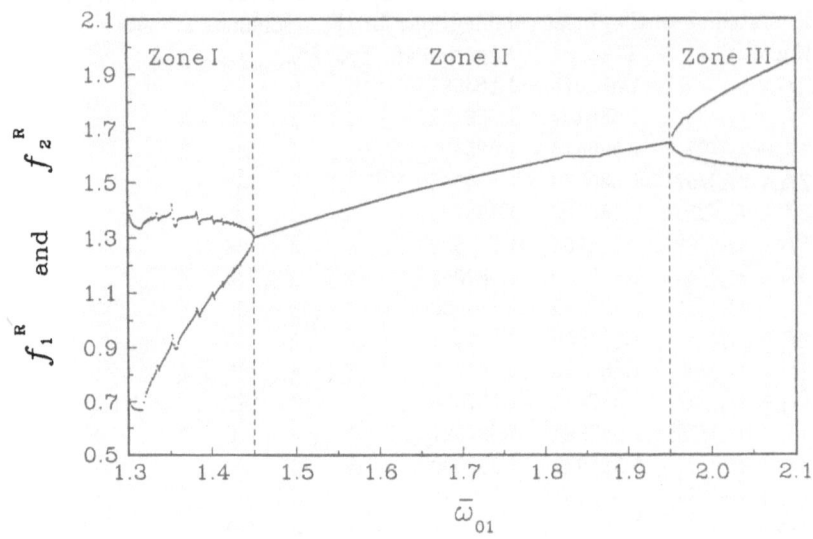

Figure 2: Detail of Fig. 1 (Zone II). The straight lines correspond to intervals where frequency responses of both oscillators are constant when the central frequency of oscillator one is varied.

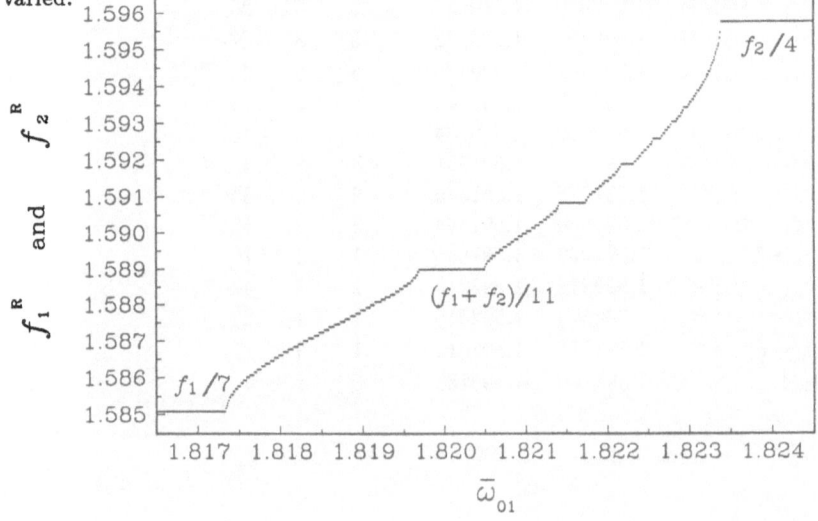

SHAPE TRANSITION IN A TWO-DIMENSION SYSTEM OF BUBBLES

V. A. Kuz

Instituto de Física de Líquidos y Sistemas Biológicos.IFLYSIB.
(U.N.L.P. - CONICET - CIC)c.c 565 .
1900 La Plata. Argentina

ABSTRACT : A two dimensional (2-D) system of bubbles in a lipid monolayer is maintained at a fixed temperature and area in the liquid-gas coexistence region. From the Einstein's fluctuation formula it is shown that the mean square relative deviation from the mean area $\lambda = (< A^2 > - < A >^2) \div < A >^2$ is proportional to the surface compressibility of this system. In the case of correlated particles, this quantity is also proportional to the integral of the particles correlation function over the area. The present model predicts a linear growth of λ with time, but a constant value when the correlations are absent. From the comparison of this result with the recent experiment of Berge et al. (Phys.Rev.A.41,6893(1990)) it can be inferred that the macroscopic transition from regular to irregular bubble pattern is due to the microscopic interaction of the surfactant particles. Changes in concentration increases the interparticles correlation, which in turn induces the experimentally observed shape transition.

1. Introduction

The great variety of 2-D cellular structures in nature are well known (1-4). Their similarity may suggest a universal behavior (5). Attention here will be concentrated on the monolayers of amphiphilic molecules spread on the air-water interface. Fluorescence microscopy technique have been used to visualize the coexistence phases in the monolayer (2-4). In a binary mixture of L-α-phosphatidylcholine /cholesterol, Rice et al observed , at the critical composition, a shape transition critical point induced by pressure. This transition can be distinguished experimentally from the standard phase transition critical point. Berge at al., also using this technique, have studied the temporal evolution of a 2-D bubble pattern in the L-G coexistence region of the monolayers. The gas bubbles, in the liquid phase, grow and transform into polygons. By changing the initial concentration of the film, the circular bubbles pattern is converted into an irregular polygonal network. A transition to a critical state is observed. It is clear that there are two phase transitions. One corresponds to a phase separation, the other is a transition of shape from a regular to an irregular polygonal network. They have also found (2) that λ remains constant in the former case while it grows linearly with time for the latter (the shape of the distribution does not reach a stable form and all the modes are present). It must also be mentioned that this type of "polygonal" phase changes can also be induced chemically (6). The Einstein fluctuation formalism in 2-D is used to study here this problem. It must be said that the system is at each time in a non-steady equilibrium. the bubbles' mean area, in

E. Tirapegui and W. Zeller (eds.), Instabilities and Nonequilibrium Structures V, 335–338.

the regular as well as in the irregular pattern, grows with time (2,7). It is assumed here that there is no correlation between the time scales of the fluctuations of the bubble's area and the time evolution of the system as a whole. This fact allowed us to apply the Einstein's fluctuation formula for the present situation (8). A relation between the mean square relative deviation from the mean area λ and the surface compressibility is derived here . By expressing this last quantity in terms of the two particles correlation function, λ results to be a linear function of time. Finally some comments are made after comparing the model with the experiment.

2. Two-Dimensional Fluctuations

Einstein's approach for treating the fluctuations of the area is discussed below. A mathematical surface of 2-D in the sense of Gibbs will represent the monolayer. The probability p that a fluctuation around the equilibrium state may occur will be given by $p \propto \exp(\Delta S/k)$, where ΔS is the change of the entropy in the fluctuation. By expanding the entropy around its value at equilibrium and considering the system isolated, the probability p will be a Gaussian distribution of the independent thermo-dynamics variables (8). The deviation of S from its value is related to the minimal work required to change the thermodynamic state of the system (9). If T, σ denote the temperature and the surface tension of the film, then the minimal work R_{min} is given by

$$R_{min} = \Delta U - T\Delta S - 2\sigma\Delta A \qquad (1)$$

where $\Delta U(= T\Delta S + 2\sigma\Delta A)$, ΔS and ΔA are the variations from the mean value in the surface energy (10), surface entropy and area of the subsystem. The surface tension σ is multiplied by a factor two because of the two side of the film. Expanding ΔU up to the second order, choosing T and A as independent thermodynamic variables, remembering the $(\partial U/\partial S)_A = T$ and $(\partial U/\partial A)_S = 2\sigma$ and expressing ΔS and $\Delta \sigma$ in terms of ΔT and ΔA ; the resulting equation will be a Gaussian distribution in these variables. The mean square fluctuation of the bubble's area is

$$< A^2 > = \frac{kT}{2}(\partial A/\partial \sigma)_T \qquad (2)$$

Then the mean square relative deviation from the mean area λ is

$$\lambda = \frac{kT}{2A^2}(\partial A/\partial \sigma)_T \qquad (3)$$

The fluctuation is proportional to the surface compressibility. If $(\sigma_o - \sigma) = NkT$, being σ_o the surface tension in the absence of the film, the right-hand side of equation (3) becomes $1/N$. This will be the order of magnitude for any slope $(\partial \sigma/\partial A)_T$ in the surface pressure-area representation of the monolayer except in the two-phase region , where the fluctuations becomes very large. In the two-phase region it has been shown (11) that $< (\Delta A)^2 >$ will be of the order of $< A >^2$ and the right-hand side of equation (3) will be of order unity. Then while ordinarily $(\partial\sigma/\partial A)_T = O(NkT/A^2)$, in the two-phase region $(\partial\sigma/\partial A)_T = O(kT/A^2)$ which is essentially zero, relative to $O(NkT/A^2)$. Keeping this analysis in mind the following relation (11) holds if the particles interaction is considered:

$$\frac{kT}{A^2}(\partial A/\partial\sigma)_T = 1 + \frac{1}{N^2}\int_A\int \left[\rho^{(2)}(r_1,r_2) - \rho^{(1)}(r_1)\,\rho^{(2)}(r_2)\right] dr_1 dr_2 \qquad (4)$$

where $\rho^{(2)}(r_1,r_2)dr_1 dr_2$ is the probability of observing particles in dr_1, dr_2 at r_1, r_2, irrespective of N and $\rho^{(i)}(r_i)$ is the probability of observing a particle at dr_i, at r_i. The integration is over the area A. The substitution of equation (3) into (4) gives :

$$\lambda = \frac{1}{2} + \frac{1}{2(\Gamma A)^2} \times \int_A\int \left[\rho^{(2)}(r_1,r_2) - \rho^{(1)}(r_1)\,\rho^{(2)}(r_2)\right] dr_1 dr_2 \qquad (5)$$

where $\Gamma = N/A$ is the density of the film and the constant A the total surface area. The film gets thinner with time. A bubble in expansion, immersed in a liquid or a gas, changes its surface density with time due to the process of surfactant dilution happened at the bubble's film. It has been shown elsewhere (12) that

$$\Gamma^2(t) = a - bt \qquad (6)$$

diminishes linearly with time, a and b are two constants. For values of t such $(bt/a) <$ 1, equation (5) becomes :

$$\lambda = \frac{1}{2} + \frac{1 + (b/a)t}{2aA^2} \times \int_A\int \left[\rho^{(2)}(r_1,r_2) - \rho^{(1)}(r_1)\,\rho^{(2)}(r_2)\right] dr_1 dr_2 \qquad (7)$$

If there is not correlation between particles λ remains constant and equal to 1/2.

3. Comments

From a statistical analysis of the bubbles in the monolayer, Berge et al. found that the mean square relative deviation from the mean area has two different behavior. At low surfactant concentrations it remains constant and equal to 1/2 while at higher concentrations it grows linearly with time. The theoretical predictions are in agreement with the experiment. The microscopic interaction of the surfactant particles induces an uneven growth of the bubbles in the monolayer.

Acknowledgments: I would like to thank to A.E.Rodriguez, to F.Vericat and M.Silbert for useful discussions.

References

[1] C.M.Knoble; Science 249,870(1990). B.G. Moore, C.M.Knoble, S. Akamasu and F.Rondelez; J.Phys.Chem.94,4588(1991).

[2] B.Berge, A.J.Simon and A.Libchaber; Phys.Rev.A.41,6893,(1990)

[3] K.J.Stine, S.A.Rauseo and B.G.Moore; Phys.Rev.A.41,6884(1990)

[4] P.A.Rice,and H.M. MacConnell; Proc.Natl.Acad.Sci.USA. 86, 6445 (1989). S.Subramanian and H.M.MacConell; J.Chem.Phys. 91, 1715(1987).

[5] D.Weaire and N.Rivier; Contemp.Phys.25,59(1984).

[6] J.Lucassen, S. Akamasu and F.Rondelez; Jour.Colloid and Interf.Sci.144, 434(1991).

[7] V.A.Kuz; Phys.Rev.A.44,8414(1991).

[8] P.Glansdorff, and I.Prigogine; Thermodynamic Theory of Stucture, Stability and Fluctuations; (1978), J.Wiley & Sons London, New York.

[9] L.D.Landau and E.M.Lifshitz; Statical Physics, Pergamon Press (1979), Oxford New York.

[10] A.W. Adamson; Physical Chemestry of Surfaces; J.Wiley & Sons (1982)

[11] T.L.Hill;Statiscal Mechanics Principles and Selected Applications; MacGraw-Hill Book Company, New York, London (1956) and J.Phys.Chem.; 57,324,(1953).

[12] V.A.Kuz; "Aging of a Liquid Drop" accepted for publication in Langmuir.

NONLINEAR BEHAVIOR AND TRAPPED DYNAMICS IN THE MICROMASER

E. LAZO[(a)], J. C. RETAMAL[(b)] and C. SAAVEDRA[(a)]

(a) Departamento de Física, Facultad de Ciencias
Universidad de Tarapacá
Casilla 7-D, Arica, Chile.
(b) Departamento de Física,
Universidad de Santiago
Casilla 307, Correo 2, Santiago, Chile.

ABSTRACT. We study the semiclassical limit of the quantum dynamics of a micromaser which provides a return map of the intracavity electromagnetic field. We consider atoms injected inside the microwave cavity in a coherent superposition. We found bifurcations, chaos, and transitions between bifurcation sequences in a broad range of the parameter space. In particular, we study the map as a function of the coherence parameter which allows the field to reach chaotic behavior in a lossless cavity. In addition, we propose an experimental setup to detect some nonlinear features of the field by simulating a measurement process of the atomic level populations when the atoms leave the cavity.

1 Introduction

In recent works interesting results have been found in the semiclassical limit of the micromaser theory [1]. In this limit the field intensity exhibits a chaotic behavior in terms of the parameter $\alpha = \exp(-\gamma t_p)$, where γ is the cavity decay constant and t_p is the average time passage between consecutive atoms. In addition, the nonlinear features of a micromaser can be observed in terms of another control parameters[2]. On the other hand, cooperative effects, that is, more than one atom simultaneously in the micromaser cavity, have been recently studied [3], showing the acceleration in the onset of chaos. We assume that the atoms are being injected in a coherent superposition. This coherence can determine the existence of the chaotic behavior in a lossless cavity, by exhibiting an inverse relation between the initial atomic coherence and chaos in the field, and under the same assumption, the system exhibits also trapped dynamics. Even when at this stage the problem seems to have only a theoretical interest, it is important to study a possible mea-

339

E. Tirapegui and W. Zeller (eds.), Instabilities and Nonequilibrium Structures V, 339–344.

surement scheme to see how the chaotic features would be observed outside the cavity.

The electromagnetic field in a micromaser is driven by two relevant processes, these are: the atom-field interaction during a time τ and the decay of the electromagnetic field through the cavity walls. The discrete nature of the micromaser dynamics arises due to passage of successive atoms through the cavity in such a way that only one atom interacts with the internal field during an interaction time τ. If the atoms are injected with an average injection rate r, the time between consecutive atoms is $r^{-1} = t_p + \tau$, so that the cavity remains without atoms for a time t_p which under current experimental conditions is $t_p \simeq r^{-1}$[4] . Considering that the cavity losses affect the internal field mainly during the time t_p, the field density matrix, after the k-th atom, can be written as $\rho^{(k+1)} = e^{Lt_p} tr_{\text{atom}} \left(U(\tau)\rho_{\text{atom}} \otimes \rho^{(k)} U^{\dagger}(\tau) \right)$, where $U(\tau)$ is the Jaynes-Cummings evolution operator [5] , and L represents the cavity losses [6] during the time t_p. The previous density matrix allows us to study the field dynamics as a discrete process, in particular, we are interested in observing the evolution of the average photon number. The atomic phase couples with the first off-diagonal of the field density matrix. Actually, when the atoms are injected with $\theta_{ab} = \pi/2$ the phase of the off-diagonal matrix elements rapidly locks to zero[7], so that the phase difference is $\theta_{ab} - \theta = \pi/2$. If we denote the semiclassical limit of the average photon number by the variable ψ, we can write the following return map

$$\psi_{k+1} = \alpha\{\psi_k - \frac{1}{2}R\sin(\mu - 2\phi\sqrt{\psi_k}) + \Omega\}, \tag{1}$$

where $\alpha = e^{-\gamma t_p}$, $R = \sqrt{A^2 + 4B^2}$, $\mu = \arctan\left(\frac{A}{2B}\right)$, $\Omega = \frac{R}{2}\sin(\mu)$, and $\phi = g\tau$. The parameters A and B are, respectively, the initial difference of atomic populations $A = \rho_{aa} - \rho_{bb}$ and the initial coherence $B = |\rho_{ab}|$. In order to study how the field evolves from a given initial condition to the stationary regime, it is important to analyze the fixed points structure of the return map $\psi_{k+1} = f(\psi_k)$. A simple calculation shows

$$\psi_q^* = q^2 \pi^2 \phi^{-2}, \qquad q = 0, 1, 2, \cdots, \quad \text{unstable}, \tag{2}$$

$$\psi_p^* = [p\pi - arctg(\frac{2B}{A})]^2 \phi^{-2}, \qquad p = 0, 1, 2, \cdots, \quad \text{stable}, \tag{3}$$

where the stability of fixed points is given through the condition $|f'(\psi^*)| < 1$.

2 Trapped dynamics and chaos

Let us consider the effects of the injected atomic coherence, B, in the dynamics of the return map (1) when there is no dissipation, i.e., $\alpha = 1$. The position of the unstable fixed points (2) is not affected by the coherence, because it only depends on the reduced time ϕ, but the stable fixed points (3) moves to the origin as B increases, separating from the unstable ones. In addition, the slope of $f(\psi_p^*)$ changes, and eventually ψ_p^* becomes unstable when $|f'(\psi^*)| > 1$. This determines a rich nonlinear behavior of the return map characterized by a period-doubling sequence and chaos as a function of B. The dynamics of the map is restricted to a finite region between two or more unstable fixed points, giving rise to a trapped dynamics with or without mixing between the basins of attraction. Mixing occurs when the maximum of $f(\psi)$ surpasses the unstable fixed point to the right of the initial point. As a consequence, there will be a lot of initial points which will arrive to the maximum after many iterations of the map, going to the right of the unstable fixed point, thus allowing the field to evolve to the next stable fixed point. However, there is an upper bound for the first block of mixed basins of attraction which separate the mixed region from the region of disconnected basins of attraction. In this way, depending on the value of A, ϕ and B, the field enters in chaos or experience bifurcations depending on which is the initial value ψ_0. In fig. (1) we show the return map and the attractors for the coherent case $B = 0.31$, for $A = 0.2$, and $\phi = 17$. There we can see the chaotic attractor for the trapped dynamics between the first and the fifth unstable fixed points, and we can observe some bifurcations for higher initial field values. The observed behavior clearly shows that the field evolves to a chaotic scenario via period doubling. Fig. (2) shows the fixed point structure of the map as a function of B for the same parameter values, for initial field condition $\psi_0 = 2.8$. There we can observe that there is an inverse relation between the appearance of bifurcations and chaos with the value of the injected atomic coherence, B, because the field evolves from a single valued stationary state for $B = 0$ (there is no chaos [2]), to a state exhibiting bifurcations and chaos for $B \to B_{\max}$.

In the incoherent atomic pump case, $B \to 0$, the function $f(\psi)$ becomes tangent to the diagonal $\psi_k = \psi_{k+1}$, and the fixed points $\psi_p^* = p^2 \pi^2 \phi^{-2}$, $p = 0, 1, 2, \ldots$ are marginally stable. In this case there is no bifurcation or chaos, but we obtain trapped dynamics and mixing of the basins of attraction [2].

3 Measurement simulation

As is well known the micromaser is a closed system, so that no direct measurement can be implemented to know in what state the internal field is. Experimentally, the field properties are indirectly measured by testing the atomic populations when the atoms get outside the cavity, by using an atomic ionization process. In a quantum description[6], the atomic variables are obtained by taking the trace with respect to the field, so that after the interaction with the cavity during an interaction time τ the atomic density matrix is $\rho_{\text{atom}}^{(k)} = tr_{\text{field}} \left(U(\tau)\rho_{\text{atom}} \otimes \rho^{(k)} U^\dagger(\tau) \right)$. Considering now the semiclassical limit in the atomic variables in the same way as we did to obtain the Eq. (1), we get for the upper level population

$$\rho_{aa}^{(k)} = \frac{1}{2} \left[1 + R\sin\left(\mu - 2\phi\sqrt{\psi_k} \right) \right] \tag{4}$$

The information about the field state that is carried out by the atoms leaving the cavity is collected outside by means of a set of N_d detectors, which measures the probability that the atom be in its upper or lower state. We assume that each detector has unitary efficiency, in this way we ensure that all atoms are detected during the measurement. The detection process begins after a transient time, typically $N_{\text{atom}} \sim 10^4$, so that the measurement gives information about the fixed point structure of the internal field. Each detector is associated with an ionizing field, and they are connected in sequence. After the transient, the j-th detector measure the set of atoms with label $j + N_d m$, with m an integer. We simulate the detection process for each atom which passes through the detector arrangement as a standard accepting-rejecting MonteCarlo process comparing the theoretical value of $\rho_{aa}^{(k)}$, given by Eq. (4), with a random number ϵ, namely, if $\rho_{aa}^{(k)} > \epsilon$ we accept the value of the measure as $\rho_{aa}^{(k)} = 1$, that is, the atom is in the upper level. In this way each detector has memory about two values, the upper and the lower level population. There are, in principle, as many values of ρ_{aa} as the number of channels in the detector arrangement. In Fig. (3.) we plot the average value of the upper level population as a function of ϕ for $A = 0.2$, $B = B_{\text{max}} = \sqrt{\rho_{aa}\rho_{bb}}$, $\alpha = 0.99$ and $\psi_0 = 2.8$ for a number of 8 channels of detection. From this figure it is evident that the measurement process mimics the nonlinear behavior of the internal field. The bifurcation sequence coincides with that of the direct calculation from Eq.(1). Given that we have considered a detector arrangement of only 8 channels, we can observe a maximum of 8 branches. This fact is reflected in Figure (3) where we

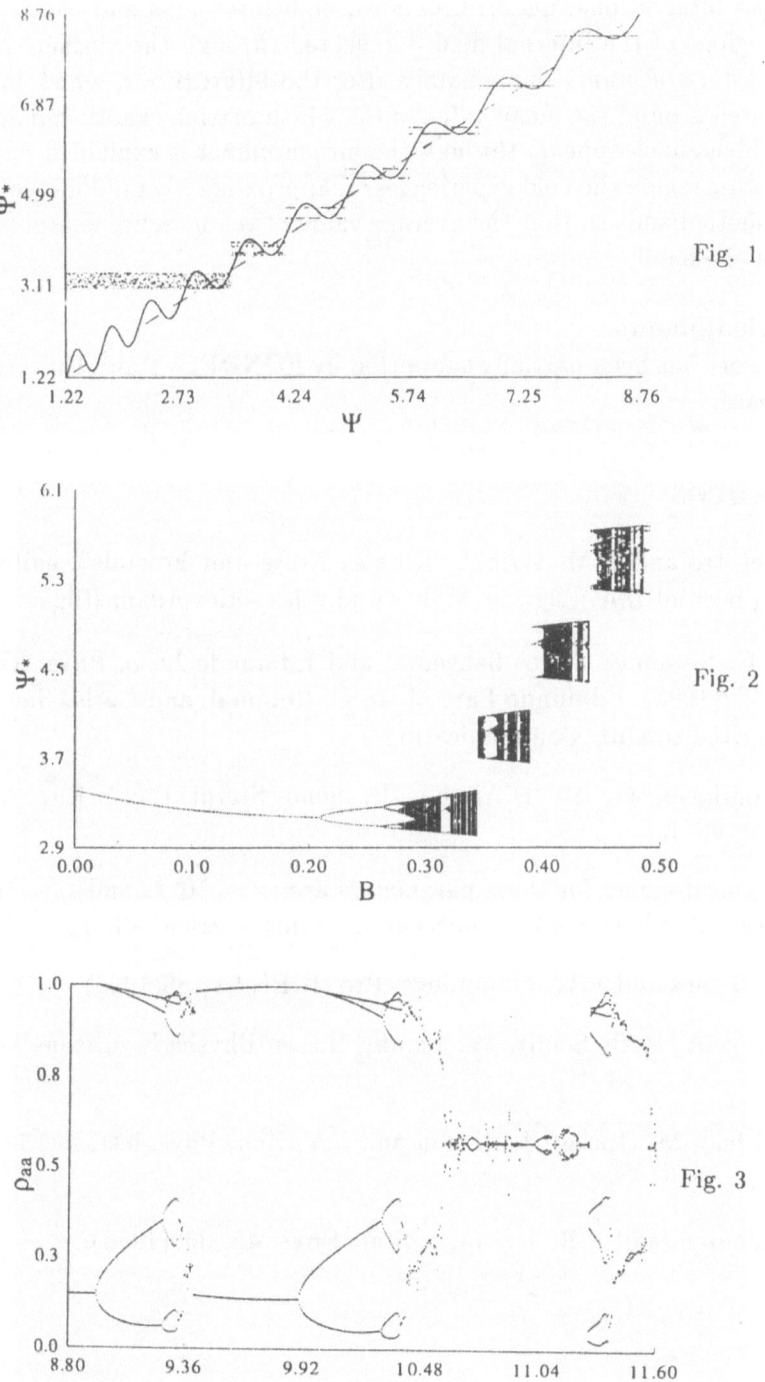

Fig. 1

Fig. 2

Fig. 3

distinguish bifurcations, quadrifurcations, eightfurcations and chaos. The chaotic regions of the internal field is reflected through the measure of ρ_{aa} by the clusters of points immediately after the bifurcations, which appear concentrated around the mean value of the width of each chaotic band. The way in which chaos appears through the measurement is explained because in the chaotic region the field experiences an approximately random sequence within chaotic bands so that the average value of the measure is around the middle of the band.

Acknowledgments

This work has been partially supported by FONDECyT and Universidad de Tarapacá.

References

[1] P. Meystre and E.M. Wright, "Chaos, Noise and Fractals", edited by E.R. Pike and L.A. Lugiato. Malvern physics series Adam Hilguer 1987.

[2] Juan C. Retamal, Carlos Saavedra, and Edmundo Lazo, Phys. Rev. A **48**, 2482(1993). Edmundo Lazo, Juan C. Retamal, and Carlos Saavedra (submitted to Opt. Communications)

[3] R. Bonifacio, G. M. D'Ariano, R. Seno Sterpi Phys. Rev A **47**, R2464(1993).

[4] The typical values for these parameters are: $\tau \sim, 10^{-5}$s and $t_p \sim 10^{-3}$s, thus the time between two consecutive atoms is essentially t_p.

[5] E. T. Jaynes and F.W. Cummings, Pro. IEEE **51,** 89(1963).

[6] M. Sargent, M.O. Scully, W. Lamb, "Laser Physics", Adison Wesley, 1970

[7] F.X. Zhao, M. Orszag, J. Bergou and S.Y. Zhu, Phys. lett. A **137**, 471 (1989).

[8] R.V. Jensen, and E.R. Jessup, J. Stat. Phys. **43**, 369(1986).

ANALITICAL STUDY OF THE CODIMENSION TWO BIFURCATIONS OF THE NEW LORENZ SYSTEM

A.C. SICARDI SCHIFINO and C. MASOLLER
Instituto de Física, Facultad de Ciencias
Tristan Narvaja 1674 Montevideo, Uruguay and
Instituto de Física, Facultad de Ingenieria
Herrera y Reissig 565 Montevideo, Uruguay.

ABSTRACT. The nonlinear dynamics of the New Lorenz System is analytically investigated. A saddle-node-Hopf bifurcation and a double saddle-node one are found when the external-heating contrast and the heating contrast between oceans and continents are varied. The normal form of the saddle-node-Hopf bifurcation is obtained, and it is shown that the codimension two bifurcations govern the structural changes in the evolution of the attractors.

In recent numerical studies [1-3] we have shown that the New Lorenz System [4,5] has a very rich and complex dynamical behavior. In this contribution we present an indepth analytical investigation of the bifurcations that occur in this model. Local analysis of the equations demonstrates the existence of saddle-node and Hopf bifurcation curves with codimension two saddle-node-Hopf and double saddle-node bifurcation points. Using the external heating contrast F and the heating contrast between oceans and continents G as control parameters, a bifurcation diagram in the (F,G) parameter space is given.

The model equations are

$$dx / dt = -y^2 - z^2 - ax + aF, \tag{1}$$

$$dy / dt = xy - bxz - y + G, \tag{2}$$

$$dz / dt = bxy + xz - z, \tag{3}$$

where the variable X represents the strength of a large-scale westerly-wind current, while Y and Z are the strengths of the cosine and sine phases of a chain of superposed waves. The parameter F represents the external-heating contrast while G represents the heating contrast between oceans and continents.

The coordinate X of the fixed points of the system verify

$$a(X - F)\left[b^2 X^2 + (1 - X)^2\right] + G^2 = 0, \tag{4}$$

while the coordinates Y and Z verify

$$Y^2 + Z^2 = G^2 / \left[b^2 X^2 + (1 - X)^2\right]. \tag{5}$$

Equation (4) has one real root if $q^2 + p^3 > 0$, two if $q^2 + p^3 = 0$ and three if $q^2 + p^3 < 0$, where

$$p = (2F + 1) / \left[3(b^2 + 1)\right] - \left[F + 2(b^2 + 1)^{-1}\right]^2 / 9, \tag{6}$$

345

E. Tirapegui and W. Zeller (eds.), Instabilities and Nonequilibrium Structures V, 345–348.

$$q = \frac{G^2 - aF}{2a(b^2+1)} + \frac{1}{6}\frac{2F+1}{b^2+1}\left[F + 2(b^2+1)^{-1}\right] - \frac{1}{27}\left[F + 2(b^2+1)^{-1}\right]^3. \tag{7}$$

The saddle-node bifurcation set is given by the set of points (F,G) for which Eq. (4) has a degenerate zero, i.e., the saddle-node curves G=G(F) are obtained from Eqs. (6)-(7) with the condition $q^2+p^3=0$. The results for a=1/4 and b=4 are shown in Fig. 1a. Notice that the bifurcation set consists of two codimension one curves on which saddle-node bifurcations occur, and the codimension two point

$$F = (1+\sqrt{3}b)/(1+b^2) \approx 0.466365, \qquad G = 2\sqrt{2\sqrt{3}ab^3}/\left[3(b^2+1)\right] \approx 0.291954 \tag{8}$$

at which we have the doubly degenerated zero of Eq.(4). In the curve labeled "a" a direct saddle-node bifurcation occurs, and two fixed points are born. In the curve labeled "b" an inverse saddle-node bifurcation occurs, and two fixed points collide and disappear. Thus, in the region I of the parameter space (F,G) the system has only one fixed point (which we shall call fixed point 1), while in the region II it has three fixed points (point 1, and the fixed points 2 and 3, which are created after the saddle-node bifurcation).

The stability of the fixed points is determined by the eigenvalues λ of the linealization of Eqs. (1)-(3) at (X,Y,Z), that satisfy

$$\lambda^3 + \alpha\lambda^2 + \beta\lambda + \gamma = 0, \tag{9}$$

where (Eq. (5) has been used to eliminate Y and Z)

$$\alpha = 2 + a - 2X, \tag{10}$$

$$\beta = 2G^2/\left[b^2X^2 + (1-X)^2\right] + b^2X^2 + (1-X)^2 + 2a(1-X), \tag{11}$$

$$\gamma = a\left[b^2X^2 + (1-X)^2\right] - 2G^2\left[(b^2+1)X - 1\right]/\left[b^2X^2 + (1-X)^2\right]. \tag{12}$$

It is interesting to notice that Eqs. (4)-(5) and (10)-(12) are invariant under the transformation (F,G)→(F,-G), and as a consequence, the bifurcation diagram is axisymmetric about the F axis in the (F,G) space. Also, notice that the saddle-node curves have $\lambda_1 = 0$ and thus satisfice γ=0.

The eigenvalues of the linealization at the saddle-node bifurcating point are (0,ρ±iω). From Eq. (9) we obtain that ρ and ω verify ρ=-α/2; $\rho^2+\omega^2=\beta$. For the values of a=1/4 and b=4 we numerically calculated α and β. In curve labeled "a" of Fig. 1a, α>0 if F<1.684 and α<0 if F>1.684 while β>0 ∀ F. Thus, in curve "a" if F<1.684 a saddle focus and a stable focus are created, while if F>1.684 a saddle focus and a source focus are created. In curve labeled "b" α, β>0 ∀ F and therefore a saddle focus and a stable focus collide and vanish.

Let us now investigate the Hopf bifurcations. At the Hopf bifurcating point the eigenvalues of the linealization are (ρ, iω,-iω) and thus, from Eq. (9), the conditions αβ=γ, $\beta=\omega^2 \geq 0$ must be satisfied. Substituting Eqs. (10)-(12) and eliminating G using Eq. (4) gives the following equation for the coordinate X of the fixed points where a Hopf bifurcation occurs

$$(b^2+1)X^3 + \left[ab^2 - b^2 - 3(a+1)\right]X^2 + \left[aF(1-b^2) + 3 + 5a + 2a^2\right]X - (a+1)(aF+a+1) = 0. \tag{13}$$

The Hopf bifurcation set is found in the following way: for a given value of F, the coordinate X of the fixed points where a Hopf bifurcation occurs is found using Eq. (13). Then, the value of G is found from Eq. (4) and thus we obtain the curves G=G(F) where a Hopf bifurcation occurs. These curves are shown in Fig. 1b, for a=1/4 and b=4. Although Eq. (13) has at least one real solution, if F<1 the fixed points do not undergo a Hopf bifurcation (the value of G^2

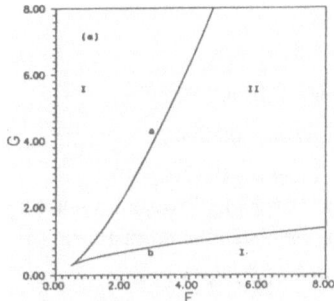 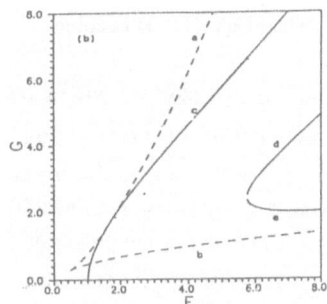

Fig. 1 (a) Saddle-node bifurcation set, for a=1/4 and b=4. Curves "a" and "b" are tangent at the codimension two bifurcating point (0.466, 0.292). (b) Hopf bifurcation set (solid line) and saddle-node bifurcation set (dashed line), for a=1/4 and b=4. Notice that curves "a" and "c" are tangent at the saddle-node-Hopf bifurcating point (1.684, 1.683).

obtained from Eq. (4) is negative). At $(F,G)=(1,0)$ the fixed point $(1,0,0)$ undergoes the Hopf bifurcation studied in [1], it becomes unstable and the "weak" cycle appears. In the region $1 \leq F \leq 5.76$ just one fixed point undergoes Hopf bifurcation, but in the region $F > 5.76$ the three fixed points undergo Hopf bifurcation (Eq. (13) has three real roots that lead to real values of G^2).

At the bifurcating point the eigenvalues of the linealization are $\lambda_1 = -\alpha$ and $\lambda_{2,3}=\pm i\sqrt{\beta}$. In the curve labeled "c" of Fig.1b, $\alpha \geq 0$ if $F \leq 1.684$ and $\alpha \leq 0$ if $F \geq 1.684$ while in the curves labeled "d" and "e" α is always positive. Thus, in curves "d" and "e" a sink focus undergoes Hopf bifurcation, while in curve "c" a sink focus undergoes Hopf bifurcation if $F \leq 1.684$ and a source focus undergoes Hopf bifurcation if $F \geq 1.684$.

The interaction of a saddle-node and a Hopf bifurcation has been studied recently [6-8]. A saddle-node-Hopf bifurcation occurs when the linealization of the vector field at a fixed point has eigenvalues $(0, i\omega, -i\omega)$ and thus the conditions $\alpha = \gamma = 0$ must be satisfied. From Eqs. (10), (12) we obtain that the coordinate X of the fixed point where the saddle-node-Hopf bifurcation occurs is $X = 1 + a/2$, while the values of the parameters at which this bifurcation occurs are

$$F = \left[3b^2(2+a)^2 + 4a + 3a^2\right]/4(2b^2 + ab^2 + a), \quad G^2 = a\left[a^2 + b^2(2+a)^2\right]^2/16(2b^2 + ab^2 + a). \quad (14)$$

For a=1/4 and b=4 the point where the saddle-node-Hopf bifurcation occurs is $(F,G) \approx (1.68405, 1.68297)$. Notice that curves "a" and "c" in Fig. 1b are tangent at this point. In spite of the fact that curves "b" and "c" intersect at one point in the parameter space (F,G), at that point it does not occur a saddle-node-Hopf bifurcation. At that point in parameter space, two of the fixed points coalese and vanish while the third fixed point undergoes a Hopf bifurcation.

For a=1/4 and b=4 the fixed point where the saddle-node-Hopf bifurcation occurs has coordinates $(X_0, Y_0, Z_0) \approx (1.125, -0.0104, 0.3737)$. We will turn to calculate the normal form for this bifurcation. The eigenvalues of the linealization at the bifurcating point are $(0, i\omega, -i\omega)$ where

$$\omega^2 = b^2 X_o^2 + (1 - X_o)^2 + 2(Y_o^2 + Z_o^2) - a^2 \approx (4.5258)^2 \quad (15)$$

To put the vector field (1)-(3) into standard form, we apply two transformations: we first apply a transformation that brings the fixed point to the origin, and then we implement the linear change of variables $(u,v,w)=S^{-1}(x,y,z)$, where $S \equiv (v_r, v_i, v_1)$, v_1 being the eigenvector of the real eigenvalue and $v_{2,3}=v_r \pm iv_i$ the eigenvectors of the pair of complex conjugated eigenvalues. Then Eqs. (1)-(3) become

$$du / dt = -0.31u^2 - 1.02v^2 - 1.20w^2 - 0.96uv + 0.96uw + 0.01vw,$$

$$dv / dt = \omega w - 0.27u^2 + 0.17v^2 - 0.20w^2 + 1.22uv + 3.67uw - 0.58vw, \qquad (16)$$

$$dw / dt = -\omega v - 1.22u^2 + 0.27v^2 - 0.32w^2 - 3.70uv + 1.29uw + 0.39vw.$$

Now, applying the following nonlinear transformation $u = \phi + h_3(\xi, \psi, \phi)$, $v = \psi + h_2(\xi, \psi, \phi)$, $w = \xi + h_1(\xi, \psi, \phi)$, (where h_i are polynomial expresions), we obtain the normal form [6].

$$d\xi / dt = -\omega\psi + a\phi\xi, \qquad d\psi / dt = \omega\xi + a\phi\psi, \qquad d\phi / dt = b(\xi^2 + \psi^2) + c\phi^2, \qquad (17)$$

where $a = 1.255$, $b = -1.11$ and $c = -0.31$. A new change of variables $r = \sqrt{bc(\xi^2 + \psi^2)}, s = |c|\phi$, reduces the system (17) of three nonlinear coupled equations to the following form

$$dr / dt = \alpha rs, \quad ds / dt = -r^2 - s^2, \qquad (18)$$

with $\alpha = -a/c$. The bifurcation set and the phase portrait of Eqs. (18) are analysed in [6-8]. In particular, if we remove the saddle-node bifurcation but not the Hopf bifurcation, varing F and G in a way that

$$dr / dt = \alpha rs, \quad ds / dt = \mu - r^2 - s^2, \qquad (19)$$

with $\mu > 0$, it is possible to show that there is a family of quasiperiodic orbits limiting on a homoclinic cycle conecting the saddle points. Then, complex secundary bifurcation phenomena appears [6] such as homoclinic orbits, horseshoes, global bifurcations, etc. This study is in progress and will be reported elsewhere.

In conclusion, we have done an analytical analysis of the bifurcations that occur in Lorenz' model of general circulation of the atmosphere. A bifurcation diagram in the (F,G) parameter space is presented and in this diagram, the saddle-node and Hopf bifurcation curves can be regarded as "organizing" the diagram. The complicated sequence of bifurcations found in [1-3], is originated after the inverse saddle-node bifurcations of curve "b". The normal form of the saddle-node-Hopf bifurcation was studied in detail. We find that the interaction of a saddle-node and a Hopf bifurcation in this system exhibits surprising complexity. The unfolding of this bifurcation and the study of the double saddle node bifurcation found is in progress and will be reported elsewhere.

ACKNOWLEDGMENT. This work was supported in part by the "Proyecto Científico Tecnológico CONICYT-BID" of the Consejo Nacional de Investigaciones Científicas y Técnicas (CONICYT, URUGUAY); by PEDECIBA (Project URU/84/002 - UNDP) and Comisión Sectorial de Investigación Científica (CSIC, URUGUAY).

REFERENCES

1. C. Masoller, A. Sicardi Schifino and L. Romanelli, *Physics Letters A* 167 (1992) 185.
2. C. Masoller, A. Sicardi Schifino and L. Romanelli, to appear in *Chaos, Solitons and Fractals*.
3. C. Masoller, A. Sicardi Schifino and L. Romanelli, in preparation.
4. E. N. Lorenz, *Tellus* 36A (1984) 98.
5. E. N. Lorenz, *Tellus* 42A (1990) 378.
6. J. Guckenheimer and P. Holmes, Nonlinear Oscillations, Dynamical Systems, and Bifurcations of Vector Fields (Springer-Verlag, New York, 1983).
7. F. Takens, Singularities of vector fields, Publ. Math. IHES, 43 (1974) 47.
8. V. Kirk, Phys. Lett. A 154 (1991) 243.

A NEW BASIS OF WAVELETS FOR STUDYING THREE PERIODIC ATTRACTORS

A. Figliola* and E. Serrano [†]

* Departamento de Física, Facultad de Ciencias Exactas y Naturales, (UBA)
Pab. I Ciudad Universitaria. (1428) Buenos Aires. Argentina.
e-mail: fig@dfuba.edu.ar
[†] Departamento de Matemática, Facultad de Ciencias Exactas y Naturales, (UBA)
Pab. I Ciudad Universitaria. (1428) Buenos Aires. Argentina.
e-mail: eduser@mate.edu.ar

Abstract:A new basis of orthogonal wavelets specially designed for the study of three period attractors, is here presented. The Logistic Map in the zone close to the inttermitency, is analyzed with this technique. Wavelets are very sensitive in order to detect the turbulent bursts. In a second step, the map of the wavelet coefficients is constructed and it leads us to caracterize the difference between the inttermitency and the chaotic behavior.

1. Introduction

Given a signal $f(t)$, in many applications one is interested in to determine its local spectrum, that is, to correlate in a suitable representation time with frequencies. This is the case when f can be considered as a time-varying frequency signal. For example, we can mention a music or a speech signal whose spectrum evolves over time in a significantly way. As it is well known, Fourier analysis is quite inappropriate for non stationary signal processing. In order to overcome this problem numerous methods to achieving time dependent frequency analysis have been proposed in the literature [1],[2],[3]. The Wavelet Transform, recently developed, give us an interesting alternative for this purpose. An extensive bibliography about this theory can be found by the reader [4]-[9]. In this paper, we construct a new base of orthogonal wavelets adapted to the study of three period attractors and we applied it to the Logistic Map [10] in the period-3 window, after the intermittency zone [11]-[12].

We will give an introduction to the Wavelet Transform and the to diadic and triadic structures in the next section. Section 3 is a brief review of intermittency in the Logistic Map and in the last section we present the results.

2. The Wavelet Transform

The Wavelet Transform is based on dilatations (or compressions) and translations of an appropriate analyzing window function ψ, called *wavelet*. It can be

E. Tirapegui and W. Zeller (eds.), Instabilities and Nonequilibrium Structures V, 349–357.

considered as a local Fourier analysis with variable precision. Depending of the type of application on hand, different wavelets can be chosen. Typically, ψ must be a smooth oscillating function, with zero mean and well localized as a band-pass filter. Then, give an admissible wavelet ψ, the continuous wavelet transform associated is defined by the map:

$$W_\psi^c f(a,b) = |a|^{-\frac{1}{2}} \int_{-\infty}^{\infty} f(t)\psi(\frac{t-b}{a})dt \qquad a,b \in \Re \quad a \neq 0 \tag{1}$$

where f is and arbitrary finite energy signal. Note that, because the action of the parameters a and b, higher frequencies are correlates with small scales; while low frequencies correspond to large scales. In signal processing applications one must restrict the parameters a, b to an appropriate discrete sublattice. Then we have the discrete family of wavelets:

$$\psi_{j,k}(t) = a_0^{-j/2}\psi(a_0^{-j}t - kb_0) \qquad j,k \in Z \tag{2}$$

where $a_0 > 0, b_0 > 0$ are fixed numbers, and the discrete transform is given by

$$W_\psi^d f(j,k) = \int_{-\infty}^{\infty} f(t)\psi_{j,k}(t)dt \tag{3}$$

2.1. The Diadic Structure

The discretization on a diadic grid, that is for $a_0 = 1/2$, $b_0 = 1$ is of particular interest for applications, and for very special choices of ψ, the diadic wavelets:

$$\psi_{j,k}(t) = 2^{j/2}\psi(2^j t - k) \qquad j,k \in Z \tag{4}$$

constitutes an orthonormal basis for the space of finite energy signals, $L^2(\Re)$. A well knows example is the Haar basis:

$$\psi_{j,k}(t) = \begin{cases} 2^{j/2}, & \text{if } 2^{-j}k \leq t < 2^{-j}(\frac{1}{2} + k); \\ -2^{j/2}, & \text{if } 2^{-j}(\frac{1}{2} + k) \leq t < 2^{-j}(1 + k); \\ 0, & \text{otherwise.} \end{cases} \tag{5}$$

Orthonormal wavelet analysis provides an important tool in signal processing. Given a signal f, its analysis consists in a successive decomposition in a hierarchical scheme of approximations and differences, called *Multiresolution Analysis*. It consists in a collection of closed subspaces $V_j, j \in Z$, such that:

. $V_j \subset V_{j+1}$

. the intersection of all V_j is 0 and its union give us the full space $L^2(\Re)$

. $f(t) \in V_j \leftrightarrow f(2t) \in V_{j+1}$

. there exists a function $\phi(t)$ such that the collection $\phi(t - k)$, $k \in Z$ is an orthonormal basis for V_0. Moreover, ϕ should be a *low pass* filter.

Denoting by W_j the orthogonal complement of V_j in V_{j+1}, we have:

. $V_{j+1} = V_j \oplus W_j$

.there is a unique admissible wavelet ψ such that the family $\psi(t - k), k \in Z$ is an orthonormal basis for W_0. From there the full family $2^{j/2}\psi(2^j t - k), j, k \in Z$ is an orthonormal basis of $L^2(\Re)$.

For example, in the Haar case, V_j are the subspaces of piecewise constant functions in each interval $2^{-j}k \le t < 2^{-j}(1 + k)$ and ϕ is the characteristic function of the interval $[0,1)$. If we denote by P_j and Q_j the orthogonal projectors onto V_j and W_j respectively, the implementation of the Multiresolution Analysis consists in the successive decomposition of any signal in the form:

$$P_{j+1}f = P_jf \oplus Q_jf \tag{6}$$

then $\lim_{j \to \infty} P_jf = f$ for all $f \in L^2(\Re)$. Note that in each step, low and high frequencies in $P_{j+1}f$ are separated. Now we can rewrite (6) as:

$$\sum_k f_{j+1}(k)\phi_{j+1,k}(t) = \sum_k f_j(k)\phi_{j,k}(t) + \sum_k c_j(k)\psi_{j,k}(t) \tag{7}$$

where:

$$f_{j+1}(k) = < f, \phi_{j+1,k} > \tag{8}$$

$$f_j(k) = < f, \phi_{j,k} > = < P_{j+1}f, \phi_{j,k} > \tag{9}$$

$$c_j(k) = < f, \psi_{j,k} > = < P_{j+1}f, \psi_{j,k} > \tag{10}$$

and using the properties and (8-10) we derive the recursive formulas:

$$f_j(k) = \sum_k f_{j+1}(n)h(n - 2k) \tag{11}$$

$$c_j(k) = \sum_k f_{j+1}(n)g(n - 2k) \tag{12}$$

where

$$h(n) = \sqrt{2} < \phi(t), \phi(2t - n) > \tag{13}$$

$$g(n) = \sqrt{2} < \psi(t), \phi(2t - n) > \tag{14}$$

The coefficients $f_j(k)$ and $c_j(k)$ can be regarded as *discrete representations* of the signal f in V_j and W_j respectively. In practice, one compute, from an initial discrete representation in V_0, the successive coefficients $f_j(k)$, $c_j(k)$, $j < 0$, by discrete convolutions with the pair of *conjugate filters* defined by (11-12).

For example, in the Haar case these filters are given by:

$$H(z) = \frac{1}{\sqrt{2}}(1 + z) \tag{15}$$

$$G(z) = \frac{1}{\sqrt{2}}(1 - z) \tag{16}$$

If we have a discrete and bounded signal $f(n), n \in Z$, using appropriate filters we can characterize its topological properties. Particularly, we can detect periodic patterns or transient phenomena. Diadic filters correspond in a natural way with diadic structures or 2^n periodic orbits.

2.2 The Triadic Structure

As it well known, some dynamical systems present 3^n periodic orbits. The most famous example is the Logistic Map [10], which evolves to the 3^n cycles after the appear of inttermitency. Hence, to detect and characterize such structures is desirable, but diadic filters disagree with these purposes. The natural idea is to use discrete filters associated with triadic wavelets, in analogous with the diadic case. Here, we propose a family of triadic wavelets, corresponding with the Haar wavelets, but this scheme can be easily extended for any orthogonal wavelet basis. Define the basic functions:

$$\phi(t) = \chi_{[0,1)}(t) \tag{17}$$

$$\psi^{(1)}(t) = \sqrt{\frac{3}{2}}(\phi(3t) - \phi(3t - 2)) \tag{18}$$

$$\psi^{(2)}(t) = -\frac{1}{\sqrt{2}}(\phi(t) - 2\phi(3t - 1) + \phi(3t - 2)) \tag{19}$$

and the orthonormal families

$$\phi_{(j,k)}(t) = 3^{j/2}(\phi(3^j t) - k) \tag{20}$$

$$\psi^{(i)}_{(j,k)}(t) = 3^{j/2}(\psi^{(i)}(3^j t) - k) \qquad i = 1, 2 \quad j, k \in Z \tag{21}$$

For each $j \in Z$ the sets $\phi_{(j,k)}(t)$, $\psi^{(1)}_{(j,k)}(t)$, $\psi(t)^{(2)}_{(j,k)}$ are orthonormal basis of the respective subspaces V_j, $W_j^{(1)}$ and $W_j^{(2)}$, generated by the parameter k running over Z. Now we have:

$$V_{j+1} = V_j \oplus W_j^{(1)} \oplus W_j^{(2)} \tag{22}$$

$$P_{j+1}f = P_j f \oplus Q_j^{(1)} f \oplus Q_j^{(2)} f \tag{23}$$

where P_j, $Q_j^{(1)}$ and $Q_j^{(2)}$ are respective orthogonal projectors and the associated *triadic* conjugate filters are defined by:

$$H(z) = \frac{1}{\sqrt{3}}(1 + z + z^2) \tag{24}$$

$$G^{(1)}(z) = \frac{1}{\sqrt{2}}(1 - z^2) \tag{25}$$

$$G^{(2)}(z) = -\frac{1}{\sqrt{6}}(1 - 2z + z^2) \tag{26}$$

Then we can write:

$$\sum_k f_{j+1}(k)\phi_{j+1,k}(t) = \sum_k f_j(k)\phi_{j,k}(t) \oplus \sum_k c_{j,k}^{(1)}(k)\psi_{j,k}^{(1)}(t) \oplus \sum_k c_{j,k}^{(2)}(k)\psi_{j,k}^{(2)}(t) \tag{27}$$

$$f_j(k) = \sum_n f_{j+1}(n)h(n - 3k) \tag{28}$$

$$c_j^i(k) = \sum_n f_{j+1}(n)g^{(i)}(n - 3k) \qquad i = 1, 2 \tag{29}$$

Given a discrete signal $f_0(n)$, the recursive applications of the triadic conjugate filters (24-26), using formulas (27-28), give us:

$$f_j(k) = \frac{f_{j+1}(3k) + f_{j+1}(3k + 1) + f_{j+1}(3k + 2)}{\sqrt{3}} \tag{30}$$

$$c_j^{(1)}(k) = \frac{f_{j+1}(3k) - f_{j+1}(3k + 2)}{\sqrt{2}} \tag{31}$$

$$c_j^{(2)}(k) = -\frac{f_{j+1}(3k) + 2f_{j+1}(3k + 1) - f_{j+1}(3k + 2)}{\sqrt{6}} \tag{32}$$

Note that a 3 periodic pattern implies a stationary point in the tridimensional map: $(f_j(k), c_j^{(1)}(k), c_j^{(2)}(k))$ for j=-n.

3. Analysis of Type-I Intermittency in the Logistic Map

As it is well known, intermittency is the alternation of long regular or laminar phases with short irregular or turbulent burst that appear as noise. The Logistic Map [10]:

$$x_{k+1} = \lambda x_k(1 - x_k) \tag{33}$$

present a three cycle period behavior for λ just above $\lambda_c = 1 + \sqrt{8}$. Just below this critical point (λ_c), the system exhibits an intermittent behavior. This route to chaos was suggest by Pomeau and Manneville [12]. The chaotic bursts are separated by windows of 3-cycles as we show in fig. 1 (a). When λ decreases below

λ_c the intermittency route to chaotic behavior take place. We applied the triadic transformation gives in (30-33) to the Logistic Map. The laminar region is caracterized by a well-defined spectral line characteristic of the three period regions and all wavelet coefficients $c_{1,k}^{(1)}$ and $c_{1,k}^{(2)}$ are constant for all k. However, this constant value might change, before and after the turbulence zone, remaining constant in the regular zones. When λ decreases, the coefficients of the Triadic Wavelet Transform show the changes among the regular zones and the turbulent bursts, and they are very sensitive for detect the starting point of the bursts. Figure 1 (b and c) shows the $c_{1,k}^{(1)}$ and $c_{1,k}^{(2)}$ coefficients, for λ such as $\lambda_c - \lambda = 10^6$. Note that they reduce the 3-cycle to a constant value.

The graphics $c_{1,k}^{(1)}$ vs $c_{1,k}^{(2)}$ can characterize different states of the logistic map (33). For the three cycles ($\lambda > \lambda_c$), presents an only point, and when the bursts are more frequent, appear a typical curve (figure 2 (a)). For all range of the parameter λ studied, into the intermittency phenomena ($\lambda_c \leq \lambda < \lambda_c - 10^{-1}$), this curve is the same. The points distribution in the curve is more uniform when the appearance of the chaotic bursts are more frequent. Only when the system has a chaotic behavior, we obtain the closed curve of the figure 2 (b). Also, we can recognize different situations: the intersection point ($c_{1,k}^{(1)} = 0$, $c_{1,k}^{(2)} = 0$) corresponds at the fixed point $x^* = 1 - \frac{1}{\lambda}$ of the first bifurcation; and the two points of the curve $c_{1,k}^{(1)} = 0$, $c_{1,k}^{(2)} \neq 0$ correspond at the period-two attractors.

4. Conclusions

The Triadic Wavelet Transform is very usefull to analize dynamical systems, specially for non stationary signals. There are remarkable advantages with this technique: the information can be represented through graphic patterns that can be easly and quickly interpretated; is very sensible to detect the start of the turbulence and the difference with the chaotic behavior and the transformation can be easely implemented using fast algorithms.

Acknowledgements: *This work was partially supported for PID 3-344200/92 CONICET, Argentina*

References

1. Y. Grenier; (1986) *Traitment du signal-Representations temps-frequence*, ENST Paris.

2. P. Flandrin; (1988) 'Non Destructive Evaluation in Time Frequency Domain by means of Wigner-Ville Distribution', NATO ASI Series, **F44**, pp 109-116.

3. P. Flandrin; (1987) 'Some Aspects of Non-Stationary Signal Processing with

Emphasis on Time-Frequency and Time-scale Methods' in *Wavelets (Time-Frequency Methods and Phase Space)*, Berlin: Springer-Verlag, pp 68-69.

4. Y. Meyer; (1992) *Ondelettes et Operateurs*, Tomes I, II, III, Paris: Hermann

5. Y. Meyer; (1990) 'Wavelets and Applications', Procedding of the International Congress of Mathematicians, Kioto Japan, pp 1621-1626.

6. Y. Meyer; (1993) *Wavelets, Algorihms & Applications*, Philadelphia: SIAM.

7. I. Daubechies; (1992) *Ten Lectures on Wavelets*, Philadelphia: SIAM.

8. C. E. Heil, D. F. Walnut; (1989) 'Continuous and Discrete Wavelet Transforms', SIAM Review, **31**, 4, pp 628-666.

9. Ch. K. Chui; (1992) *Wavelets: A Tutorial in Theory and Applications* San Diego, Academic Press Inc..

10. M. Feigenbaum; (1979) 'The Universal Properties of Nonlinear Transformations' J. Stat. Phys.,**21**, 6, pp 669-705.

11. J. E. Hirsch, B. A. Huberman, D. J. Scalapino; (1982) Theory of Intermittence, Phys. Rev. A,**25**, 1, pp 243-257.

12. Y. Pomeau and P. Manneville; (1980) 'Intermittent Transition to Turbulence in Dissipative Dynamical Systems' Comun. Math. Phys.,**74**, pp 189-197.

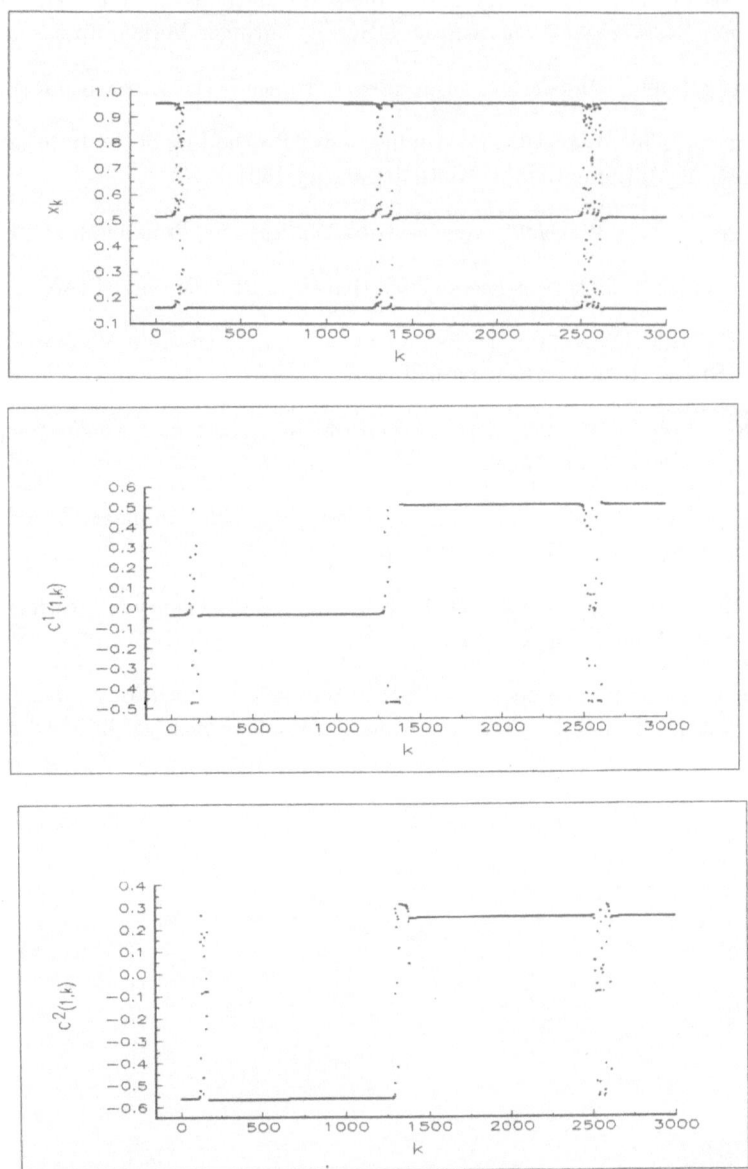

Figure 1: (a) Intermittent behavior for $\lambda_c - \lambda = 10^{-6}$. Note that turbulent bursts are evident. (b and c) Triadic Wavelet coefficients $c^{(1)}(1, k)$ and $c^{(2)}(1, k)$ as a function of k por x_k series shown in (a)

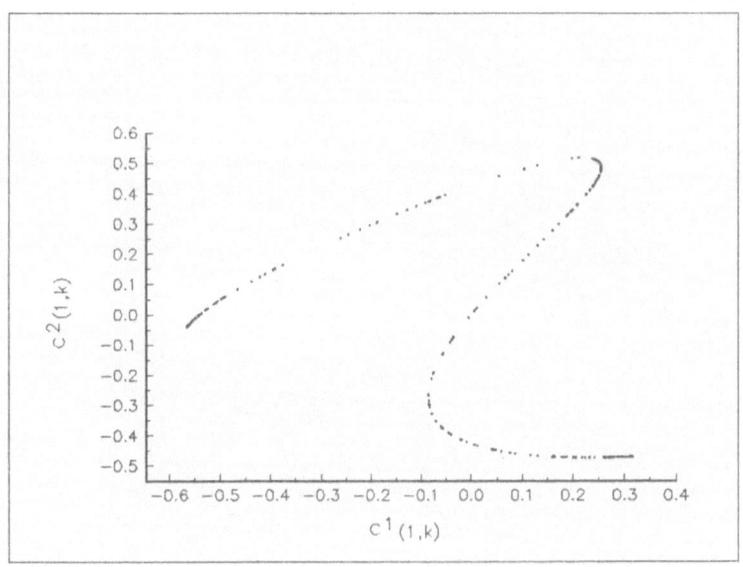

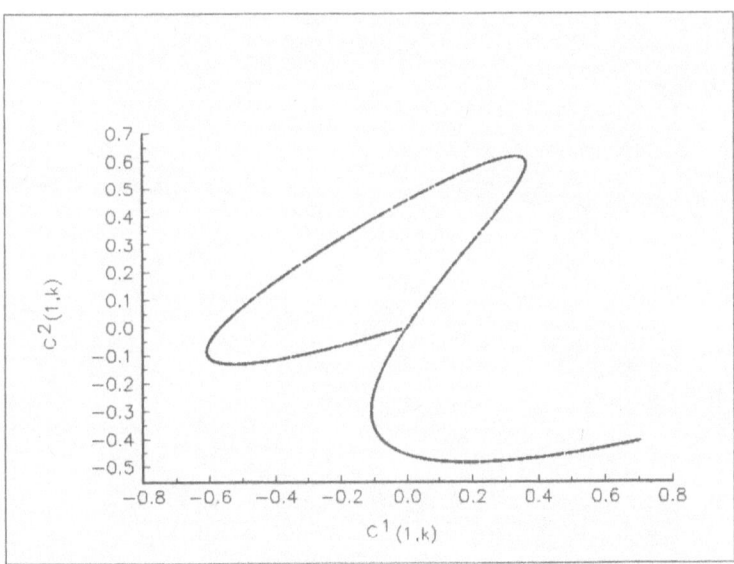

Figure 2: Typical map of the Triadic Wavelet coefficients $c^{(1)}(1,k)$ vs $c^{(2)}(1,k)$ for: (a) the intermittency zone ($\lambda_c - \lambda = 10^{-6}$) and (b) chaotic behavior ($\lambda - 3.75$)

TREHALOSE IN SOLUTION AND ITS INTERACTION WITH WATER AND BIOLOGICAL STRUCTURES. MOLECULAR DYNAMICS SIMULATION

M.C. DONNAMARIA AND J.R. GRIGERA

Instituto de Fisica de Liquidos y Sistemas Biologicos (IFLYSIB), CONICET, UNLP and Departamento de Ciencias Biologicas, Facultad de Ciencias Exactas, Universidad Nacional de La Plata, C.C. 565, 1900 La Plata, Argentina

ABSTRACT. Molecular simulations of an aqueous solution of trehalose have been done to contribute to the understanding of the effect of trehalose as protecting agent against water stress in biological systems. The hydrogen bond network and water dynamics were found to be only slightly altered when compared with pure water (SPC/E model). Some internal hydrogen bonds in trehalose contribute to stabilize the conformation. Results support the view that the protective effect of trehalose against water stress is due to direct contribution to the stabilization of biological structures and not through the modification of water properties.

1. Introduction

Trehalose (α-D-glucopyranosil (1-1) α-D-glucopyranose) is a disaccharide well known as a natural protectant against water stress [1]. This effect of trehalose may be due either to modifications of the water structure and dynamics or to direct stability action on the biological structures to be protected. Its actual mechanism is still unclear. By analysis of Molecular Dynamics Simulation, which has been proved to be highly reliable to predict carbohydrate properties in solution [2-5], we have been able to state several aspects of the characteristic dynamic behaviour, structure, and conformation of the water trehalose system. Some preliminary data of molecular dynamics and molecular mechanics results were already published [6]

2. Method

We run the Molecular Dynamics simulation using the GROMOS package in an IBM RS/6000 32H. Results were analyzed in a VAX 11/750 and in a 486 personal computer. All runs were done at constant pressure and constant temperature, keeping them at 1.013×10^5 Pa and 300 K, respectively. Integration time step was 0.001 ps and simulation continues for 100 ps after the equilibrium was reached. The criteria for equilibrium were the stability of box length and potential energy. One trehalose molecule was initially located in an octahedral box of $1.786 \times 1.813 \times 2.283$ nm surrounded by 270 water molecules.

E. Tirapegui and W. Zeller (eds.), Instabilities and Nonequilibrium Structures V, 359–364.
© 1996 *Kluwer Academic Publishers.*

2.1 MODEL

In trehalose two hexapyranose rings are connected *via* 1-1 glycosidic linkage. In the model used, both rings were kept rigid in the 4C1 conformation by applying improper torsion potentials that avoid transitions between other possible conformations (Fig.1). All atoms were explicitly included and tetrahedral geometry of carbon atoms was maintained using improper torsion potentials. Bond lengths were kept rigid using SHAKE procedure and bond angles treated as having harmonic potentials. There were not applied torsion potentials on glycosidic linkage, in order to avoid a bias in the calculated conformations. [3,5,7]. Force field parameters were used as given in GROMOS package. The charges were computed with semi-empirical quantum calculation: a modified version of AM1 in which the solvation effect is incorporated *ad hoc*. The model used for water was the SPC/E [8]

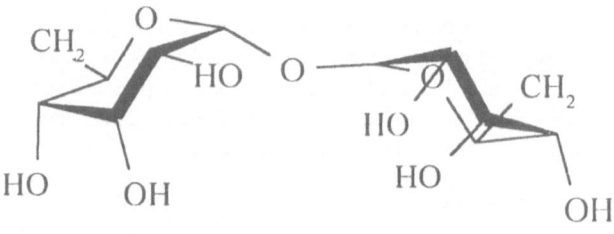

Figure 1. The trehalose molecule.

3. Results

3.1 FLEXIBILITY

The relative mobility of both trehalose rings has not any imposed restriction than mutual atom interaction. Figure 2 shows the trajectories of glycosidic dihedral angles ϕ(Ca2-Ca1-Ogl-Cb1) and ψ(Ca1-Ogl-Cb1-Cb2) recorded during the spanned 100 ps. The averages values are: $\phi = 216°$ (RMS=13.13), and $\psi =215°$ (RMS=11.98). Both angles fluctuate around almost the same mean value; and only one conformation was accessed. These data agree with those obtained [9] using optical rotation method although our results are closer to those observed in the crystal.

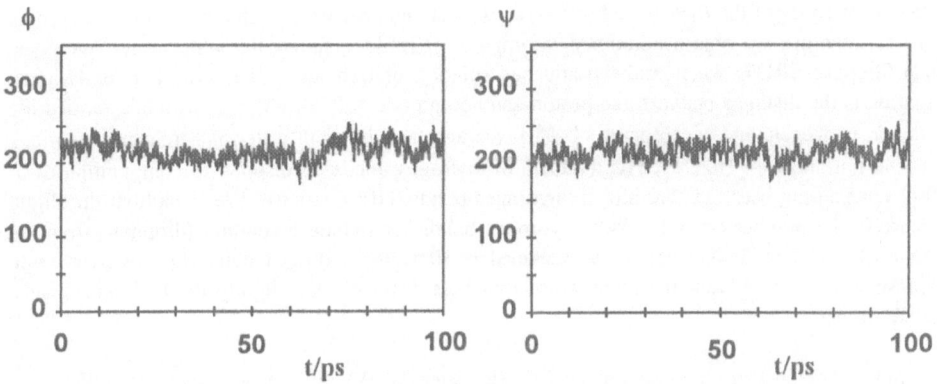

Figure 2 Time trajectories of dihedral angles of glycosidic linkage of trehalose.

3.2 HYDRATION

Hydration has been studied through the radial distribution functions RDF of water around each atom of trehalose. As expected, equivalent atoms corresponding to different rings have the same behaviour. The RDF of water around oxygens indicates that ring and glycosidic oxygens have a non-polar behaviour. The other oxygens show a polar feature, effect that is clearer for O6. Fig. 3 shows a sample of typical RDF. For the "hydrophilic" oxygens the first peak is around 0.35 nm. As rule, the peaks are low when compared with similar situation on other carbohydrates. [2-4]

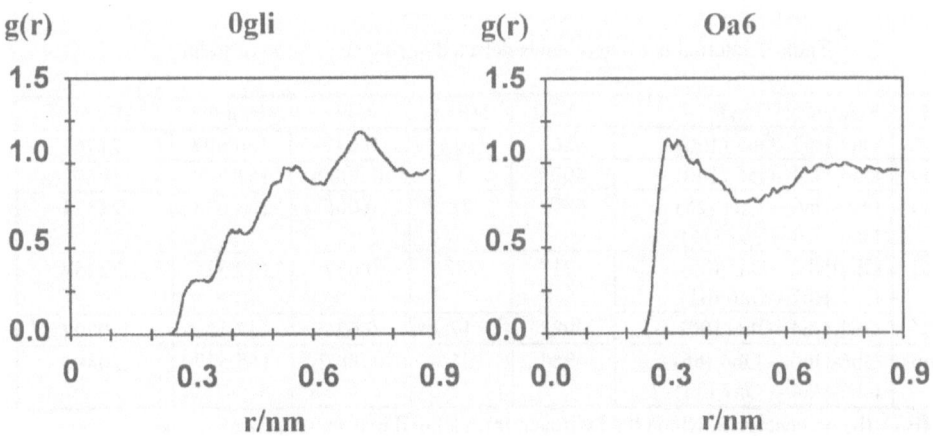

Figure 3- Radial Distribution Functions of water molecules around trehalose oxygens

By the analysis of the *hydrogen bond network*, including the water-trehalose, water-water, and trehalose-trehalose we obtain additional information. Table 1 shows the water-water hydrogen bonds for pure SPC/E water and the aqueous solution of trehalose. The criterion for H-bond formation is the distance between the proton and acceptor ($d < 0.24$ nm) and the angle formed by the donor, hydrogen, and acceptor ($\varphi > 145°$). Averages are, for both water and trehalose solution, $d = 0.186$ nm and $\varphi = 162.1°$. The numbers of hydrogen bonds correspond to computation over 100 ps considering between zero and six hydrogen bonds (HB=i, i=1-6). We found that the slight decrease in the number of water-water hydrogen bonds for trehalose solution (dropping from an average of 3.183 to 3.141) could be explained if 10 water hydrogen bonds are involved with trehalose. We conclude that the presence of trehalose disturbs very slightly the hydrogen bonds network of water.

Table 1-Water-water hydrogen bonds in pure water (W-W) and trehalose solution (T-W).

	HB=0	HB=1	HB=2	HB=3	HB=4	HB=5	HB=6	$<N^o>$	$<A/^o>$	$<D/nm>$
W-W	2792	44447	234074	531405	496929	40106	248	3.183	162.077	.18641
T-W	3151	49127	248865	540472	470699	37438	247	3.141	162.069	.18631

The average number of *internal hydrogen bond* of trehalose is very small (Table 2). There are some number of hydrogen bonds between O6 and O2 of different sugar rings. The existence of these H-bonds has been also suggested by Rees et al.[9] These H-bonds will contribute to inhibit large movements of the glycosidic dihedral angles. A few hydrogen bonds in which the ring oxygen, O5, was involved have been detected, which is indeed striking due to its non-polar character. However, we note that the small number of H-bonds formed are not enough to modify the overall non-polar character of the ring oxygen, although may have some contribution to the stabilization of the molecular conformation.

Table 2.Internal hydrogen bonds detected during simulation of trehalose.

H	Sequence(%HB)	HB=0	HB=1	$<N^o>$	$<Ang/^o>$	$<Dist/nm>$
Ha2	Oa2-Ha2 -Ob5 (100)	4803	197	0.039	160.697	.21762
Ha4	Oa4-Ha4--Oa6 (100)	4999	1	0.0002	151.739	.19699
Ha6	Oa6-Ha6 -- Oa4 (25) Oa6-Ha6 --Ob2 (75)	4972	28	0.006	163.033	.21524
Hb2	Ob2-Hb2 --Oa5 (99) Ob2-Hb2 --Oa6 (01)	4717	283	0.057	160.315	.22259
Hb4	Ob4-Hb4--Ob6(100)	4988	12	0.024	147.100	.19902
Hb6	Ob6-Hb6 -- Ob4 (66) Ob6-Ha4 -- Oa2 (33)	4964	36	0.0072	155.517	.20896

%HB is the percentage in which the hydrogen form a bond in each sequence.

3.3 WATER DIFFUSION COEFFICIENT

To evaluate the effect of trehalose in the dynamics properties of water, we have computed the water-water diffusion coefficient, from the atomic mean square displacement [10] The values for pure SPC/E water, computed for the same box geometry and size, is $D_w = 2.211 \times 10^{-5}$ cm^{2}s^{-1}, and the one for simulated aqueous trehalose solution, $D_t = 2.163 \times \times 10^{-5}$ cm^{2}s^{-1}. It becomes that, for concentrations to up to at least 0.2 mol/l, the trehalose does not alter significantly the water diffusion coefficient.

4. Conclusions

Working with a model of two rigid rings, but allowing them to move free around the involved glycosidic angles, it was possible to observe a moderate flexibility of trehalose. Since we have not applied torsion potentials on glycosidic linkage, the conformation determined through the dihedrals comes only from the atom-atom interaction and solvent effects and it is not biased by any pre-defined structure. The relative rigidity of trehalose seems to be due to the existence of a small number of internal hydrogen bonds that do not allow large fluctuation around glycosidic linkage. The presence of trehalose seems to disturb very slightly the H-bond network of water as well as its dynamics. So, we cannot consider the changes in the hydrogen bonding pattern of the water as an explanation for the protective effect of trehalose.

We note that the partial folding between the two rings, along the glycosidic linkage, leads to a special structure, where the HO(2) of one residue is close to HO(6) of the other, increasing the availability of hydrogen bonding. Also unsuspected O5--H2 hydrogen bonding could contribute to keep this structure stable. Such a conformation produces a spatial arrangement of hydroxyl groups that maximize hydrogen bonding interactions with a putative tridymite structure of water. The spacing of hydroxyl groups in trehalose matches to tetrahedral coordinated oxygen atoms hydrogen bonded into the dynamic network array for water (Fig. 4). As is seen on the model, most of OH groups capable to form H-bonds with water are oriented toward one surface while the more hydrophilic surface fits into the water structure. It is sensible to suppose that water structure also matches on hydrophilic surfaces of biological structures. Therefore, we may assume that hydrogen bonds of trehalose will match onto biological structures [11]

Under the circumstances we propose that for a system in which the water plays a role in the stabilization of the structure, as in biological systems, the trehalose may replace a number of water molecules bonded to the structure. In this sense the action of trehalose as protectant under water stress, corresponds to a direct action on the biological structures to be protected, rather than a modification of the water structure and dynamics.

Acknowledgments. This work has been partially funded by the Consejo Nacional de Investigaciones Cientificas y Tecnicas (CONICET) of Argentina. J.R.G. and M.C.D are members of Research Career of CONICET and Comision de Investigaciones Cientificas of Province of Buenos Aires (CIC) respectively. The authors wish to thank Dr. H. Villar for this valuable work in calculating the charges of trehalose and are also indebted to Juan Grigera for his help with technical support.

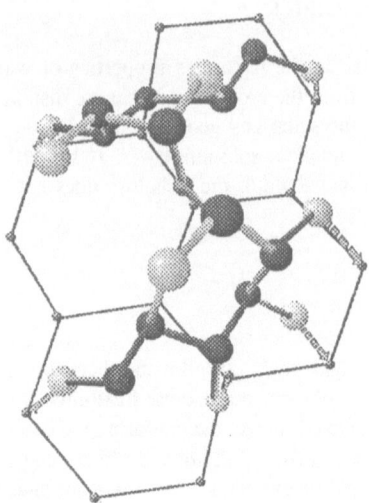

Figure 4 Scheme of trehalose molecule over an expanding ice structure.

5. References

[1]. Franks F. *Biophysics and biochemistry a low temperatures.* Cambridge University Press. Cambridge 1985.

[2] J.W. Brady. *Curr. Op. Struct. Biol.* 1990, **1**, 711.

[3] F. Franks Dadok J. and Kay R. and J.R. Grigera. *J. Chem. Soc. Faraday Trans.* 1991, **87**, 579.

[4] E.I. Howard and J.R. Grigera, *J. Chem. Soc. Faraday Trans*, 1992, **88**, 437.

[5] J.R. Grigera, *Hydration of Carbohydrates as seen by Computer Simulation.* In: *Adv. Comp. Biol.* Vol 1, 203-229, JAI Press, New York 1994

[6] J.R. Grigera, M.C. Donnamaria and E.I. Howard. *Flexibility of polysaccharides using Molecular Dynamics.* In: *Condensed Matter Theories.* Vol 8,L. Blum , Ed.   Plenum Press, New York 1993.

[7] F. Franks and J.R. Grigera *Solution properties of low molecular weight polyhydroxy compounds.* In: *Water Science Review* Vol 5. F. Franks ed. Cambridge Univ. Press. Cambridge 1990.

[8]. H.J.C. Berendsen, J.R. Grigera and T. Straastma. *J. Phys .Chem.* 1887, **91**, 6269.

[9]. D. A. Rees and D. Thom, *J.S. Perkin II*, 1977, 191.

[10] J.P. Hansen and I. R. McDonald, *Theory of Simple Liquids*, Academic London, 1976.

[11]. H.J.C Berendsen. *Water Structure.* In: *Theoretical and Experimental Biophysics.* A. Cole, editor. Marcel Dekker, New York 1967.